W0071755

Science and American Foreign Relations since World War II

The sciences played a critical role in American foreign policy after World War II. From atomic energy and satellites to the Green Revolution, scientific advances were central to American diplomacy in the early Cold War, as the United States leveraged its scientific and technical preeminence to secure alliances and markets. The growth of applied research in the 1970s, exemplified by the biotech industry, led the United States to promote global intellectual property rights. Priorities shifted with the collapse of the Soviet Union, as attention turned to information technology and environmental sciences. Today, international relations take place within a scientific and technical framework, whether in the headlines on global warming and the war on terror or in the fine print of intellectual property rights. *Science and American Foreign Relations since World War II* provides the historical background necessary to understand the contemporary geopolitics of science.

GREG WHITESIDES is Assistant Professor of History at the University of Colorado, Denver.

Cambridge Studies in US Foreign Relations

Edited by

Paul Thomas Chamberlin, Columbia University
Lien-Hang T. Nguyen, Columbia University

This series showcases cutting-edge scholarship in US foreign relations that employs dynamic new methodological approaches and archives from the colonial era to the present. The series will be guided by the ethos of transnationalism, focusing on the history of American foreign relations in a global context rather than privileging the US as the dominant actor on the world stage.

Also in the Series

Science and American Foreign
Relations since World War II

GREG WHITESIDES

University of Colorado Denver

CAMBRIDGE
UNIVERSITY PRESS

CAMBRIDGE
UNIVERSITY PRESS

University Printing House, Cambridge CB2 8BS, United Kingdom

One Liberty Plaza, 20th Floor, New York, NY 10006, USA

477 Williamstown Road, Port Melbourne, VIC 3207, Australia

314-321, 3rd Floor, Plot 3, Splendor Forum, Jasola District Centre, New Delhi - 110025, India

79 Anson Road, #06-04/06, Singapore 079906

Cambridge University Press is part of the University of Cambridge.

It furthers the University's mission by disseminating knowledge in the pursuit of education, learning and research at the highest international levels of excellence.

www.cambridge.org
Information on this title: www.cambridge.org/9781108409919
DOI: 10.1017/9781108303965

First published 2019
First paperback edition 2020

A catalogue record for this publication is available from the British Library

Library of Congress Cataloging in Publication data
NAMES: Whitesides, Greg, author.
TITLE: Science and American foreign relations since World War II /
Greg Whitesides, University of Colorado, Denver.
DESCRIPTION: Cambridge, United Kingdom ; New York, NY : Cambridge
University Press, 2019. | Series: Cambridge studies in US foreign relations |
Includes bibliographical references and index.
IDENTIFIERS: LCCN 2018034473 | ISBN 9781108420440 (hardback : alk. paper) |
ISBN 9781108409919 (paperback : alk. paper)
SUBJECTS: LCSH: Science and international relations–United States–History–20th century. |
Science and international relations–United States–History–21st century. | United States–
Foreign relations–1945-1989. | United States–Foreign relations–1989–
CLASSIFICATION: LCC JZ1254 .W45 2019 | DDC 327.73–DC23
LC record available at https://lccn.loc.gov/2018034473

ISBN 978-1-108-42044-0 Hardback
ISBN 978-1-108-40991-9 Paperback

Dedicated to my family and Lee

Contents

Figures

Acknowledgments

I would like to thank Lionel M. Jensen, Lawrence Badash, and Fredrik Logevall for their inspiration and tutelage. Additionally, my editors at Cambridge University Press, especially Deborah Gershenowitz, Paul Thomas Chamberlin, and Natasha Whelan, deserve special thanks for their patient assistance in guiding this book to publication.

Abbreviations

AAAS	American Association for the Advancement of Science
ABCC	Atomic Bomb Casualty Commission
AEC	Atomic Energy Commission
AFP	Atoms for Peace (Chapter 2)
AFP	Alliance for Progress (Chapter 3)
ARPA	Advanced Research Projects Agency
ASTP	Apollo–Soyuz Test Project
BSF	Binational Science Foundation (the United States & Israel)
CAS	Chinese Academy of Sciences
CBD	Convention on Biological Diversity
CCF	Congress for Cultural Freedom
CERC	Clean Energy Research Center
CERN	*Conseil European pour la Recherche Nucléaire*
CFCs	Chlorofluorocarbons
COCOM	Coordinating Committee for Multilateral Export Controls
COMSAT	Communications Satellite Consortium
COPUOS	Committee on the Peaceful Uses of Outer Space
EPA	Environmental Protection Agency
ESA	European Space Agency
FAO	Food and Agriculture Organization
FAS	Federation of American Scientists
FDA	Food and Drug Administration
G7	Group of 7
G77	Group of 77
GAO	General Accounting Office
GARP	Global Atmospheric Research Program

GATT	General Agreement on Tariffs and Trade
GEF	Global Environment Facility
HGP	Human Genome Project
IAEA	International Atomic Energy Agency
IBP	International Biological Program
ICAO	International Civil Aviation Organization
ICGEB	International Center for Genetic Engineering and Biotechnology
ICSU	International Council of Scientific Unions
IGY	International Geophysical Year
IIAA	Institute for Inter-American Affairs
IICST	Iraqi International Center for Science and Technology
IIST	International Institute of Science and Technology
IPCC	Intergovernmental Panel on Climate Change
IRI	Industrial Research Institute
IRRI	International Rice Research Institute
ISS	International Space Station
ISTC	Institute for Scientific and Technical Cooperation (Chapter 4)
ISTC	International Science and Technology Center (Chapter 6)
ITER	International Thermonuclear Reactor
ITU	International Telecommunication Union
JECOR	Joint Economic Cooperation Office Riyadh
MAP	Mexican Agricultural Program
MEP	Malaria Eradication Program
NACA	National Advisory Committee for Aeronautics
NAS	National Academy of Sciences
NASA	National Aeronautics and Space Administration
NATO	North Atlantic Treaty Organization
NBS	National Bureau of Standards
NDRC	National Defense Research Committee
NGO	Non-Governmental Organization
NIH	National Institutes of Health
NIST	National Institute of Standards and Technology
NOAA	National Oceanic and Atmospheric Administration
NRC	National Research Council
NRDC	Natural Resources Defense Council
NSF	National Science Foundation
OES	Bureau of Oceans and International Environmental and Scientific Affairs
OMB	Office of Management and Budget

ONR   Office of Naval Research
OSRD   Office of Scientific Research and Development
OSTP   Office of Science and Technology Policy
PSAC   President's Science Advisory Committee
SITE   Satellite Instructional Television Experiment
SDI   Strategic Defense Initiative
TRIPS   Trade Related aspects of Intellectual Property rightS
TVA   Tennessee Valley Authority
TWAS   The World Academy of Sciences
UNAEC   United Nations Atomic Energy Committee
UNEP   United Nations Environmental Program
UNESCO   United Nations Educational Scientific and Cultural
           Organization
UNFCC   United Nations Framework on Climate Change
USAID   United States Agency for International Development
USDA   United States Department of Agriculture
USGS   United States Geological Survey
USPTO   United States Patent and Trade Office
WHO   World Health Organization
WIPO   World Intellectual Property Organization
WMO   World Meteorological Organization
WTO   World Trade Organization

Introduction

Commissioned to find a northwest passage by the British crown, Captain James Cook sailed into political headwinds – part of the empire was in rebellion. Thankfully, Benjamin Franklin, patron of "useful knowledge" and a prominent colonist, asked the colonial navy to "not consider the HMS *Resolution* [Cook's ship] as an enemy."[1] Mapping the Arctic and Pacific oceans was in the greater interest of mankind; such a scientific voyage was above the fray. Joseph Banks, president of the British Royal Society, returned the favor, thanking Franklin for being a "Friend of disinterested Discovery."[2] But Franklin's plea fell upon deaf ears: The rebellious Continental Congress thought the *Resolution* should be captured if possible.[3] The story raises a key question: Is science bound by national interest? The answer was academic at the time, but that would change.

In January of 1939, Germans Otto Hahn and Fritz Strassmann published a paper outlining nuclear fission. The revelation sent shockwaves around the world: four American teams, as well as Frederic Joliot-Curie's

---

[1] Franklin's original proposal for the American Philosophical Society in 1743 suggested "the promoting of Useful Knowledge," and his letter is reprinted in A. Kippis, *A Narrative of the Voyages Round the World Performed by Captain James Cook, with an Account of His Life during the Previous and Intervening Periods* (Philadelphia: Porter & Coates, [N.D.]), 391–392. The narrative was first published in 1788.

[2] Banks quoted in Deborah Allen and Deborah J. Allen, "Acquiring 'Knowledge of Our Continent': Geopolitics, Science and Jeffersonian Geography, 1783–1803," *Journal of American Studies* 40 (August 2006): 205–232, quote on 208.

[3] On the Congressional reaction, see Kippis, *A Narrative*, 392–393; J. C. Beaglehole, *The Life of Captain James Cook* (Stanford: Stanford University Press, 1974), n.685; Alan Villiers, *Captain James Cook* (New York: Charles Scribner's Sons, 1967), 231.

lab in Paris, verified the paper before the end of the month.[4] Nazi scientists alerted the War Office in the spring; a German newspaper asked whether nuclear energy could be put to "practical uses" in June.[5] The openness of the period is revealing in hindsight: stories about fission and possible chain reactions appeared frequently, including in more than 100 scientific papers.[6] Worried about possible military implications, scientists tried to impose a voluntary ban on publications but Joliot-Curie's team ignored it. Europeans imposed secrecy only after German tanks rolled into Poland in September; a self-imposed censorship became effective in the United States the following year.[7]

The conversion of knowledge into power during World War II altered international relations. The US government funded research critical to victory and the destruction of Hiroshima and Nagasaki testified to the effectiveness of the approach. After the conflict, nations sought the benefits of applied research, but research remained concentrated in a few countries. Indeed, World War II exacerbated imbalances by advancing American scientific capability while reducing the capabilities of European and Asian rivals. The United States had been a leading scientific power before the war; the country occupied a singular position after.

Science became central to American diplomacy. Although tensions over nuclear, biological, and chemical weapons are familiar, the larger story remains unknown. *Science and American Foreign Relations since World War II* provides the first history of science in American foreign relations alongside analysis of science as a tool of American statecraft. Agricultural research, export controls, and genetics, for example, played key roles in postwar diplomacy, as the United States leveraged its scientific and technical preeminence to secure alliances and markets. American funding underwrote international scientific undertakings and diplomats offered developmental assistance to curry favor: Middle Eastern nations, for example, benefited from extensive scientific and technical aid, whether under the Shah of Iran (1950–1979), through the JECOR initiative with

---

[4] Lawrence Badash, *Scientists and the Development of Nuclear Weapons* (Atlantic Highlands, NJ: Humanities Press, 1995), esp. 22–24.

[5] Regarding the German scientists, see Mark Walker, *German National Socialism and the Quest for Nuclear Power, 1939–49* (Cambridge: Cambridge University Press, 1989). Regarding the newspaper, see L. Badash, et al., "Nuclear Fission: Reaction to the Discovery in 1939," *Proceedings of the American Philosophical Society* 130 (June 1986): 196–231, quote on 215.

[6] Badash, *Scientists and the Development of Nuclear Weapons*, 29.

[7] Badash, "Nuclear Fission," 196–231.

Saudi Arabia (1975–2000), or via multiple programs with Israel (1950 to the present). This assistance boosted American soft power and prestige overseas: Polls indicate foreigners view American science and technology more favorably than other aspects of American society.[8] The growth of commercial research led the United States to promote intellectual property rights, while environmental sciences, the rise of Asian competitors, and collapse of the Soviet Union reshaped American priorities. Science remains essential to American foreign relations today, whether in cooperative activities such as weather prediction and disease prevention or in geopolitical disputes over climate change, genetically modified organisms and rogue nuclear programs.

Science and American Foreign Relations since World War II builds upon a growing body of work on science in international relations.[9] In 2008, the American Association for the Advancement of Science (AAAS) established a Center for Science Diplomacy and suggested a three-tiered framework to separate different approaches:[10]

Diplomacy for science – using diplomacy to advance a scientific goal or project
Science for diplomacy – using science to build international relations (aka "science diplomacy")
Science in diplomacy – using science to inform and shape diplomacy

For illustration, consider the US/Saudi solar energy project in the 1980s: First, American politicians secured support for solar energy research in Saudi Arabia (*diplomacy for science*); the resulting project benefited the

---

[8] Regarding polls, see NSF Director Arden L. Bement, "Prepared Statement of Arden L. Bement," in Committee on Science and Technology, House of Representatives, 110th U.S. Congress, 2nd Session, April 2, 2008, *International Science and Technology Cooperation* (Washington: GPO, 2008), 19.

[9] Recent overviews include National Research Council, *U.S. and International Perspectives on Global Science Policy and Science Diplomacy: Report of a Workshop* (Washington: National Academies Press, 2011) and Lloyd S. Davis and Robert G. Patman, *Science Diplomacy: New Day or False Dawn?* (Hackensack, NJ: World Scientific Publishing, 2015). See also British Royal Society, *New Frontiers in Science Diplomacy: Navigating the Changing Balance of Power* (London: Science Policy Centre, 2010). For an overview of different national approaches, see Tim Fink and Ulrich Schreiterer, "Science Diplomacy at the Intersection of S&T Policies and Foreign Affairs: Toward a Typology of National Approaches," *Science and Public Policy* 37 (November 2010): 665–677.

[10] I have modified the AAAS framework: The AAAS framework stresses large projects like the International Space Station or International Thermonuclear Experimental Reactor under "Diplomacy for Science," while I include much smaller projects as well. See the Center's website for a more detailed explanation: www.aaas.org/program/center-science-diplomacy.

Saudi people and government, bolstering US–Saudi relations (*science for diplomacy*); finally, the data guided US international energy policy (*science in diplomacy*). In this case, a single project – Saudi solar energy research – involved all three approaches. Or not. Rather than learn from the project, the Reagan administration minimized the potential of solar research at home while advertising its potential abroad, leading to a fourth consideration: the politicization of science. History shows data alone are rarely determinative of policy or cooperation.

Instead, international scientific relations are shaped by a variety of actors and considerations, such as cost/benefit analyses, geopolitics (including national security, economic competitiveness, and diplomacy), and scientific merit (often the last consideration). The motivations of various participants, whether nation-states, scientists, scientific societies, international and nongovernmental organizations, or industry represen-tatives, are instrumental and each has their own agenda. No country is a "disinterested" patron. Since World War II, for example, the American government has engaged in significant "science for diplomacy," but has demonstrated less interest in supporting international science projects (*diplomacy for science*) or in allowing science to shape foreign policy (*science in diplomacy*) contrary to American interests. Finally, in addition to the approaches above, this study also considers American diplomacy to limit access to science, whether through export controls, intellectual property rights or non-cooperation.

A complete picture of science and American foreign relations requires an inclusive definition of "science," incorporating commercial and med-ical research, engineering and advanced technology. Disputes over access to American industrial research and Soviet cancer research, for example, chilled US–Soviet relations in the early Cold War and the United States established an export control system to deny advanced technology and technical know-how to the Soviet bloc. Research with national-security or commercial applications – "applied" science – became a focus of Ameri-can intelligence and diplomacy. "Basic" science, defined by the govern-ment as systematic study "without specific applications towards processes or products in mind," was of interest, but less concern.[11] However, these

---

[11] In general, the United States government separates "science" into "basic research," "applied research," and "development." For an overview of the various definitions and federal regulations, see National Science Foundation, "Definitions of Research and Development: An Annotated Compilation of Official Sources," available at: www.nsf .gov/statistics/randdef/fedgov.cfm.

categories have been accepted as man-made and fluid since their creation: in the 1950s and 1960s, the Department of Defense provided both justifications for the same research and fields like genetic engineering defy easy classification.[12] As such, whenever possible, I will specify a scientific field rather than use "science."

It is important to be field-specific because American diplomacy is field-specific. After the development of genetic engineering and an American biotech industry, for example, the CIA began tracking research and foreign proficiency in the field. Access to genetic material became part of Cold War diplomacy, as the United States hoped to limit Soviet programs while offering genetic aid to entice China toward normalization. The profitability of the field led the United States to undermine a UN center for genetic engineering in the 1970s and refuse cooperation with G7 partners the following decade. Nor were genetic engineering and biotechnology alone; the importance of commercial research to American diplomacy increased throughout the postwar period.

Commercial research and related technologies have a long history in American diplomacy. American acquisition of enemy patents began during World War I; the occupations of Germany and Japan after World War II provided the United States with thousands of commercially valuable properties. When newly independent nations asked for access to advanced science and technology, the United States helped introduce intellectual property rights to protect commercial research. As domestic R&D became more market-driven, American diplomacy for science became more demanding of legal protection and less cooperative. Although biotech is the foremost example studied here, other similar fields include pharmaceuticals, nanotechnology and computer sciences or information technology.

The shift from public to private funding of domestic R&D (Figure I.1) shaped American diplomacy and the text's organization. The first three chapters chronicle the period from World War II to the 1970s, when the government was the dominant source of funding. In the early Cold War, federal funding guaranteed influence over research at home and abroad as the United States played the dominant role in international scientific endeavors (Chapters 1 and 2). At the same time, anti-communism ensured "science for development" a prominent place in American diplomacy

---

[12] Regarding the DOD's "two-title" policy, see Deborah Shapley, "Defense Research: The Names Are Changed to Protect the Innocent," *Science* 175 (February 25, 1972): 866–868.

U.S. total R&D expenditures, by source of funds: 1953–2015

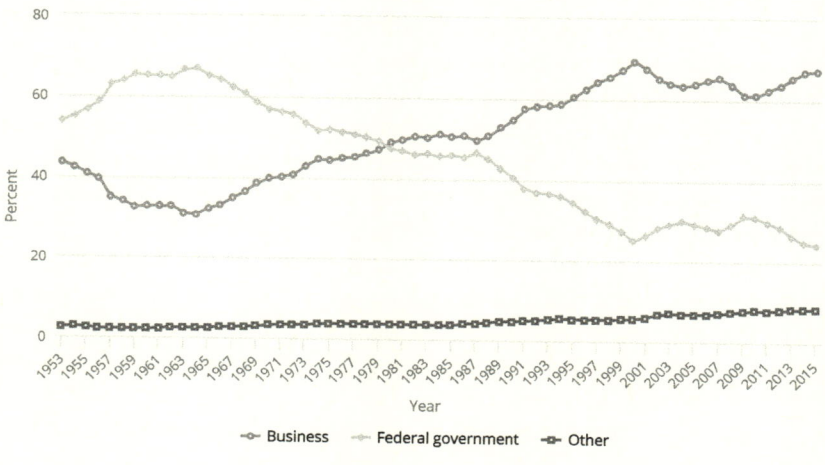

Note(s)

Data for 2015 are preliminary and may later be revised. The other category includes nonfederal government, higher education, and other nonprofit organizations.

Source(s)

National Science Foundation, National Center for Science and Engineering Statistics, National Patterns of R&D Resources (annual series).

Science and Engineering Indicators 2018

FIGURE 1.1 US total R&D expenditures, by source of funds: 1953–2015.

(Chapter 3). Chapter 4, titled "The Crossing Point," covers the transition to more private funding of R&D. During this critical era, the intersection of ecology and geopolitics led to fears of industrial regulation, while the Vietnam War sparked campus protests over defense research (Chapter 4). In the 1970s, genetic engineering exemplified the boom in commercial research which led American diplomats to refuse allied cooperation in biotechnology and advance global patent protections instead. The final three chapters bring the story into the present, when private industry became the primary source of domestic R&D funding. As policy-makers came to view research and development as critical to national prosperity (the "knowledge economy"), American diplomacy focused on implementing intellectual property rights (Chapters 4 and 6) and keeping pace with rivals (Chapters 6 and 7). Participation in international scientific projects declined (Chapters 4 and 5) while domestic politics complicated relations with allies and international institutions (Chapter 7).

A handful of blocs and nations – NATO allies, the Soviet Union and affiliates, China, Japan, Iran, Saudi Arabia, and Israel – played prominent roles in postwar American science diplomacy. Iran, for example, became one of the first and largest recipients of scientific and technical aid and the United States encouraged the Iranian nuclear program until the Islamic revolution. Both Saudi Arabia and Iran received assistance to recycle American payments for high oil prices back into the American economy. The United States also provided unique scientific support for Israel – including access to top-secret research and shielding the Israeli nuclear program from international oversight – while President Reagan funneled hundreds of millions of dollars in research contracts to the country. But relations with China may be the most remarkable: The United States attempted to limit Chinese access to advanced science and technology until the 1970s; a decade later the People's Republic was America's largest bilateral science partner. New State Department and intelligence documents highlight the central role scientific exchanges played in normalizing Chinese–American relations. The relationship grew throughout Reagan's presidency and expanded after the Tiananmen protests and China's membership in the World Trade Organization. In 2011, amid worries about corporate espionage and satellite warfare, the House of Representatives held hearings on "Efforts to Transfer America's Leading-Edge Science to China."[13] One benefit of the current approach is the ability to see, for the first time, the arc of US–Chinese scientific relations from World War II to the present.

American diplomacy reflected the geopolitical realities unique to each era and field. The Soviet Union, for example, was America's primary scientific rival after World War II, leading to concerns of espionage and export controls (Chapters 1 and 2), while Japan became a commercial competitor in the 1970s (Chapter 6) and the People's Republic of China in the twenty-first century (Chapter 7). Although nuclear physics was one of the earliest and most influential examples of the relationship between science and state power (Chapters 1 and 2), the field evolved so far past application the United States lost interest in maintaining world-class

---

[13] Committee on Foreign Affairs, House of Representatives, 112th Cong., 1st sess.s, November 2, 2011, *Efforts to Transfer America's Leading-Edge Science to China* (Washington: Government Printing Office, 2011). On concern over corporate espionage and satellite warfare, see Caroline S. Wagner, Lutz Bornmann, and Loet Leydesdorff, "Recent Developments in China–U.S. Cooperation in Science," *Minerva* 53 (2015): 199–214.

research facilities after the Cold War (Chapter 6). Environmental sciences and biotechnology played only a minor role in American diplomacy until the 1970s (Chapter 4), while meteorology first became controversial with weather modification (Chapter 5) and global warming only became problematic in the late 1980s (Chapter 7).

A final theme is the relationship between scientific universalism and American diplomacy in three critical areas: national security, commercial research and environmental sciences. Aspects of scientific universalism – especially the free sharing of knowledge and the apolitical evaluation of research – conflict with American interests. In national security-related fields, for example, the United States accepts global oversight on its own terms. In commercial research, the United States supports a global system to enforce intellectual property rights. Environmental sciences provoke the most conflict, because they underlie global regulations threatening to American sovereignty and industry, whether over climate change or genetic engineering. International organizations, including the United Nations, scientific societies, and NGOs, have challenged American positions on environmental grounds. However, before discussing the postwar period, we must briefly revisit Captain Cook's time for three key histories: the relationship between scientific universalism, privilege and geopolitics; the American focus on applied research, commerce and agriculture; and the impact of World War I and the interwar period on science worldwide.

SCIENTIFIC UNIVERSALISM, PRIVILEGE, AND NATIONAL SERVICE

The idea science is universal has a long and distinguished pedigree. Rule number two of Isaac Newton's *Mathematical Principles of Natural Philosophy* (1687) stresses universality, requiring the same cause – gravity – control "the descent of stones in both Europe and in America."[14] Newton aspired to uncover universal knowledge, i.e. knowledge able to transcend politics, religion, and culture, and the evolution of shared practices, journals and societies across borders eventually created a global community working toward the same ideal. The experimental method, historian

---

[14] Isaac Newton, *Mathematical Principles of Natural Philosophy* (1687), Book III, Rule II.

John Henry points out, is "a means for generating and maintaining consensus in a self-ordering community without any arbitrary authority."[15] But intellectual autonomy complicates state relations: Benjamin Franklin's American Philosophical Society (1745), for example, predated the existence of an independent "America" and welcomed British members during the Revolution.[16] Indeed, Franklin's letter supporting Cook speaks to the potential tension between American nationalism and scientific internationalism since the country's founding.

Researchers asked for and often received special privileges. Scientists, specimens and equipment, including Harvard astronomer John Winthrop and his telescope, crossed battle lines without interruption during the French and Indian War and Thomas Jefferson suggested maintaining the tradition, writing: "These [scientific] societies are always at peace, however their nations may be at war ... their correspondence is never interrupted by any civilized nation."[17] International interaction was common, whether for visiting medical and laboratory facilities or collaborating on the periodic chart of elements and a universal system of weights and measures.[18] Throughout the nineteenth century, the British admiralty launched a series of protected scientific voyages, instructing its captains: "You are to refrain from any act of aggression towards a vessel or settlement of any nation with which we may be at war, as expeditions employed on behalf of discovery and science have always been considered by all civilized communities as acting under a general safeguard."[19] From 1872 to 1876, the HMS *Challenger* traveled 69,000 nautical miles, cataloging new species and measuring the oceans (depth, temperature,

---

[15] John Henry, *The Scientific Revolution and the Origins of Modern Science* (New York: Palgrave Macmillan, 2008), 53.

[16] Gilbert Chinard, "The American Philosophical Society and the World of Science (1768–1800)," *Proceedings of the American Philosophical Society* 87 (July 14, 1943): 1–11.

[17] On scientists in the French and Indian War, see Gavin de Beer, *The Sciences Were Never at War* (London: Thomas Nelson & Sons, 1960) and Badash, *Scientists*, 8. Thomas Jefferson's letter of 1809 is available at: www.let.rug.nl/usa/presidents/thomas-jefferson/letters-of-thomas-jefferson/jefl190.php.

[18] Maurice Crosland, "Relationships between the Royal Society and the *Academie des Sciences* in the Late Eighteenth Century," *Notes and Records of the Royal Society of London* 59 (January 22, 2005): 25–34.

[19] The Admiralty quoted in A. V. Hill, "The International Status and Obligations of Science," *Science* 38 (February 1934): 146–156, quote on 146.

current, etc.).[20] The Royal Society published frequent accounts, as the *Challenger* captured international attention and helped launch the field of oceanography (the United States paid homage a century later with its namesake space shuttle). Scientists soon inaugurated the First International Polar Year (1882–1883), when researchers from twenty nations studied the high northern latitudes, focusing on surface meteorology, geomagnetism, and the aurora borealis.[21] The community also policed itself: After the eminent German bacteriologist Robert Koch stated bovine tuberculosis presented no human health risk via milk consumption, American and British scientists tested Koch's claim, eventually proving a link to children's tubercular disease and helping educate the public about the benefits of pasteurization.[22]

Yet science served national interests as well. Early state support for research focused on practical concerns like astronomy to aid navigation and mineralogy to exploit natural resources. Scientists and engineers participated in imperial crusades from Napoleon's invasion of Egypt (1798) to the establishment of British and French rubber plantations a century later. Biology provided a system for the acquisition of raw materials for commerce. "Botanical knowledge concerning useful plants," historian Lucile Brockway argued, "was a counterpart of today's academic-industrial research ... as important in furthering the national welfare as our modern research laboratories today."[23] Advances in science and technology shaped European attitudes, seemingly testifying to European superiority as well as the backwardness of the non-European "other."[24] The imperial powers collaborated when it was to their benefit, such as coordinating weather observations at sea, and often cloaked colonial expeditions in a scientific purpose, lending an air of danger to nineteenth-century research: On Cook's return trip to the newly

---

[20] Anon., "The Exploring Voyage of the Challenger," *Science* 3 (May 9, 1884): 576–580.

[21] US House of Representatives, *Science, Technology and Diplomacy in the Age of Interdependence* (Washington: Government Printing Office, 1976), 46.

[22] Susan D. Jones, "Mapping a Zoonotic Disease: Anglo-American Efforts to Control Bovine Tuberculosis before World War I," *Osiris* 19 (2004): 133–148.

[23] Lucile Brockway, *Science and Colonial Expansion: The Role of the British Royal Botanic Garden* (New Haven: Yale University Press, 2002), 7. See also Zaheer Baber, *The Science of Empire: Scientific Knowledge, Civilization and Colonial Rule in India* (New York: State University of New York Press, 1996), 158.

[24] Michael Adas, *Machines as the Measure of Men: Science, Technology and the Ideologies of Western Dominance* (Ithaca, NY: Cornell University Press, 1989).

christened "Sandwich Islands," for example, locals killed him. European explorations in North America looked eerily familiar.

Thomas Jefferson, an outspoken advocate for science, worried about European intentions.[25] In 1781, the future president wrote to George Rogers Clark that England was "exploring the country from Mississippi to California. They pretend it is only to promote knowledge. I am afraid they have thoughts of colonizing that quarter."[26] Four years later, he feared a French expedition carrying "men of eminence in different branches of science" foreshadowed "colonizing on the Western coast of America."[27] By the time of his "secret" message to Congress (1803), President Jefferson described the relationship between geography, nationalism, and commerce rather cynically: "While other nations have encountered great expense to enlarge the boundaries of knowledge... our nation seems to owe the same object, as well as its own interest, to explore [the Missouri River], the only line of easy communication across the continent ... the interests of commerce place the principle object within the constitutional powers and care of congress, and that it should incidentally advance the geographical knowledge of our own continent can not but be an additional gratification."[28] The United States purchased Louisiana a few months later and the US Army Corps of Discovery (the "Lewis and Clark" expedition) set off the following May. Commerce, science and American diplomacy went hand-in-hand for the rest of the century.

APPLIED SCIENCES, COMMERCE, AND AMERICAN FOREIGN RELATIONS TO WORLD WAR I

Science was part of the American identity from the beginning. To support an experiment in self-rule, American revolutionaries argued the political world could be as rationally understood and constructed as the physical

[25] See Silvio A. Bedini, *Thomas Jefferson: Statesman of Science* (New York: Macmillan, 1990). For an excellent examination of the episode discussed, see Allen and Allen, "Acquiring 'Knowledge of Our Continent:' Geopolitics, Science and Jeffersonian Geography, 1783–1803," *Journal of American Studies* 40 (August 2006): 205–232.

[26] The British crown paid for a group of explorers in 1781. Jefferson quoted in Allen and Allen, "Acquiring 'Knowledge of Our Continent'," 214.

[27] Ibid., 216.

[28] Thomas Jefferson, *Jefferson's Secret Message to Congress Regarding the Lewis & Clark Expedition* (1803).

world.[29] "Natural laws" provided the intellectual firepower behind rebellion and individual rights. Appropriately, the US Constitution committed the federal government to promoting the "progress of science and the useful arts," a clear indication of the founders' faith in science and its ability to progress.[30] But the "useful arts" were more imperative. Arguably the first federal research took place in the 1790s at Springfield and Harper's Ferry armories under the US Ordinance Department (eventually contributing to interchangeable parts) while Hamiltonian tariffs for domestic industrial protection dominated early economic policy.[31] Foreigners noticed the American tendency; Alexis de Tocqueville's famed masterpiece asked "Why the Americans are More Concerned with Applications than with the Theory of Science."[32] Driven by practical needs such as national defense and the recovery of untapped resources, the government's "manifest destiny" tilted it toward the naturalist tradition in the nineteenth century – i.e. frontier sciences. Geology, biology, and paleontology surged as America introduced European naturalists to a plethora of new species and practices: Alexander von Humboldt spent ten days with Jefferson examining everything from mammoth teeth to slave conditions (he later apologized for being so "carried away by my devotion to the cause of the blacks").[33] As the west opened for settlement, civilian scientists gradually displaced the army as the primary agent of continental exploration.[34]

Scientific surveys provided a rationale for American exploration and trade abroad. The Wilkes expedition surveyed more than 250 Pacific islands from 1838 to 1842 and undertook astronomical and meteorological

---

[29] Henry Steele Commager, *The Empire of Reason: How Europe Imagined and America Realized the Enlightenment* (Garden City, NY: Anchor Press/Doubleday, 1977).

[30] U.S. Constitution, Article I, Section 8.

[31] Sylvia Kraemer, *Science and Technology Policy in the United States: Open Systems in Action* (New Brunswick, NJ: Rutgers University Press, 2006), 18–21.

[32] Alexis de Tocqueville, *Democracy in America* (New York: Harper & Row, 1966). *Volume II, Book I, Chapter I* is entitled "Why the Americans Are More Concerned with Applications Than with the Theory of Science."

[33] For an overview of the first trip see Leonard G. Wilson, "Geology on the Eve of Charles Lyell's First Visit to America, 1941," in *Proceedings of the American Philosophical Society* 124 (June 30, 1980): 168–202. For the second trip, see Charles Lyell, *A Second Visit to the United States of America* (New York: Harper and Brothers, 1849). Regarding Humboldt, see Helmut de Terra, "Alexander von Humboldt's Correspondence with Jefferson, Madison and Gallatin," *Proceedings of the American Philosophical Society* 103 (December 15, 1959): 783–806, quote on 790.

[34] William H. Goetzmann, *Exploration and Empire: The Explorer and the Scientist in the Winning of the American West* (New York: Alfred A. Knopf, 1966).

observations in the Sandwich Islands.[35] In 1847, the Navy sent the Lynch mission to the Dead Sea and the west coast of Africa for geological, zoological, and botanical data collection.[36] The Chilean expedition of 1849–1852, aided by the American Philosophical Society, conducted research in astronomy, earthquakes, and terrestrial magnetism; a later mission focused on solar parallax and the climate of Argentina.[37] The Herndon/Gibbon Naval Expedition of 1851, partially inspired by a Southern desire to ship slaves to restricted Brazilian lands, explored the Amazon and local resources. As trade expanded within the hemisphere and with Asia, the Navy sent specialists to locate Pacific coaling stations and possible sites for a Central American canal, leading to extensive information on Nicaraguan topography, rainfall, and temperature.[38]

Nationalism spurred domestic science. Against the backdrop of the Civil War, proponents pitched both the National Academy of Sciences (NAS, 1863) and the US Coastal Survey as examples of a single nation amid the dissension and bloodshed.[39] A. D. Bache, leader of the survey, wanted the geographic assessment to demonstrate America's great power status.[40] After the war, the desire to equal European advances inspired the lab-based research at the new Johns Hopkins University (1876) as well as the Marine Biological Laboratory at Woods Hole (1888).[41] Although

---

[35] Doris Esch Borthwick, "Outfitting the United States Exploring Expedition: Lieutenant Charles Wilkes' Expedition Assignment, August–November, 1836," *Proceedings of the American Philosophical Society* 109 (June 15, 1965): 159–172.

[36] Hunter Dupree, *Science in the Federal Government: A History of Policies and Activities* (Baltimore: Johns Hopkins University Press, 1986), 95–97.

[37] Geoffrey Sutton Smith, "The Navy before Darwinism: Science, Exploration, and Diplomacy in Antebellum America," *American Quarterly* 28 (Spring 1978): 41–55. See also, Merl Curti, *Prelude to Point Four: American Technical Missions Overseas, 1838–1938* (Madison, WI: University of Wisconsin Press, 1954), 12.

[38] Bailey Willes, "Work of the US Geological Survey," *Science* 10 (August 18, 1899): 203–213.

[39] Lillian B. Miller, *The Lazzaroni: Science and Scientists in Mid-Nineteenth Century America* (Washington, DC: Smithsonian Institution Press, 1972) and Dupree, *Science in the Federal Government*, 135–148.

[40] Hugh Richard Slotten, *Patronage, Practice and the Culture of American Science: Alexander Dallas Bache and the U.S. Coast Survey* (Cambridge: Cambridge University Press, 1994), 96.

[41] Anon., "The Marine Biological Laboratory," *Science* 13 (April 19, 1889): 303–304. See also Philip Pauly, "Summer Resort and Scientific Discipline: Woods Hole and the Structure of American Biology, 1882–1925," and Jane Maienschein, "Whitman at Chicago: Establishing a Chicago Style of Biology?" in Ron Rainger, et al., eds., *The American Development of Biology* (Philadelphia: University of Pennsylvania Press, 1990), 121–150 and 151–184 respectively.

many students still went abroad for advanced training, American physics education and research was world-class by the turn of the century.[42] As a sign of maturity, the federal government established the National Bureau of Standards (NBS, 1901) to set American standards to the levels of accuracy required by international trade. Physicist Edward B. Rosa, first head of the NBS, remembered, "to be obliged to ask the German imperial or other foreign laboratories to do our testing for us, because we lacked a well-equipped laboratory, was clearly a situation that ought to be corrected."[43] Yet scientific independence proved difficult to secure: when American industries established laboratories – whether GE, Du Pont, AT&T, or Eastman Kodak – many were staffed by foreigners or Americans with foreign degrees.[44]

Foreign competition spurred the American system for agricultural research. In 1887, the Hatch Act increased funding for state agricultural stations, creating an unprecedented network of federal-state research partnerships through the Department of Agriculture.[45] Faced with wary farmers, the department promised the stations would make "the fruit of foreign research and experience" available for their use.[46] Among the successes: improved irrigation in Colorado; increased wine-making in California; the introduction of more than seventy varieties of cane to sugar-making in Louisiana; and the discovery of fibrin in milk (critical to making cream). In 1892, the Nelson agricultural mission to Mexico, co-sponsored by the Department of Agriculture and the Department of State, ushered in a fourteen-year biological and botanical study, an early synthesis of agricultural science and diplomacy.

---

[42] Paul Forman, et al., *Physics circa 1900* (Historical Studies in the Physical Sciences) (Princeton, NJ: Princeton University Press, 1975), 5.

[43] Edward B. Rosa., "The National Bureau of Standards and Its Relation to Scientific and Technical Laboratories," *Science* 21 (February 3, 1905): 161–174, quote on 162.

[44] For example, many of the primary physicists at GE were either German or German-trained: Charles Steinmetz (who worked on alternating current), Willis Whitney and Irving Langmuir. See Daniel Kevles, *The Physicists: The History of a Scientific Community in Modern America* (Cambridge, MA: Harvard University Press, 1996), 99–101.

[45] Anon., "Organization of the Agricultural Experiment Stations," *Science* 14 (August 23, 1889): 132–133. The Stations' scientific roster was diverse: by 1902, they employed 146 chemists, 49 botanists, 48 entomologists, six zoologists, 14 meteorologists, seven biologists, five physicists, five geologists, 21 bacteriologists, etc. See A. C. True, "Work of the Agricultural Experiment Stations," *Science* 15 (June 13, 1902): 939–943, figures on 942.

[46] Farmer's Bulletin No. 1, produced by W. O. Atwater, director of the Office of Experimental Stations, Department of Agriculture, reprinted as "The What and Why of Agricultural Experiment Stations," *Science* 14 (August 9, 1889): 96–100, quote on 98.

Agriculture became America's first major scientific export. In his inaugural Message to Congress, President Roosevelt boasted the Department of Agriculture was "searching the world" to aid American farming and stock-growing interests.[47] The following year, the Director of the Office of Experiment Stations proudly noted Germany was moving "toward the extension of the so-called 'American system' of field experiments, conducted on a large scale and in a more practical way."[48] Backed by the American government, the system spread across the globe: By 1908, the Department's Bureau of Plant Industry oversaw experimental stations in the American possessions of Cuba and the Philippines and maintained representatives in eight countries; an American officer was Director of Agriculture in Mysore State, India.[49] Thus the United States acquired a reputation for applied science, especially experimental agriculture, before World War I and had incorporated the field into American diplomacy.

SCIENCE, WORLD WAR I, AND INTERWAR MODERNITY

World War I stimulated government-directed research, injecting nationalism and division into the global scientific community. All sides explored poisonous gasses before the conflict; duplication among allies in World War I was one reason for the improved coordination of scientific efforts in World War II.[50] Barely a month into hostilities, the German army destroyed the famous fifteenth-century library in Louvain (the library sheltered snipers). One of many incidents in the well-publicized "rape" of Belgium, the destruction sparked international outrage and a counter manifesto signed by ninety-three German intellectuals, including

---

[47] Theodore Roosevelt, "Annual Message to Congress," reprinted in *Science* 14 (December 13, 1901): 907–912.

[48] True, "Work of the Agricultural Experiment Stations," 943.

[49] The eight countries were Colombia, Brazil, Guiana, China, Siam, East Africa, South Africa, and Argentina. See the following: Curti, *Prelude to Point Four*, 36; F. S. Earle, "Botany at the Cuban Experimental Station," *Science* 20 (September 30, 1904): 444–445; Anon., "Experimental Stations in Hawaii and Porto Rico," *Science* 12 (October 5, 1900): 531–532.

[50] See, for example, L. F. Haber, *The Poisonous Cloud: Chemical Warfare in the First World War* (Oxford: Clarendon Press, 1986); Harry W. Paul, *The Sorcerer's Apprentice: The French Scientist's Image of German Science, 1814–1919* (Gainesville: University of Florida Press, 1972); and Jeffrey Allan Johnson, *The Kaiser's Chemists: Science and Modernization in Imperial Germany* (Chapel Hill: University of North Carolina Press, 1990).

numerous Nobel prize-winning scientists.[51] The willingness of German scientists to take sides shocked their brethren: the British removed Germans from scientific societies; across the Atlantic, George Hale, a prominent American astronomer, campaigned to remove Germany from all international scientific associations.[52] Others argued against exclusion of the central powers, stating it divided scientists into "hostile camps."[53] But internationalism is hard to sustain during wartime.

The European conflict shaped American institutions and diplomacy. Convinced of the need to keep abreast of advancements in aviation, Congress established the National Advisory Committee for Aeronautics (NACA) in 1915. Support for the field rose as airpower demonstrated its potential throughout the war. Meanwhile, scientists lobbied politicians for increased funding: Willis Whitney, a leading chemist at GE, pointed to German accomplishments in the fixation of atmospheric nitrogen, arguing research was a "national duty."[54] In 1916, the National Defense Act funded research into nitrate fixation for munitions and fertilizer, other legislation established the National Research Council (NRC) as the experimental arm of the NAS. George Hale, appointed first NRC director, went to Europe to aid the war effort; the detection of submarines became one of the first military concerns to result in scientific cooperation among allies.

The demands of war influenced domestic research. Army and Navy research occurred at more than forty universities, leading to the first campus security regulations.[55] In order to determine the location of enemy artillery, for example, General John "Blackjack" Pershing

---

[51] The letter is available at http://wwi.lib.byu.edu/index.php/Manifesto_of_the_Ninety-Three_German_Intellectuals. The scientists included Max Planck, Fritz Haber, Philip Lenard, and Wilhelm Roentgen. For Planck, the manifesto represented a lifelong dilemma between nationalism and his membership in an international community, see John L. Heilbron, *The Dilemmas of an Upright Man: Max Planck as Spokesman for German Science* (Berkeley: University of California Press, 1986). See also Matthew Stanley, "'An Expedition to Heal the Wounds of War:' The 1919 Eclipse and Eddington as Quaker Adventurer," *Isis* 94 (March 2003): 57–89.

[52] Lawrence Badash, "British and American Views of the German Menace in World War I," *Notes and Records of the Royal Society of London* 34 (July 1979): 91–121. Anon., "The Brussels Meeting of the International Research Council," *Science* 50 (September 5, 1919): 226. See also E. M. Washburn, "The New International Union of Pure and Applied Chemistry," *Science* 50 (October 3, 1919): 319–323.

[53] Kevles, "'Into Hostile Camps,' The Reorganization of International Science in World War I" *Isis* 62 (Spring 1971): 47–60, quote on 59.

[54] Willis R. Whitney, "Research as a National Duty," *Science* 43 (May 5, 1916): 629–637.

[55] Kevles, *The Physicists*, 138.

requested physicists work on sound-ranging.[56] Chemistry received special attention: the Chemical Warfare Service at American University employed more than 1,200 researchers while the Bureau of Mines farmed out work to over twenty university and industry laboratories.[57] The war also altered industrial research practice: during both world wars the federal government, under threat of patent seizure, directed firms to suspend litigation and form patent pools so all could benefit.[58]

Science symbolized modernity and progress after the war. Both Soviets and Americans assumed science and technology underlay a better standard of living; Bolshevik banners proclaimed: "Communism equals Soviet power plus electrification of the whole country." The May 4th independence movement in China adopted "Mr. Science" and "Mr. Democracy" as its namesakes; the revolutionary Sun Yat-Sen admonished his countrymen, "All these new inventions and weapons have come since the development of science ... so now if we want to learn from Europe we should learn what we ourselves lack – science – but not political philosophy."[59] Science was one of the few "Western" artifacts unblemished by imperialism and cooperation embodied the new international spirit: the League of Nations organized to standardize vaccines and serums, setting up epidemiological centers in Geneva, Singapore, and Melbourne.[60] Scientific exchanges became one of the few contacts between the United States and the nascent Soviet Union: during the Russian civil war, the NRC provided materials to Russian scientists, eventually delivering some twelve tons of material through the American Relief Administration.[61] Scientific specialization encouraged international collaboration, since differences in focus, personnel and funding shaped national strengths.

The American "model" relied on links between the government and commercial and academic research. The need for hydroelectric power, for example, stimulated "big science" partnerships between the state,

---

[56] Ibid., 126–131.
[57] Gilbert Whittemore, "World War I, Poison Gas Research, and the Ideals of American Chemists," *Social Studies of Science* 5 (May 1975): 135–163.
[58] Sylvia Kraemer, *Science & Technology Policy in the United States: Open Systems in Action* (New Brunswick, NJ: Rutgers University Press, 2006), 66–67.
[59] Sun Yat-Sen, "*San Min Chu I* (The Three Principles of the People)," in *Decolonization: Perspectives from Now and Then* (New York: Routledge, 2004), 21–28, quote on 28.
[60] US House of Representatives, *Science, Technology and Diplomacy*, 108.
[61] Vernon Kellogg, "Work of the National Research Council," *Science* 58 (November 2, 1923): 337–341.

industry, and academia in California, while MIT created a Division of Industrial Cooperation and Research in 1921 to make the school's "scientific and industrial experience" available for a fee.[62] The NRC helped found the Industrial Research Institute (IRI), a conglomerate of private laboratories, and legislation spurred corporate research: more than 2,000 companies employed a total laboratory staff of 70,000 by the end of the Depression.[63] Finally, the Tennessee Valley Authority, the largest project of the New Deal, symbolized American modernization. A massive program of hydroelectric dams and ecological education, the TVA became an international tourist destination, with official delegations from China and India journeying to Knoxville.[64] But there was competition: Mussolini promised to drain the Agro Pontino marshes and end malaria; Stalin had five-year plans, the steel city of Magnitagorsk and the Dneprostroi hydroelectric complex.

In many ways, a new framework for science and international relations emerged before World War II. World War I necessitated government-directed research and introduced scientific cooperation among allies. A dynamic arose that would become familiar during World War II and after: wartime needs required specialized research and research transformed foreign relations. When hostilities ended, the League of Nations championed internationalism, scientific exchanges preserved US–Russian relations and the global scientific community came together for a Second International Polar Year in 1932–1933. Yet international competition intensified as nations strove to demonstrate scientific and technical superiority. American legislation supported industrial research during the

[62] Stanford, Caltech, and Berkeley were among the schools involved, see Peter Galison, "The Many Faces of Big Science," in Peter Galison and Bruce Hevly, eds., *Big Science: The Growth of Large-Scale Research* (Stanford: Stanford University Press, 1992): 1–20, quotes on 3. See also S. S. Schweber, "Big Science in Context: Cornell & MIT," in Peter Galison and Bruce Hevly, eds., *Big Science*, 149–183, quote on 151.

[63] Regarding the IRI, see Eric S. Hintz., "Selling the Research Idea: The Genesis of the Industrial Research Institute, 1916–1945," *Research-Technology Management* 56 (November/December 2013): 46–50. Legislation included the Plant Patent Act (1930) to commercialize the results of scientific agriculture and the Revenue Act (1936) which made corporate research expenditures tax deductible. Kendall Birr, "Industrial Research Laboratories," in Nathan Reingold, ed., *The Sciences in the American Context: New Perspectives* (Washington: Smithsonian Institution, 1979): 193–208, figure on 199.

[64] David Ekbladh, *The Great American Mission: Modernization & the Construction of an American World Order* (Princeton: Princeton University Press, 2010), 72. See also Baber, *The Science of Empire*, 232.

Depression and the country advertised its developmental model abroad. Of course, the US position differed before World War II: The United States was not the preeminent scientific nation, science played only a limited role in American diplomacy, and the country was not the primary supporter of international scientific cooperation. Instead, other countries, such as Germany, were scientific powers.

The ascension of Nazism in one of the leading scientific nations upset international relations. Germany was the epicenter of atomic physics before a right-wing movement labeled the theory of relativity "Jewish" physics.[65] Scientists from a variety of fields were quick to defend the principle of universalism. A. V. Hill, a Nobel prize-winning British physiologist, opined, "Those who talk, for example, of Aryan and non-Aryan physics or of proletarian and capitalist genetics, as though they were different, simply make themselves ridiculous."[66] Sociologist Robert K. Merton emphasized, "The criteria of validity of claims to scientific knowledge are not matters of national taste and culture. Sooner or later, competing claims to validity are settled by universalistic criteria."[67] Once in power, Nazi racial policies purged Jewish scientists from the civil service, the Kaiser Wilhelm Society and German Academy of Science; more than 100 physicists, including Albert Einstein, fled.[68] The backlash was swift. International societies canceled conferences in Germany, while academics formed rescue committees to aid the fleeing scientists.[69] Abraham Flexner, director of the Princeton Institute for Advanced Studies,

[65] There were seven Nobel prizewinners in physics in Berlin. Weiner, "A New Site for the Seminar," 200–202. See also Paul Forman, "Scientific Internationalism and the Weimar Physicists: The Ideology and Its Manipulation in Germany after World War I," *Isis* 64 (1973): 151–180, esp. 157. See also Mark Walker, *Nazi Science: Myth, Truth, and the German Atomic Bomb Project* (New York: Plenum, 1995).

[66] A. V. Hill, "The International Status and Obligations of Science," *Science* 38 (February 1934): 146–156, quotes on 146.

[67] Merton quoted in Mark Walker, "Introduction," in Mark Walker, ed., *Science and Ideology: A Comparative History* (New York: Routledge, 2003), 3.

[68] Mark Walker, *German National Socialism and the Quest for Nuclear Power, 1939–1949* (Cambridge: Cambridge University Press, 1989). See also Mark Gordin, et al., "Ideologically Correct Science," in Mark Walker, ed., *Science and Ideology: A Comparative History* (New York: Routledge, 2003): 35–65, esp. 44.

[69] For an introduction to European reactions, see Greta Jones, *Science, Politics and the Cold War* (London: Routledge, 1988), 1–4. See also, Doel, "National States and International Science," 51–59.

joked that Hitler was his "best friend" because "he shakes the tree and I collect the apples [Einstein among them]."[70] Undeterred, the German state funded two large research centers to marry aeronautical education and domestic industry.[71] As Nazi forces occupied Czechoslovakia, German physicists learned how to split the atom.

[70] Flexner quoted in Laura Fermi, *Illustrious Immigrants: The Intellectual Migration from Europe, 1930–1941* (Chicago: University of Chicago Press, 1968), 78. The "illustrious immigrants" included, among others, Edward Teller, Eugene Wigner, Leo Szilard, Jon Neumann, Michael Polanyi, Enrico Fermi, George Gamow, Hans Bethe, Victor Weiskopf, and Niels Bohr.

[71] The two centers were the Army Experimental Institute in Peenemünde Ost and Aeronautics Research Center in Völkenrode. See Burghard Ciesla and Helmuth Trischler, "Legitimation through Use: Rocket and Aeronautic Research in the Third Reich and the U.S.A.," in Mark Walker, ed., *Science and* Ideology, 156–185.

I

The Battle of the Laboratories

As Allied soldiers poured into France, members of a secret American scientific mission – codenamed ALSOS – followed the front; University of Michigan physicist Samuel Goudsmit's team was among the first into Paris.[1] Charged with determining the progress of the Nazi nuclear program, Goudsmit and crew quickly searched Joliot-Curie's lab before continuing on to Germany. Time was of the essence; the war still raged and the Soviet scientific exploitation mission had a head start.[2] Simultaneously, US Department of Commerce teams scoured German industrial plants (especially the chemical giant I. G. Farben), acquiring thousands of patents.[3] Ultimately, ALSOS teams captured ten leading German physicists and investigated dozens of targets (including bioweapons facilities, university labs, and the medical experimentation sites of four concentration camps).[4] When news of Hiroshima's destruction filtered down to Hitler's "Uranium club" interned in England, the captured German physicists debated

---

[1] Samuel A. Goudsmit, *ALSOS* (Woodbury, New York: American Institute of Physics Press, 1996), 34–37. Goudsmit's text was originally published in 1946.

[2] Soviet scientific exploitation teams, including about three dozen Soviet scientists, ultimately captured numerous scientists and engineers and secured 240–340 tons of Uranium oxide from German and Czechoslovakia. David Holloway, *Stalin and the Bomb: The Soviet Union and Atomic Energy, 1939–46* (New Haven: Yale University Press, 1994), 109–111.

[3] The Commerce department teams were known as FIAT teams, for Field Intelligence Agency – Technical. Clarence G. Lasby, *Project Paperclip: German Scientists and the Cold War* (New York: Atheneum, 1971), 26.

[4] John D. Hart, "The ALSOS Mission, 1943–45: A Secret U.S. Scientific Intelligence Unit," *International Journal of Intelligence and Counterintelligence* 18 (2005): 508–537, lists of visited sites on 510.

whether to go "East" (to the Soviet Union) or "West" (to the United States).[5] Such questions, posed while American and Soviet soldiers confiscated uranium ores at gunpoint, shaped international relations for decades.

World War II undermined the previous world order and required the construction of a new one. Science, so prominent in the conclusion to the conflict, would be central. Nations, especially the United States and Soviet Union, recognized the necessity of funding and controlling research, while the colonized world and nascent international community – represented by the United Nations and affiliated institutions – considered unfettered access to science and technology critical to global peace and development. As the Cold War tied science to national security and international blocs, the United States used its dominant scientific position to advance various foreign policy goals, whether the promotion of allied solidarity and anti-communism or the acquisition of foreign materials and markets.

The first section of this chapter, "Nation Bound," considers how World War II embedded science in American diplomacy. Collaborative research provided an important tie with allies and the United States offered medical and technical aid to enlist Latin American nations. Atomic research, however, tested the limits of cooperation and the United States began restricting allied access before the end of the conflict. At home, wartime needs required the United States expand its role in funding domestic research, often using a contract system to create partnerships between the government, academia, and industry. As the United States undertook a broader role in overseeing science at home, questions arose about the control of science abroad.

The world was "At Loose Ends" after the war. Anchored by the UN, a suite of international institutions arose championing global governance, whether through agreement on scientific and technical norms or international cooperation in healthcare, agriculture, and education. But the Cold War intruded: disputes between the United States and Soviet Union occurred frequently, whether over nuclear energy, the role of new global institutions, or cures for cancer. Maintaining control of atomic energy, for example, was a central concern for American policy-makers; as such, the United States refused to cede oversight to the UN Atomic Energy Commission and pressured the French government to remove Left-leaning

---

[5] The British interned the German physicists at a country estate – Farm Hall – where each room was carefully "bugged" for the benefit of American and British intelligence services. See Jeremy Bernstein, *Hitler's Uranium Club: The Secret Recordings at Farm Hall* (Woodbury, NY: AIP, 1995).

scientists from positions of influence. Additionally, the American preference for national or private research limited support for UN laboratories while anti-communists in Congress required the World Health Organization renounce "social medicine" before receiving US support.

The final section, "Rebound," considers American influence on science overseas. The occupations of Germany and Japan provided unique access to research, researchers, and intellectual properties, although the Soviet Union disputed American patent policies and American employment of Nazi scientists briefly caused embarrassment. Multiple government authorities took on new responsibilities: the Central Intelligence Agency began tracking European research; the State Department limited access to Marshall plan funds for Left-leaning researchers; and the Commerce Department required allies adopt export controls to deny advanced technology to the Soviet bloc. Even medicine became caught in the Cold War: the United States offered radioisotopes as diplomatic favors and clashed with allies over the patenting of penicillin.

NATION BOUND

We Soviet scientists are employing all our knowledge and all our endeavor to secure the early defeat of Hitler's hordes. The scientists of the world must devote all their energies and all their knowledge to the fight against the most horrible tyranny history has ever known, against Hitlerism.

– Appeal of 20 world-famous Soviet scientists (1941)[6]

Bob Wilson came in and said that he had been funded to do a job that was secret . . . a bomb . . . I said I didn't want to do it. . . .So I went back to work on my thesis – for about three minutes. Then I began to pace the floor and think about this thing. The Germans had Hitler and the possibility of developing an atomic bomb was obvious, and the possibility that they would develop it before we did was very much of a fright. So I decided to go to the meeting at three o'clock. By four o'clock I already had a desk in a room and was trying to calculate whether this particular method was limited by the total amount of current that you get in an ion beam, and so on. I won't go into the details. But I had a desk, and I had paper, and I was working as hard as I could and as fast as I could . . .

– American physicist Richard Feynman, reminiscing about 1943 (1980)[7]

---

6 The appeal is quoted in Lasby, *Project Paperclip*, 11.
7 Richard Feynman, "Los Alamos from Below," in Lawrence Badash, et al., *Reminiscences of Los Alamos, 1943–45* (Boston: D. Reidel Publishing Company, 1980), 105–132, quotes on 105.

Scientific expertise trumped politics and nationality in wartime. Throughout World War II, the major participants enlisted the sciences for the battlefront and the home front. Scientific exchanges helped cement allied relations, although tensions were quickly evident over the war's greatest prize – the atomic bomb – which emerged as the dominant symbol of the fusion of science and national power. The bomb also represented the most significant science and engineering challenge to date, requiring nations rethink their relations with industry and academia.

National politics influenced the mobilization of science. The fascist powers – Nazi Germany and Imperial Japan – simply took over domestic scientific establishments, while researchers in the Soviet Union already labored under state control.[8] In all three cases, state authorities scrutinized scientists' political loyalties, but did not allow ideological incorrectness to hamper research.[9] Nor were approaches equally effective: the Japanese, for example, passed a National Mobilization Law in 1938 and protected militarily significant industries, but never effectively coordinated civilian and military research, leading to few wartime achievements. The Germans, on the other hand, effectively mobilized industrial resources to avoid duplication and replaced scarce metals, oils, and rubber with synthetics. And though Nazi Germany and Imperial Japan worked on the same uranium separation method, they never sought to cooperate; the Allies would not make the same mistake.

Science and American Foreign Relations before World War II

The American scientific enterprise in World War II began cautiously and at the urging of individual scientists. As Nazi Germany defied international pressures in the late 1930s, numerous scientists feared the United States was unprepared for a possible war. Vannevar Bush, head of the

[8] See, for example, Alan D. Beyerchen, *Scientists under Hitler: Politics and the Physics Community in the Third Reich* (New Haven, CT: Yale University Press, 1977); Mark Walker, *Nazi Science* and Mark Walker, ed., *Science and Ideology*; Holloway, *Stalin and the Bomb*; and Walter E. Grunden, et al., "Laying the Foundation for Wartime Research: A Comparative Overview of Science Mobilization in National Socialist Germany, Japan and the Soviet Union," *Osiris* 20 (2005): 79–106. See also Walter E. Grunden, et al., "Wartime Nuclear Weapons Research in Germany and Japan," *Osiris* 20 (2005): 107–130.

[9] See Mark Gordin, et al., "Ideologically Correct Science," in Walker, ed., *Science and Ideology*, 35–65, esp. 44–47.

Carnegie Institute and former head of NACA, worried about advances in German aerospace technology. He began offering informal advice to President Roosevelt early in 1939 and convinced the president to create a National Defense Research Committee (NDRC) by June. Neither Bush nor FDR knew what the committee would do, but Bush believed scientists should play key roles; he later observed, "there were those who protested that the action of setting up NDRC was an end run; a grab by which a small company of scientists and engineers, acting outside established channels, got hold of the authority and money for the program of developing new weapons. That, in fact, is exactly what it was."[10] At the same time, two refugee Hungarian physicists, Leo Szilard and Eugene Wigner, persuaded Albert Einstein to write the president a letter on the potential explosiveness of German discoveries with uranium, ominously observing, "Germany has actually stopped the sale of uranium from the Czechoslovakian mines which she has taken over."[11] Unimpressed, the letter resulted in a few thousand dollars from the Navy to investigate nuclear propulsion for submarines; it took a foreign intervention to stimulate the Americans.

The British, fearing German airpower, hoped for American collaboration on critical wartime research (especially radar). In May 1940, A. V. Hill, the renowned physiologist and Member of Parliament for Cambridge University, met with American political leaders. Months later, as the Luftwaffe pounded the English coasts, the British Technical and Scientific Mission – often called the "Tizard mission" after its leader Sir Henry Tizard – arrived in the United States.[12] The Crown authorized the mission to offer secret technology in exchange for American assistance. In particular, Sir Henry brought a cavity magnetron, a British invention capable of generating high-energy microwaves. Impressed, Bush and others suggested collaboration while the NDRC contracted with MIT to develop a radar system. In October, the "Rad Lab" was born.

---

[10] Vannevar Bush, *Pieces of the Action* (New York: Morrow, 1970), 31–32.
[11] Albert Einstein, "Letter to Franklin D. Roosevelt" (August 2, 1939) reprinted in Philip L. Cantelon, et al., eds., *The American Atom: A Documentary History of Nuclear Policies from the Discovery of Fission to the Present*, 2nd edn. (Philadelphia: University of Pennsylvania Press, 1991), 10.
[12] James Phelps, *The Tizard Mission: The Top-Secret Operation That Changed the Course of World War II* (London: Westholme Publishing, 2010). See also David Zimmerman, *Top-Secret Mission: The Tizard Mission and the Scientific War* (Montreal: McGill-Queen's University Press, 1996) and James Phinney Baxter 3rd., *Scientists against Time* (Cambridge, MA: Massachusetts Institute of Technology Press, 1946), 119–120.

The "Radiation Laboratory" – or "Rad Lab" – expanded the previous state-directed fusion of industry and academia. Bush, the head of the NDRC and a former Dean at MIT, argued for the Institute's centrality; industrial partners included Bell Labs, GE, Westinghouse, RCA, and Sperry – all organized and funded by the federal government. The Rad Lab at MIT foreshadowed the later Manhattan Project, with a staff of thousands and more than a billion dollars in industrial contracts.[13] New labs sprouted as needed: whether the Confidential Instruments Development Laboratory (for gyroscopic gunsights); the Gas Turbine Laboratory; the Naval Supersonic Laboratory; or the Aeroelastic and Structures Laboratory. Academic partnerships also sprouted as needed; the Rad Lab alone housed representatives from over 69 different academic institutions.[14] As the MIT "model" spread to other universities and industries within the United States, the NDRC solidified partnerships overseas.

Scientific cooperation was part of American foreign policy before Pearl Harbor. The United States, Great Britain, and Canada began coordinating research on radar and underwater detection in September of 1940; the NDRC and US Naval Research Laboratory set up London offices in early 1941. Working with Harvard and observatories in Peru, Australia, and Alaska, the allies created the Interservice Radio Propagation Laboratory to provide communications throughout the global conflict.[15] By the time the shared research expanded to include nuclear physics, allied programs had already made progress with the proximity fuse, rockets, and radar. Thus scientific cooperation helped cement the "special relationship" across the Atlantic.

Science diplomacy also fortified alliances to the south. As part of his "Good Neighbor" policies, FDR created the Interdepartmental Committee on Scientific and Cultural Cooperation to coordinate US technical assistance to Latin America. Its focus was practical and political – agriculture, geology, public health, and meteorology to cement relations south of the border.[16] A month before Pearl Harbor, Nelson Rockefeller, whose business interests included Venezuelan oil, approached the War department

---

[13] Stuart R. Leslie, *The Cold War and American Science: The Military-Industrial-Academic Complex at MIT and Stanford* (New York: Columbia University Press, 1993), 21.

[14] Baxter, *Scientists against Time*, 22.

[15] See "Radio Science and World War II" in Needell, *Science, Cold War and the American State*, 67–96.

[16] Miller, "An Effective Instrument of Peace," 136.

about a public works program in South America. An alliance was easy to arrange; plans were already in motion to construct transportation and communication facilities throughout the hemisphere (Brazil's proximity to Africa, for example, was ideal for trans-Atlantic airfields).[17] The government quickly established the Institute for Inter-American Affairs (IIAA) to provide sanitation and health care to militarily relevant areas.

The IIAA served the American military first and the host population as necessary. With Rockefeller as chairman and overseen by the US military, the IIAA operated through *servicios* – offices jointly funded and staffed, and often run through local Ministries of Education, Health, or Agriculture.[18] In 1942 the United States signed a bilateral agreement with Brazil, a key strategic country and rubber producer; eventually, the program provided limited sanitation, anti-malarials (atabrine and DDT), and water treatment. Within the year eighteen countries signed similar bilateral agreements. The IIAA established a model for incorporating science and technology into diplomacy with the developing world: Paraguay, Peru, and Haiti, for example, received medical facilities, including penicillin to wipe out a yaws epidemic (the Haitian program convinced Rockefeller aid campaigns created regional good will toward the United States).[19] By the time the IIAA merged with other aid programs in the late 1950s, the Institute had spent less than $50 million and left behind over 1,500 individual projects, including hospitals, health centers, water, and sewage systems, etc.[20] In the words of historian Claude Erb, the IIAA gave "top priority to cleaning up areas that might become bases for US troops or sources of strategic materials. From the outset this aid program demonstrated a mixture of idealism and self-interest."[21] A similar mixture of idealism and self-interest shaped US assistance for decades.

American entrance into the conflict led to a reexamination of the relationship between science and the state among all participants. The Japanese, Germans, and Soviets, for example, increased funding for military research and upgraded their nuclear programs (the Soviets also

---

[17] Claude C. Erb, "Prelude to Point Four: The Institute of Inter-American Affairs," *Diplomatic History* 9 (July 1985): 249–269, esp. 24–25.
[18] For an early history of the IIAA, see "Learning by Doing – In Latin America," in Jonathan B. Bingham, *Shirt-Sleeve Diplomacy: Point 4 in Action* (New York: The John Day Company, 1954), 16–23.
[19] Bingham, *Shirt-Sleeve Diplomacy*, 20.
[20] Erb, "Prelude to Point Four," 268–269.
[21] Ibid., 256.

increased attempts to penetrate British and American programs).[22] In the United States, the federal government reconstituted the NDRC as the Office of Scientific Research and Development (OSRD) and initiated the Manhattan Project.

OSRD and the Manhattan Project

OSRD expanded the federal role in directing and funding American science. Rather than centralize research, OSRD adopted a contract model, which carried responsibility but not "subservience" or "paternalism" and maintained a free-market approach.[23] Operating with only a small staff, the Office eventually offered over two thousand contracts to hundreds of industries, universities and nonprofit research institutes.[24] For example, Columbia and Harvard worked on underwater explosives; Princeton specialized in ballistics; the University of Rochester in optics.[25] OSRD also created a committee to fund medical research. Among its successes: the isolation of gamma globulin and creation of new blood substitutes, the mass production of penicillin and the synthesis of atabrine after the Japanese seized the primary sources of quinine (an anti-malarial).[26] When new technologies required on-site technicians, OSRD established an Office of Field Services to provide scientists and engineers to battlefield commanders.[27] Not everything operated smoothly. The state/industry partnerships could be sources of tension; DuPont, GM, and others chafed at government patent regulations and clauses.[28] The Industrial Research

---

[22] Walter E. Grunden, et al., "Laying the Foundation for Wartime Research;" see also Holloway, *Stalin and the Bomb*, 82–83.

[23] This phrasing comes from Larry Owens, "The Counterproductive Management of Science in the Second World War: Vannevar Bush and the Office of Scientific Research and Development," *The Business History Review* 68 (Winter 1994): 515–576, esp. 525. OSRD followed precedents set by foundations and industry in the use of contracts, see Daniel Lee Kleinman, *Politics on the Endless Frontier: Postwar Research Policy in the United States* (Durham, NC: Duke University Press, 1995), 51.

[24] See Owens, "The Counterproductive Management of Science in the Second World War," which lists more than 2,300 contracts with 321 industrial and 142 academic and non-profit institutions in an appendix. See also Baxter, *Scientists against Time*.

[25] Ibid., 23.

[26] Paul Starr, *The Social Transformation of American Medicine* (New York: Basic Books, 1982), esp. 338–345.

[27] Roy Macleod, "Combat Science: OSRD's Postscript in the Pacific," *Boston Studies in the Philosophy of Science* 207 (2000): 13–26.

[28] Owens, "The Counterproductive Management of Science in the Second World War," 515–576.

Institute, representing fifty-five industrial concerns, met to discuss "patent problems" with the American government, specifying cooperative research in aeronautics.[29] The top-secret atomic program only increased the tension over government control.

The American bomb program, like the Rad Lab, had roots across the pond. In 1939, German physicists hypothesized uranium – the heaviest naturally occurring element – could be split and the fission could cause a nuclear chain reaction releasing enormous energy. However, no one knew how much uranium would be needed, which isotope of uranium (U^{235} or U^{238}) would react or if the reaction could be initiated and controlled. At first, only the British were seriously interested. In 1940, two refugee physicists in Britain worked out the theoretical basis for the bomb; a year later, the British "Maud" committee suggested the amount of uranium required for a chain reaction. Hoping to stimulate the Americans, the British sent a copy of the Maud committee report to their allies in the summer of 1941.[30] After much internal wrangling, and only one day before Pearl Harbor, FDR authorized a significant expansion of nuclear research, providing funds for nuclear research at Berkeley, Princeton, Columbia, and the University of Chicago.[31] Six months later, the "Manhattan Project" was born.

The Manhattan Project was a network of scattered university, industry, and government facilities.[32] Early work on initiating and controlling a chain reaction was done at the University of Chicago, while production of fissionable material – U^{235} and plutonium – took place in Oak Ridge, Tennessee, and Hanford, Washington, respectively. Final engineering and construction of the bombs took place in Los Alamos, New Mexico. The project involved dozens of additional sites and facilities, as it consumed significant national and industrial resources. Just separating the extremely rare U^{235} from the more common U^{238} was an enormous undertaking; Chrysler built gaseous diffusion machines while GE and others worked on electromagnetic separation. As the United States harnessed its industrial

---

[29] See Anon., "The Industrial Research Institute," *Science* 97 (June 11, 1943): 527–28 and Anon., "The Industrial Research Institute," *Science* 99 (June 9, 1944): 464–65.

[30] An excellent account of this period is Richard Rhodes, *The Making of the Atomic Bomb* (New York: Simon & Schuster, 1986), 367–373. The Maud Committee report is reprinted in Cantelon, *The American Atom*, 16–20.

[31] Badash, *Scientists and the Development of Nuclear Weapons*, 33.

[32] Two good inroads into the literature are Rhodes, *The Making of the Atomic Bomb* and Stephane Groueff, *Manhattan Project: The Untold Story of the Making of the Atomic Bomb* (Boston: Little, Brown & Co., 1967).

base to produce an atomic bomb, Allied concern over German and Japanese nuclear programs grew.

Industrial and scientific centers were common targets in World War II. After the war, acting Secretary of War Patterson argued science influenced the decision to strike Germany first: "But the reason that seemed to me as compelling as any was the danger of the German scientists, the risk that they would come up with new weapons of devastating destructiveness. There was no time to lose in eliminating German science from the war. There was no comparable peril from Japanese science."[33] Such talk was not mere hindsight; the Allies directed numerous special operations against German scientific targets, including "heavy water" sites (necessary for atomic research) and the V-2 rocket program at Peenemünde. In one of the more daring examples, allied intelligence recruited the science editor for German publishing giant Springer Verlag to relay information on both the Nazi bomb and V-2 rocket programs (including co-ordinates of the site at Peenemünde), often using aspiring Norwegian physicists as spies.[34] Multiple bombing runs, including Operation Crossbow targeting German scientists, forced the relocation of the German bomb program.

Although free from aerial bombardment, the American/British/Canadian atomic program was not free from inter-allied tensions. For the first two years of the war, the British were hesitant to collaborate while the United States worried about British–Soviet scientific cooperation.[35] As early as December 1942, Vannevar Bush wrote there would be "no unduly serious hindrance ... if all further interchange between the United States and Britain in this matter were to cease."[36] Nonetheless, the following year the Quebec Agreement on "tube alloys" (the codename for the bomb projects) reasoned success would "be more speedily achieved if all available British and American brains and resources are pooled."[37] The agreement required the bombs never be

[33] Patterson quoted in Baxter, *Scientists against Time*, 26.

[34] Arnold Kramish, *The Griffin: The Greatest Untold Espionage Story of World War II* (Boston: Houghton Mifflin, 1986).

[35] E. H. Beardsley, "Secrets between Friends: Applied Science Exchange between the Western Allies and the Soviet Union during World War II," *Social Studies of Science* 7 (1977): 447–473, esp. 449.

[36] Bush quoted in Spencer R. Weart, *Scientists in Power* (Cambridge, MA: Harvard University Press, 1979), 198.

[37] See "The Quebec Agreement (August 19, 1943)," in Cantelon, *The American Atom*, 31–33, quote on 32.

used against one another, nor could either pass information to a third party, like the Soviet Union, without receiving consent. The agreement also denied French patent claims on reactor design and plutonium, excluded France from atomic collaboration and limited British access to engineering and technical specifications.[38] Canada emerged as the geographic lynchpin in the trio and British physicists like Allan Nunn May joined the Montreal Laboratory/Chalk River Project, while American intelligence worried about Canadian visits to the Oak Ridge and Chicago facilities.[39] But the American program quickly eclipsed British capabilities and imbalanced the relationship, even though Roosevelt and Churchill signed a memoire at Hyde Park to continue cooperation after the war.[40] Ominously, the memoire required "enquiries" into the political activities of Danish physicist Niels Bohr.[41]

The Manhattan Project and other scientific endeavors dramatically raised the profile of elite scientists, especially physicists, providing limited political influence at home and making them targets of foreign espionage. US Army Signals Intelligence at Arlington Hall, VA (better known as the "Venona" project) determined Soviet spies penetrated the Manhattan Project from the start; Soviet agents asked physicist Robert Oppenheimer, the leading scientist at Los Alamos, to provide information in 1943.[42] Other scientists waded into political waters: Vannevar Bush suggested to FDR they "start Russia down the path of scientific and political collaboration with us," while Niels Bohr personally asked the president to internationalize the US atomic program in August of 1944 (Roosevelt preferred to keep it secret).[43] Complicating matters, a few Manhattan Project scientists quietly petitioned against the use of bomb without prior warning or demonstration, but there was no scientific consensus regarding use. The official Manhattan Project Science Panel,

---

[38] Other concerns included isotope separation methods, see Weart, *Scientists in Power*, 179.

[39] Donald H. Avery, *The Science of War: Canadian Scientists and Allied Military Technology during the Second World War* (Toronto: University of Toronto Press, 1998), esp. 67 and 213.

[40] See "Anglo-American Declaration of Trust (June 13, 1944)," in Cantelon, *The American Atom*, 34.

[41] See "Roosevelt–Churchill Hyde Park Aide-Memoire (September 19, 1944)," in ibid., 36.

[42] Nigel West, *Mortal Crimes: The Greatest Theft in History – The Soviet Penetration of the Manhattan Project* (New York: Enigma books, 2004). See also Philip M. Stern and Harold P. Green, *The Oppenheimer Case* (New York: Harper & Row, 1969), 43–45.

[43] Bush quoted in Joseph Manzione, "'Amusing and Amazing and Practical:' The Legacy of Scientific Internationalism in American Foreign Policy, 1945–1963," *Diplomatic History* 24 (Winter 2000): 21–54, quote on 34.

for example, including Oppenheimer, supported immediate use without warning or demonstration.[44]

The Atomic Bomb and Questions After

"Sixteen hours ago an American airplane dropped one bomb on Hiroshima, an important Japanese Army base. That bomb had more power than 20,000 tons of T.N.T."[45] With those words, President Truman introduced the atomic bomb to the American people. In his press release August 6, 1945, Truman provided a remarkable window into American foreign policy at the dawn of the atomic age. First, he marveled at the new role of science in warfare: "It is an atomic bomb. It is a harnessing of the basic power of the universe ... The battle of the laboratories held fateful risks for us as well as the battles of the air, land and sea, and we have now won the battle of the laboratories as we have won the other battles."[46] After crediting British contributions, the president spoke with pride about the scale and economics of the American Manhattan Project:

We now have two great plants and many lesser works devoted to the production of atomic power. Employment during peak construction numbered 125,000 ... We have spent two billion dollars on the greatest scientific gamble in history – and won. ... Both science and industry worked under the direction of the United States Army ... What has been done is the greatest achievement of organized science in history.[47]

The release then veered from triumph to threat, demonstrating the depth of wartime hostility and the headiness of new power: "We are now prepared to obliterate more rapidly and completely every productive enterprise the Japanese have above ground in any city. We shall destroy their docks, their factories, and their communications. Let there be no mistake; we shall completely destroy Japan's power to make war."[48]

Having presented the American position, Truman pondered the future in the statement's final paragraphs. In language that would become familiar over the years, he promoted the benefits of nuclear power as well

---

[44] Other scientists in the panel included A. H. Compton, E. O. Lawrence, and E. Fermi. The panel recommendations are found in Cantelon, *The American Atom*, 47–48.

[45] White House Press Release, Statement by the President of the United States (August 6, 1945), *FRUS: Diplomatic Papers of the Conference of Berlin (the Potsdam Conference)*, *1948*: 1376–1378, quote on 1376.

[46] Ibid., 1376.   [47] Ibid., 1377.   [48] Ibid.

as the need for additional research: "The fact that we can release atomic energy ushers in a new era in man's understanding of nature's forces. Atomic energy may in the future supplement the power that now comes from coal, oil, and falling water, but at present it cannot be produced on a basis to compete with them commercially. Before that comes there must be a long period of intensive research."[49] However, for the first time, not all research would be made public; nuclear physics would remain classified. It is worth quoting the final lines at length:

It has never been the habit of the scientists of this country or the policy of this Government to withhold from the world scientific knowledge. Normally, therefore, everything about the work with atomic energy would be made public. But under present circumstances it is not intended to divulge the technical processes of production or all the military applications, pending further examination of possible methods of protecting us and the rest of the world from the danger of sudden destruction.[50]

Like fruit from the mythical Tree of Knowledge, nuclear power empowered and burdened its recipient; and like Pandora's box, once "opened," threatened the world. Atomic energy needed to be controlled. But how?

World War II entangled science in American policies at home and abroad. Mobilization and the Manhattan Project forcefully raised expectations of research among politicians and the public while providing the United States a unique scientific advantage: the war left the American government in possession of a remarkable set of cutting-edge scientific institutions and an equally remarkable network of industrial and academic contacts.[51] At the same time, the war increased the importance of science in American foreign relations. Even before American entrance, for example, the British used access to science and technology – whether a cavity magnetron or specifics on nuclear fission – to entice American participation and raise alarm. The United States collaborated on necessary research and shared scientific information with allies during the conflict, but reserved the right to limit access, leading the Soviets to covertly penetrate US research facilities. Away from battle, American

---

[49] Ibid., 1378.   [50] Ibid.
[51] See Vannevar Bush, *Science: The Endless Frontier*, available at www.nsf.gov. See also Pascal Zachary, *Endless Frontier: Vannevar Bush, Engineer of the American Century* (Cambridge, MA: Massachusetts Institute of Technology Press, 1999) and Daniel Sarewitz, *Frontiers of Illusion: Science, Technology and the Politics of Progress* (Philadelphia: Temple University Press, 1996).

assistance programs provided scientific and technical aid to secure allies, supplies, and bases. Throughout World War II, the United States successfully harnessed science for American interests; however, once the war was over, new questions arose about the role of science in American diplomacy and international relations.

AT LOOSE ENDS

... they did not see it. They did not see it until the atomic bombs burst in their fumbling hands. ... Destruction was becoming so facile that any little body of malcontents could use it; it was revolutionising the problems of police and internal rule. ... For a time the war spirit defeated every effort to rally the forces of preservation and construction. ...Why should anyone give in while he can still destroy his enemies? Surrender? While there is still a chance of blowing them to dust?[52]

– H. G. Wells, *The World Set Free* (1914)

Since wars begin in the minds of men, it is in the minds of men that the defences of peace must be constructed ... The purpose of the Organization is to contribute to peace and security by promoting collaboration among the nations through education, science and culture ...

– *UNESCO Constitution* (1945)

Modernity unraveled between the two quotes above. Wells wrote his remarkably prescient warning on the eve of World War I; an international committee composed the UNESCO preamble in the aftermath of the World War II. The wars fatally punctured European imperialism and a Eurocentric model of modernity; by 1945, European economies lay in ruins and the colonial world agitated for independence. In H. G. Wells's *The World Set Free*, originally published in 1914, the destructiveness of the then-hypothetical atomic bombs compelled world government, as individual nations could not be trusted with such awesome power. A later work, *The Open Conspiracy* (1928), took the argument a step further.[53] Wells proposed that governments be run by scientists, creating intellectual aristocracies operating in transparent, rational fashion. An

[52] H.G. Wells, *The World Set Free: A Story of Mankind* (London: MacMillan & Co., 1914), reprinted in Cantelon, et al., eds., *The American Atom*, 5–7.
[53] See Gregg Herken, *Cardinal Choices: Presidential Science Advising from the Atomic Bomb to SDI* (Stanford: Stanford University Press, 2000), 4.

opponent of nationalism, he then imagined these aristocracies networked into a "Scientific World Commonweal" for the good of mankind.[54]

Unfortunately, the United States and Soviet Union struggled with scientific relations during World War II and after. One of the first technical exchanges – a mission on synthetic rubber in 1942 – undermined future cooperation; after a delay, the Soviets failed to provide technical specifications and demanded reciprocal patents from DuPont and Standard Oil (who refused).[55] The American business community turned against cooperation and American diplomats, worried about losing US trade secrets, pressured the British to cancel a planned Tizard mission to the Soviet Union.[56] Only limited US/Soviet medical exchanges remained (see below). After the defeat of Germany, Stalin invited allied scientific delegations to the 220th anniversary of the Soviet (originally "Russian") Academy of Sciences, but the United States discouraged attendance: The British government, on American advice, refused to allow eight British scientists to attend.[57] Nonetheless, more than one hundred delegates from eighteen countries heard foreign minister Molotov offer a toast to "close collaboration between Soviet and world science."[58] Within months of the destruction of Hiroshima and Nagasaki, however, the Soviet line changed: Stalin spoke about the need to "catch up and overtake" his former allies.[59] The Cold War soon clouded Wells's hopeful idealism, as the United States and Soviet Union disagreed over the control of atomic energy, access to research, and mandate of new global institutions.

Postwar Institutions and Internationalism

The United States championed internationalism during the war to maintain unity. In 1942, twenty-six nations united in the war against fascism. The following year, the Tehran Declaration, signed by FDR, Churchill,

---

[54] H. G. Wells, *The Open Conspiracy: Blue Prints for a World Revolution* (London: Albatross, 2017). The phrase is taken from the eighth chapter, entitled, "Broad Characteristics of a Scientific World Commonweal."

[55] E. H. Beardsley, "Secrets between Friends: Applied Science Exchange between the Western Allies and the Soviet Union during World War II," *Social Studies of Science* 7 (1977): 447–473, esp. 452.

[56] Ibid., 450–458.   [57] On the "missing eight," see Avery, *The Science of War*, 223.

[58] Molotov quoted in Pollock, *Stalin and the Soviet Science Wars*, 5.

[59] Stalin quoted in N. L. Krementsov, "In the Shadow of the Bomb: U.S.–Soviet Biomedical Relations in the Early Cold War, 1944–1948," *Journal of Cold War Studies* 9 (Fall 2007): 41–67, quote on 47.

and Stalin, imagined a new global order, as the three agreed to welcome all "into a world family of Democratic Nations." At the same time, the Food and Agriculture Organization (FAO) sprouted at an international conference held in Hot Springs, VA; the organization arose, in part, because of the belief the improvement of global agriculture and nutrition was a shared responsibility.[60] Formally established in Canada in 1945, the FAO was the first in a new suite of international organizations led by the United Nations (UN).

International institutions, often with a scientific or technological focus, promised to bind the wounds of World War II and prevent future conflict. Only five years after the war, the UN system included the United Nations Educational, Scientific, and Cultural Organization (UNESCO, 1945), the International Civil Aviation Organization (ICAO, 1947), the World Health Organization (WHO, 1948), and the World Meteorological Organization (WMO, 1950). Such organizations represented a new order based on rationalism, internationalism, and democratic dialogue, creating a global public sphere in which science and technology would be cooperatively managed.[61] But scientific internationalism could conflict with American interests.

UNESCO embodied the renewed scientific internationalism. Its original director general, Julian Huxley, a contributor to the modern evolutionary synthesis, dreamed of making science globally accessible through translation and abstraction of the most critical biological and medical texts.[62] The first director of the natural sciences section, Joseph Needham, worked in China during the war and proposed a global scientific corps to disseminate information from the "bright" zone to the less-developed nations.[63] He also suggested scientists deserve the equivalent of a diplomatic passport, allowing them to travel freely.[64] Although the passports never came to pass, scientific exchanges helped sell the idea of UNESCO to Secretary of State Dean Acheson, provided certain topics were off-limits.

[60] John H. Perkins, *Geopolitics and the Green Revolution: Wheat, Genes and the Cold War* (Oxford: Oxford University Press, 1997), 143.

[61] I've borrowed this idea of a public sphere from many, esp. Jessica Wang, "Scientists and the Problem of the Public in Cold War America, 1945–50," *Osiris* 17 (2002): 323–347, esp. 363–338.

[62] Eileen R. Cunningham, "UNESCO Initiates Cooperation in the Abstracting of Biological and Medical Sciences," *Science* 106 (December 19, 1947): 609–611.

[63] Thomas Mougey, "Needham at the Crossroads: History, Politics and International Science in Wartime China (1942–1946)," *BJHS* 50 (March 2017): 83–109.

[64] Dominques and Petitjean, "International Science," 45.

UNESCO often accommodated American preferences. Regarding the proposed exchanges, Acheson explained, "First, the role of scientists, scientific collaboration, and interchange of scientific knowledge should be emphasized and made more explicit. This suggestion obviously has no bearing on scientific research for military purposes."[65] It seemed clear, but distinguishing between civilian, military, and commercial research proved difficult and disagreement could imperil cooperation. Beginning in 1946, for example, France spearheaded an effort to establish a series of UN research laboratories, suggesting UNESCO, rather than scientific societies or philanthropies, serve as the basis for global scientific cooperation.[66] The proposal won support from less-advanced nations and the Soviet Union, but the United States and Great Britain disapproved, worrying about limited resources and competition with industrial and national research. France ultimately received the International Union for the Conservation of Nature (1948), while UNESCO settled for field offices across the globe. In response, the Soviet Union boycotted UNESCO, arguing the organization was an instrument of US domination. Reconciling American and Soviet positions on atomic energy proved more difficult.

Atomic Energy and Espionage

Atomic energy unleashed political battles over science at home and abroad. In 1945, the War Department supported legislation – the May/Johnson bill – which permitted active-duty military to serve in positions of authority in a new atomic agency, a move widely interpreted as foreshadowing military control of atomic energy. In a rare and controversial response, physicists at Oak Ridge, Chicago, and other Manhattan Project facilities established the Federation of American Scientists (FAS) to protest; often rallying behind the cry of "civilian control of atomic energy!" The FAS championed the competing bills, which limited military participation in the new agency. To influence debate, a "Scientists' Movement" led by the FAS published an anthology titled *One World or None* (1946),

---

[65] The Acting Secretary of State (Acheson) to the Ambassador in the United Kingdom (Winant), *FRUS: Foreign Relations, 1945* 1: 1514–1515, quote on 1515.

[66] Kirtley F. Mather, "United Nations Research Laboratories," *Science* 111 (April 21, 1950): 397–399. See also Heloisa Maria Bertol Domingues and Patrick Petitjean, "International Science, Brazil and Diplomacy in UNESCO (1946–50)," *Science, Technology & Society* 9 (2004): 29–50.

arguing, as had Wells forty years earlier, the destructiveness of the bombs necessitated international civilian control.[67]

The entrance of scientists supercharged the politics of atomic research. Both sides dug in as the bills moved through Congress. The *Bulletin of the Atomic Scientists* updated its "Doomsday" clock counting down to Armageddon. Others questioned American militarism: the *One World* anthology highlighted the "Made in the USA" stamp on bombs, while Norbert Weiner, a famous MIT mathematician, refused to provide an unpublished paper to the air force, later stating, "the scientist ends by putting unlimited power in the hands of the people he is least inclined to trust with its use."[68] Such outspokenness split the scientific community, creating a moral schizophrenia over participation in politics (Vannevar Bush and Robert Oppenheimer, for example, supported the May/Johnson bill).[69] It also alarmed the authorities. The FBI, concerned about un-American activities, began monitoring FAS meetings and members around the country, even infiltrating regional chapters.[70] As the politicians debated, media reports confirmed American fears.

Revelations of espionage undermined US/Soviet relations, strengthening the case for surveillance of scientists and American control of atomic energy. Only weeks after the Japanese surrender, the Canadian prime minister informed President Truman of the defection of Igor Goushenko, a Soviet cipher clerk claiming to have proof of a Soviet spy ring.[71] Goushenko confirmed FBI Director Hoover's fears of infiltration and his revelations attracted international attention when reported in early 1946. The prosecution of British physicist Allan Nunn May, who provided uranium samples to Soviet intelligence, began the hunt for "atomic spies," a group eventually including May, Klaus Fuchs, Bruno

---

[67] Dexter Masters and Katharine Way, eds., *One World or None: A Report to the Public on the Full Meaning of the Atomic Bomb* (New York: The New Press, 2007). Originally published in 1946. See also Alice Kimball Smith, *A Peril and a Hope: The Scientists' Movement in America, 1945–47* (Chicago: University of Chicago Press, 1965). Another good source, less positive in its assessment, is "The Atomic Scientists: From Bomb-Makers to Political Sages," in Paul Boyer, *By the Bomb's Early Light: American Thought and Culture at the Dawn of the Atomic Age* (Chapel Hill: University of North Carolina Press, 1994), 47–106.

[68] Weiner quoted in Manzione, "Amusing and Amazing and Practical," 21–54.

[69] David Dickson, *New Politics of Science* (Chicago: University of Chicago Press, 1988), 308.

[70] Jessica Wang, *American Science in an Age of Anxiety: Scientists, Anticommunism & the Cold War* (Chapel Hill: University of North Carolina Press, 1999), 59–63.

[71] Paul Dufour, "'Eggheads' and Espionage: The Gouzenko Affair in Canada," *Journal of Canadian Studies* 16 (Fall 1981): 188–198.

Pontecorvo, and Ethel and Julius Rosenberg (see below).[72] Goushenko became a frequent witness against communism in Congress, while the "Goushenko affair" and subsequent convictions provided evidence of Soviet duplicity, influencing American domestic policy and diplomacy.[73]

The passage of the Atomic Energy Act in August of 1946, only a few months after the revelation of Soviet espionage, heralded a new age of federal research. At its creation, the Atomic Energy Commission (AEC) inherited 37 installations in 19 states and Canada worth more than $2 billion, and quickly began purchasing uranium ore from South Africa, the Congo, Canada, Australia, and domestic miners (locking down national production almost immediately).[74] The Act prohibited active-duty military from serving on the commission and maintained the federal government's monopoly on fissionable material, while waiving restrictions on non-military research and according private companies limited patent rights. Additionally, federally employed researchers, including those who had worked on the Manhattan Project, were required to take loyalty oaths (another first) and lost influence in setting research priorities and funding; they were to be on tap, not on top. Having established domestic control, the Act terminated collaboration with allies (Great Britain and Canada) in nuclear physics. Establishing international control of atomic energy proved more complicated.

The Soviet Union challenged American atomic diplomacy at the UN. As the sparring over civilian control grew heated in the United States, a similar debate intensified globally. At issue was the mandate of the United Nations Atomic Energy Commission (UNAEC), a body composed of the security council plus Canada. The United States proposed continuing the American monopoly over nuclear materials and research until an effective global monitoring system was in place (the "Baruch Plan"); the Soviet Union countered with the destruction of all atomic bombs, sharing of relevant material and immediate international oversight. The FAS, already immersed in politics at home, organized to support oversight, including mailing pamphlets to foreign scientists to secure a voice for the scientific community in global affairs.[75] As the politicians negotiated, the

---

[72] May worked in the Canadan atomic energy project. See Amy Knight, *How the Cold War Began: The Goushenko Affair and the Hunt for Soviet Spies* (Toronto: McClelland & Stewart, 2005).

[73] See "Igor Goushenko," in Reg Whitaker and Steve Hewitt, *Canada and the Cold War* (Toronto: James Lorimer & Company Ltd., 2003), 13–17.

[74] Badash, *Scientists and the Development of Nuclear Weapons*, 74–75.

[75] Manzione, "Amusing," 33.

US Army conducted the appropriately named Operation Crossroads, the Bikini atoll atomic tests. Many felt the timing was suspect; journalist Raymond Gram Swing wrote at the time, "So we strive to save civilization, and we learn how to wreck it, all on the same weekend," adding, "this war game [the Bikini tests] will appear to others, as not being defensive in its ultimate meaning. It is a notice served on the world that we have the power and intend to be heeded."[76]

The UN's Atomic Energy Commission amounted to little. Opportunities to cooperate on atomic energy deteriorated alongside US/Soviet relations. Within a few years, the commission ceased operations. Thirty years later, a US Congressional inquiry acknowledged a "basic contradiction in the US negotiating position" which "partly justified" Soviet suspicions.[77] Additionally, because the accompanying scientists supported more foreign oversight, the Commission also noted conflicts "between the relationships and respective roles of the American scientists and diplomats."[78] While the intensity surrounding nuclear physics was unique, friction between national and international interests arose frequently in the UN system; Cold War politics attended the birth of the World Health Organization as well.

US–Soviet Conflict over the WHO

Organized international health efforts date to the early twentieth century. Beginning with the International Office of Public Hygiene in Paris (1907), numerous succeeding bodies and philanthropies, from the League of Nations Health Organization (1920) to the Rockefeller International Health Board (1916), Ford Foundation (1936), and the Wellcome Trust (1936), focused on the issue. Healthcare was a pressing global need after the war, whether to treat the wounded, stop the spread of disease or establish safety standards. US Surgeon General Thomas Parran explained how global health was the responsibility of the new network of international institutions: "the Food and Agriculture Organization [works] on worldwide nutrition and rural health; the International Labor Organization on industrial hygiene and social insurance; the Provisional

---

[76] Swing quoted in Boyer, *By the Bomb's Early* Light, quotes on 82–83.
[77] See "Case One – The Baruch Plan: U.S. Diplomacy Enters the Nuclear Age," in Committee on International Relations, US House of Representatives, *Science, Technology and Diplomacy in the Age of Interdependence*, 20–28, quotes on 24.
[78] Ibid., 24.

International Civil Aviation Organization, on checking the spread of disease through rapid transport; the Trusteeship Council, on the health of dependent peoples; the Narcotics Commission, in the field of habit-forming drugs."[79] The only remaining piece was a World Health Organization to co-ordinate the disparate efforts.

The United States hesitated to ratify the WHO constitution. The earliest health advocates highlighted the links between health, disease, and social conditions like sanitation, nutrition, and education. By 1945, many international health experts were natural champions of the poor.[80] But "social medicine" sounded like socialism to staunch anti-communists. President Truman publicly supported national healthcare (1945) and the WHO (1947), stating the United States could "play an important role in improving the health conditions of more backward states" because it was "far advanced in medical science."[81] However, anti-communist Republicans won the 1946 elections and free-market ideology framed the WHO debate: The American Medical Association stated, "the socio-economic aspects of medical practice should be a concern of the individual country," while Hugh S. Cummings, a former Surgeon General and the director of American public health programs in Latin America, denounced attempts to form the WHO as the work of "advanced internationalists" and communists.[82] Although Parran quickly repudiated the comments, Brock Chisholm, the WHO's first director and a Canadian, complained "ignorant and prejudiced" Republicans blocked American participation.[83]

Republican concerns over "social medicine" delayed American entrance to the WHO for two years (1946–1948). To garner Republican support, Surgeon General Parran offered a commercial motivation for participation: "Higher standards of living and greater buying power in countries with a high degree of health will have a direct bearing on fundamentals of our economy," adding WHO membership would

---

[79] Thomas W. Parran, "The World Health Organization" *Yale Journal of Biology and Medicine* 19 (March 1947): 401–410, quote on 406.

[80] Kelley Lee, *The World Health Organization (WHO)*, (New York: Routledge, 2009), 22–23.

[81] Truman quoted in John Farley, *Brock Chisholm, the World Health Organization & the Cold War* (Vancouver: University of British Columbia Press, 2008), 48.

[82] AMA statement quoted in ibid., 49. Cummings was director of the Pan-American Sanitary Board Board, which operated as the foreign arm of the US Public Health Service, see Norman Howard-Jones, *International Public Health between the Two World Wars – the Organizational Problems* (Geneva: World Health Organization, 1976).

[83] Chisholm quoted in Farley, *Brock Chisholm*, 49.

increase "demand for our skills and materials."[84] He closed by observing that the war's "most efficacious medical weapons" – penicillin (England), the sulfonamides (Germany), and DDT (Switzerland) – came from foreign laboratories.[85] Even as Americans debated, a cholera outbreak in Egypt (1947) demonstrated the organization's potential: twenty million doses were sent from the United States, USSR, and India. In an emergency, medical need still trumped politics.

Eventually opposition was overcome; most Americans supported ratification and WHO success required American funding and participation.[86] The United States received a unique opt-out clause and the organization tabled attempts to include socioeconomic justice, focusing on disease prevention and treatment. A later US Congressional commission concluded the "two-year delay of the United States in ratifying the constitution of the WHO seems to have been motivated by both medical-political and national-political considerations, including fears that the WHO would become involved in such questions as health insurance and socialized medicine in an international context rather than the problems of preventive medicine on an international scale."[87] Not that the organization was free from politics: When the United States blocked the admission of China, North Vietnam, and North Korea, the rest of the Soviet-bloc countries withdrew; the Polish letter of withdrawal argued the WHO had demonstrated "its complete surrender to the imperialistic States, and in particular the United States."[88] As US/Soviet relations deteriorated, divisions within the medical and scientific communities became official.

The KR Affair and a Scientific Iron Curtain

The collapse of US/Soviet relations upset cooperation in medical research and altered Soviet diplomacy. Limited medical exchanges had resumed during the war: Researchers established the American–Soviet Medical

---

[84] Parran, "The World Health Organization," 407 and 410.   [85] Ibid., 409.

[86] See "The World Health Organization and Its Budget," in Francis Hoole, *Politics and Budgeting in the World Health Organization* (Bloomington: Indiana University Press, 1976), 33–48.

[87] House of Representatives, *Science, Technology and American Diplomacy*, 111.

[88] The Polish Letter of Withdrawal is quoted in Anon., "Editorial Comments: Poland Withdraws from World Health Organization," *Canadian Medical Association Journal* 64 (January 1951): 75. See also "American and Soviet Participation" and "The Politics of Exclusion," in Javed Siddiqi, *World Health and World Politics: The World Health Organization and the UN System* (Columbus: University of South Carolina Press, 1995), 101–116. See also Lee, *The World Health Organization*, 22–23.

Society in 1943 and the Allies sent a medical mission including members from OSRD, the National Cancer Institute, and Howard Florey, the "father" of penicillin.[89] After the war, the American public learned of a mysterious Soviet cancer cure called "KR," leading to desperate requests for samples and official offers to finance further research and production.[90] The Soviets cooperated throughout 1946, proudly offering American delegations tours of medical facilities and an experimental sample. However, as political relations worsened, the medical exchanges came under suspicion: Stalin arrested those involved with sharing KR and blamed American intelligence for stealing Soviet secrets. A series of dramatic policy changes ensued: Soviet scientists resigned from foreign scientific societies; the Politburo classified "experimentation in all spheres of science" as state secrets, the revelation of which was punishable by a lengthy jail sentence; and Soviet scientific journals ceased publishing in foreign languages.[91] Finally, the Soviet Union released a film on the affair, in multiple translations, as part of a global propaganda campaign.

Stalin launched a "Peace Offensive" in 1948 based on access to science and technology.[92] The Marshal's offensive supported international oversight of atomic energy and the prohibition of nuclear weapons (the Soviet bomb program remained unsuccessful). At the first "Congress of Intellectuals for Peace and Science" in Wroclaw, Poland (1948), the participants proposed replacing UNESCO with a more inclusive and scientifically administered organization. Julian Huxley, UNESCO's director, attended but left early; the American chargé in Poland promptly informed the State Department, "Huxley conducted himself well throughout."[93] Nor was Huxley alone: Although the Congress welcomed numerous anti-Western regions – republican Spain, liberated China and Vietnam, Soviet-zone Germany – the United States had a larger delegation than the Soviet Union.[94] Over the next year, Stalin's offensive marched through more than ten countries, decamping in New York City for a "Scientific and

---

[89] Nikolai Krementsov, *The Cure: A Story of Cancer and Politics from the Annals of the Cold War* (Chicago: University of Chicago Press, 2002), 62–63.

[90] "KR" was an abbreviation for the discoverers' last names: Nina Kliueva and Grigorii Roskin. Ibid., 81.

[91] Ibid., 130–131.

[92] Draft Paper Prepared in the Department of State, The Soviet "Peace" Offensive (December 9, 1949), *FRUS: Foreign Relations, 1949* 5: 839–849.

[93] The Chargé in Poland (Crocker) to the Secretary of State, *FRUS: Foreign Relations, 1948* 4: 912–915, quote on 914.

[94] Ibid.

Cultural Conference for World Peace" in March 1949. To rebut the Soviet initiative, the State department suggested "stressing the failure of the Soviet Union to utilize the UN organization UNESCO."[95] Additionally, American diplomats criticized the Soviet rejection of genetics and state control over research, positions with significant support in the global scientific community (see Chapter 2).

Disagreements over science, technology, and medicine plagued US/Soviet relations after World War II. National interests limited international cooperation or oversight, whether in atomic energy or global laboratories and healthcare. In each case the United States pressured the UN to accommodate American preferences, leading the Soviet Union to protest American influence, boycott international organizations and argue for a return to wartime cooperation. But revelations of atomic espionage eroded trust and the "iron curtain" closed on the global scientific community. Although economic and political conflicts instigated the Cold War, scientific conflicts were influential as well: before the Berlin airlift, for example, there were disputes over atomic espionage, the Baruch plan, UNESCO, the WHO, and the KR affair. US/Soviet tensions often required global institutions and the scientific community swear allegiance, an act antithetical to their natural internationalism. In the Soviet Union, Stalin engaged in a series of "wars" to control Soviet research, while federal funds directed a majority of American research and anti-communism precluded hosting neutral events; few major international scientific conferences were held in the United States between 1947 and 1954 (the McCarthy period is covered in greater detail in Chapter 2).[96] As early as 1946, prominent scientists such as Edward Condon openly criticized government policies; in 1948, when the United States detained French physicist Irene Joliot-Curie – daughter of the famed researcher and wife of Frederic Joliot-Curie – her response was acidic: "Americans look with much more favor on fascism than on communism ... Americans think fascism has more respect for money."[97] In one sense, she was correct: Scientific proficiency was a valuable currency in the Cold War and the United States welcomed scientists from fascist Germany and Japan.

---

[95] Draft Paper Prepared in the Department of State, The Soviet "Peace" Offensive (December 9, 1949), *FRUS: Foreign Relations, 1949* 5: 839–849.

[96] Regarding Stalin, see Pollock, *Stalin and the Soviet Science Wars.* Regarding US scientific conferences, see Manzione, "Amusing," 40.

[97] Ibid., 242.

REBOUND

We would like you to know and to appreciate that you are here in the interest of Science and we hope that you will work with us in close harmony to further develop and expand upon various subjects of interest to ourselves as well as to you. ... Do not think of yourselves as under restrictions while here ... You are not POWs but are more in the category of employees of the U.S.A. and will therefore be accorded corresponding courtesies and privileges as far as it is possible in keeping with your own security.

from the "Instructions to German Scientists" at Wright Field,
Ohio (1945)[98]

We face a situation in which a future world war, employing atom bombs, rockets guided by radio, and many other marvels of man's perverted ingenuity, will achieve a destructiveness thousands of times greater than ever achieved before ... the scientific life of the country must not be subordinated to, or derive its chief support from, the military ... we are confronted in America with a situation in which scientists are being held very strictly under military domination, to severe detriment of our scientific development and the development of wholesome international relations. What is going on?

American physicist Edward Condon (1946)[99]

After World War I, many scientists feared their international community would break into "hostile camps." After World War II, such a split seemed eminently possible. As the Cold War heated up, elite scientists found themselves under increased scrutiny. Edward Condon, a physicist of impeccable national credentials (member of the National Academy of Sciences, contributor to MIT's Rad Lab and the Manhattan Project, and head of the postwar National Bureau of Standards) eventually left government service and private industry because of persecution by the House Un-American Activities Committee (HUAC).[100] Although HUAC found no evidence of guilt, Condon's membership in the American–Soviet Science Society and support for internationalism convicted him in the eyes of

---

[98] The "Instructions to German Scientists" are reprinted in Clarence Lasby, *Project Paperclip* (New York: Atheneum Press, 1971), 11. Written by anonymous American officers, the instructions were left on the beds of six German aeronautical specialists at Wright Field to welcome and prepare them for an extended stay in the United States.

[99] E. U. Condon, "Science and Our Future," *Science* 103 (April 5, 1946): 415–417.

[100] Note also that HUAC was able to secure $200,000 in additional funding, in part, because of its pursuit of Condon. See "HUAC and the Condon Case, 1947–1948," in Wang, *American Scientists in an Age of Anxiety*, 130–147.

American anti-communists. Nor were suspicions limited to American scientists. The international scientific community struggled to remain independent of superpower geopolitics, as both the United States and the USSR profited from postwar occupations and bound science within "blocs" to secure leverage and influence.

Foreign research became a focus of American diplomacy. German patents and scientists were a prize throughout the war and the American occupation of Germany continued the focus on acquiring useful material, regardless of origin. In Japan, the United States reorganized the national science council, acquired agricultural samples, and initiated a medical study on the effects of the atomic bombs. Aid money provided influence with European allies, but the United States clashed with Great Britain over patenting penicillin and competed with the country in the distribution of radioisotopes, two early examples of the intersection of commercial research and geopolitics.

Science in Occupied Germany

The United States seized German patents for a second time in World War II. Many of the thousands of patents confiscated during World War I related to radio, chemistry, or medicine (such as salvorsan); twenty-five years later, FDR's Executive order 9095 seized enemy patents and inventions, establishing the Office of the Alien Property Custodian (APC) to make them available to American industry. The following year, John Roe, assistant general counsel for the APC, suggested:

One of the most potent weapons we possess in our war against the Axis powers is our control over patent and patent applications which have been seized by the Alien Property Custodian, covering the most advanced developments in chemistry, electricity, meteorology, and other sciences and controlling processes that can be of tremendous value to America, both in the prosecution of the war and in the economic development to follow.[101]

Impounded properties included patents for tungsten carbide, high-strength rayon, optical heat sensors and a process for producing wood alcohol, all made available at field offices across the country.[102]

[101] Michael White, "Patents for Victory," *Science and Technology Libraries* 22 (2001): 5–22, quote on 12–13.
[102] Otto C. Sommerich, "Treatment by United States of World War I and II Enemy-Owned Patents and Copyrights," *The American Journal of Comparative Law* 4 (Autumn 1955): 587–600.

Exploitation of German science and scientists grew as the European war drew to a close. While ALSOS teams scoured the countryside, Project Overcast brought more than one hundred German scientists to the United States to help with the war against Japan (none contributed before V-J day). Later renamed Project Paperclip, the program eventually naturalized hundreds of German scientists in a variety of fields (a more coercive Soviet program repatriated a few thousand). Physicists and aerospace engineers are the most famous; the most infamous are biologists renowned for their work at Edgewood Arsenal, Maryland, an Army center for chemical warfare. Although only a minor part of the larger containment of science, this project struck a nerve with the American public.

Project Paperclip illustrated how scientific potential trumped wartime allegiances. Following the war, Secretary of Commerce Henry Wallace argued for intellectual reparations, asking why American dollars should be spent to reproduce work already done by Germans. Nonetheless, the War Department, wary of publicity, kept Paperclip secret. It became news, however, when the press discovered the project bypassed immigration laws. The Department of Commerce highlighted the program's potential economic benefit, but public reaction was negative, especially from politically active scientists. Albert Einstein denounced the program and the FAS met with the President; physicist Hans Bethe, a German refugee, asked if America wanted "science at any price?"[103] In response, the project retreated from view and opposition collapsed. Commerce quietly fielded employment applications from more than 170 companies and dozens of universities.[104] The Air Force enlisted more than 260 "paperclips" and argued they saved taxpayers more than $2 billion. German contributions to the space program will be addressed in Chapter 2, here it is sufficient to note Redstone Arsenal near Huntsville,

---

[103] Bethe quoted in Lasby, *Project Paperclip*, 202. See 185–202 for information on the promotion campaign. See also Linda Hunt, *Secret Agenda: The United States Government, Nazi Scientists, and Project Paperclip, 1945 to 1990* (New York: St. Martin's Press, 1991) and Annie Jacobsen, *Operation Paperclip: The Secret Intelligence Program That Brought Nazi Scientists to America* (New York: Back Bay Books, 2014).

[104] The universities included Yale, Michigan State, Wisconsin, Oregon State, and Minnesota, among others. The companies and institutes included Boeing, AVCO, Lockheed, Dow Chemical, Raytheon, Convair, General Electric, Bell, Northrup, RAND, Westinghouse, RCA, and many others. See ibid., 26.

Alabama housed so many Germans it was known as "Peenemünde South."[105] Unfortunately for Americans, the real Peenemünde was in Soviet hands.

German science worried US policy-makers. In 1945, Secretary of the Treasury Henry Morgenthau published *Germany is Our Problem,* which proposed destroying all research and industrial facilities to reduce the country to an agricultural state.[106] Stalin and many in the US military agreed. In May 1945, a directive from the Joint Chiefs (JCS 1067/8) closed all German laboratories and research institutions except those needed for public health. The Army occupation government confiscated research of interest and detained all personnel, ultimately maintaining files on the activities and movements of around five thousand German scientists.[107] The following year, Allied Control Council Law Number 25 banned military research and required scientists apply for permission to resume experiments. But such controls swiftly eroded. Although the program excluded former Nazis, determining previous loyalties proved difficult and economic rebuilding took precedence.[108] Reparations tested US/Soviet relations as well.

Though Truman and Stalin agreed to share German science and industry at Potsdam, both nations scavenged their respective occupation zones: the Soviets dismantled and carried off entire laboratories, while the US Navy shipped home more than 9,000 tons of material (including a large wind tunnel).[109] In one famous example, Operation LUSTY (LUftwaffe Secret TechnologY) occupied the Hermann Göering Aeronautical Research Institute, which pioneered swept-back wings.[110] A LUSTY member, George Schairer from Boeing, wrote to the company headquarters afterwards; swept wings soon graced the B-47.[111] Meanwhile, FIAT

[105] Richard Ciesla and Helmuth Trischler, "Legitimation through Use: Rocket and Aeronautic Research in the Third Reich and the U.S.A.," in Mark Walker, ed., *Science and Ideology: A Comparative History* (New York: Routledge, 2003):156–85, quote on 171. See also Monique Laney, *German Rocketeers in the Heart of Dixie: Making Sense of the Nazi Past during the Civil Rights Era* (New Have: Yale University Press, 2015).

[106] Henry Morgenthau, *Germany Is Our Problem: A Plan for Germany* (New York: Harper & Bros, 1945).

[107] Lasby, *Project Paperclip,* 76.

[108] John Krige, *American Hegemony and the Postwar Reconstruction of Science in Europe* (Cambridge, MA: Massachusetts Institute of Technology Press, 2008), 45–56.

[109] John Gimbel, *Science, Technology and Reparations: Exploitation and Plunder in Postwar Germany* (Stanford: Stanford University Press, 1990).

[110] The National Museum of the Air Force dedicated a page to Operation LUSTY, available at www.nationalmuseum.af.mil/.

[111] Lasby, *Project Paperclip,* 29.

teams paid German scientists to provide summaries of their work and translate thousands of patents from research industries like chemical giant I. G. Farben.[112] German intellectual properties were in high demand. When a Commerce Department official commented "on the immense value to the United States of these patents," the Soviet representative V. I. Molotov argued the United States was holding back.[113] Secretary of State Marshall responded, "We have used United States scientists to obtain information on German science, including patents, all of which information is being published in pamphlets and made available to the world. As a matter of fact, Amtorg, the Soviet Purchasing Agency in the United States, has been so far the biggest single purchaser of these pamphlets."[114] Masterfully, Marshall demanded equal access to patents found in the Soviet occupation zone, at which point Molotov demurred.[115] American diplomats were thankful there were no Soviet exploitation teams in Japan.

Science in Occupied Japan

As in West Germany, Americans designed scientific policies in Japan to complement a larger strategy of democratization, stabilization, and economic reconstruction.[116] Determined to secure Japan as an ally and partner in a global free-market system, the United States engaged in an unprecedented reconstruction effort, including providing loans, writing a constitution, reorganizing industry, and guaranteeing a favorable trading partner through currency manipulation. All laboratories closed except those deemed "necessary to the purposes of the occupation."[117] American personnel reorganized the Science Council of Japan while an NAS committee, led by Nobel prize winning physicist I.I. Rabi, toured the country.

[112] Richard H. Beyler and Morris F. Low, "Science Policy in Post-1945 West Germany and Japan: Between Ideology and Economics," in Mark Walker, ed., *Science and Ideology*, 97–123, esp. 100.

[113] Hearing of Representative of the Inter-Allied Reparations Agency, *FRUS: Foreign Relations, 1947* 2: 258–261, quote on 261.

[114] Ibid.   [115] Ibid.

[116] Richard H. Beyler and Morris F. Low, "Science Policy in Post-1945 West Germany and Japan: Between Ideology and Economics," in Mark Walker, ed., *Science and Ideology*, 97–123

[117] The US. directive is reprinted in Hideo Yoshikawa and Joanne Kauffman, *Science Has No National Borders: Harry C. Kelly and the Reconstruction of Science and Technology in Postwar Japan* (Cambridge, MA: Massachusetts Institute of Technology Press, 1994), quote on 4.

Impressed, the committee argued Japanese scientists and research facilities could be of use.[118] The Scientific and Technical Division of SCAP, headed by a young American physicist, played numerous roles overseeing and reconstructing Japanese science: on one hand, the division aimed to solve industrial problems and achieve economic growth; on the other, it was responsible for reopening laboratories, surveilling scientists and inspecting on-going research. The two areas could conflict.

The attempt to rebuild Japanese atomic facilities revealed the distance between scientific and military perspectives. During the war, the Army unintentionally pushed RIKEN's cyclotrons into Tokyo harbor. After the war, Lee Dubridge, director of the Rad Lab, wrote to Frank Jewett, president of the National Academy of Sciences: "Many of us believe it is a most important matter of principle that American scientists could make a very important move toward international goodwill and the principle of internationalism in science by rectifying the tragic mistake that the War Department has made."[119] Jewett forwarded the suggestion to Acting Secretary of War Kenneth Royall in hopes of rectifying the mistake. Royall's response was clear and forceful:

The destruction of the cyclotrons was a mistake, or more accurately, an accident of war. As such, it was only one of many accidents and I feel that any action to amend it now would be harmful and ill-advised ... It is unsound to even intimate that scientists are citizens of the world alone, are internationalists and not loyal to their native lands and never willing participants in the ambitions of dictators or tyrants. The evidence to the contrary is too overwhelming for the American public to accept this thesis, for modern war is scientific and technical war in toto. Without the scientist or the technical worker the terrible instruments of destruction of the present day world would not have been possible.[120]

Indeed. The desire of physicists to maintain a sense of collegial internationalism was secondary to national security and economic interests; American national interests, not scientific manners, determined American diplomacy.

The United States, for example, appropriated research from various Japanese biological and medical units, including the infamous Unit 731. During the war, the Japanese military operated more than twenty laboratories for human experimentation, primarily in China, which performed

---

[118] Ibid., 114.
[119] Exchange included in Yoshikawa and Kauffman, *Science Has No National Borders*, 10–11.
[120] Ibid.

gruesome tests, including vivisections. In return for access to research, the US occupation government granted researchers immunity from war crimes prosecution; seven directors of the Japanese National Institute of Health – an institute modeled on NIH and built with American support – conducted wartime experimentation.[121] Additionally, when Chinese and Soviet trials produced evidence of Japanese war crimes, the State Department denounced it as mere communist propaganda, a readily accepted explanation.[122] The deals were first disclosed in the 1980s, and Dr. Edwin Hill, the former Chief of Basic Sciences at Fort Detrick (home of the US biological weapons program), had argued frankly: "such information could not be obtained in our own laboratories because of scruples attached to human experimentation."[123] While the use of Japanese medical data later became controversial, studying and treating radiation poisoning was uncomfortable from the beginning.

The atomic bombs left an unexpected medical legacy. The American government did not plan to study the genetic effects of radiation. Instead, SCAP discouraged discussion of the bombs because it allowed the Japanese to portray themselves as the final victims of the war, rather than as its original aggressors.[124] Nonetheless, the AEC, working alongside the new Japanese NIH, set up an Atomic Bomb Casualty Commission (ABCC) to investigate genetic mutation and its generational influence.[125] When Stalin sent a similar team, the Commission became politicized; Hermann Müller, that year's Nobel winner in physiology and medicine (1946), delivered an anti-communist speech to Kyoto University (one of the more "radical" Japanese universities).[126] Geneticist George Beadle, an ABCC advisor, argued the Commission provided a "great opportunity to encourage science in a much needed area and a great deal of goodwill toward the USA could come along as a byproduct. This latter point is important, I feel. Many Japanese intellectuals are eyeing Red China and

---

[121] See Andrew Goliszek, *In the Name of Science: A History of Secret Programs, Medical Research and Human Experimentation* (New York: St. Martin's Press, 2003), 43–49. See also Jing-Bao Nie, "The United States Cover-up of Japanese Wartime Medical Atrocities: Complicity Committed in the National Interest and Two Proposals for Contemporary Action," *The American Journal of Bioethics* 6 (July 2006): 21–33.
[122] Regarding the Russian Khabarovsk trials of 1949, see ibid., 22.
[123] Hill quoted in ibid., 25.   [124] Beylor and Low, "Science policy," 115.
[125] M. Susan Lindee, *Suffering Made Real: American Science and the Survivors at Hiroshima* (Chicago: University of Chicago Press, 1994).
[126] John Beatty, "Scientific Collaboration, Internationalism and Diplomacy: The Case of the Atomic Bomb Casualty Commission," *Journal of the History of Biology* 26 (Summer 1993): 205–231, esp. 214.

Russia with interest and if we don't take steps soon to keep them on our side, it may be too late."[127]

The ABCC's "no treatment" policy complicated US/Japanese relations. Medical treatment could be interpreted as an admission of guilt and if the United States treated some victims, why not all? Rather than discriminate between those who deserved treatment, the ABCC chose not to treat at all. Instead, the ABCC presented their research as a partnership and an example of "international science" while censoring news of local tensions.[128] Asked to provide body fluids, information about abnormal births, and allow autopsies, survivors demonstrated against the "no treatment" policy in 1955; critics argued the Japanese were merely "guinea pigs." The policy changed two years later; after a few tumultuous decades, the ABCC mutated into the Radiation Effects Research Foundation in 1975, which still operates today.[129] However, the occupation had perhaps its greatest impact in agriculture.

The roots of the Green Revolution lie in the American occupation of Japan. Pacific collaboration in agricultural sciences had a long history: in the 1920s, the Department of Agriculture sent specialists to Japan to investigate chestnuts (oriental chestnuts are now common in the United States); the following decade, USDA botanists returned from Japan with 3,000 varieties of soybeans (which became a major oilseed crop in World War II).[130] After the war, the Natural Resources section of SCAP and the USDA collaborated on wheat research; Dr. S. D. Salmon – the USDA project leader on wheat – returned to the United States with multiple 10 gram samples of semi-dwarfing wheat. Salmon selected one strain, known as "Norin" because of its Japanese institutional home, for further study at the USDA/Washington State University Experiment station. Once crossed with a local strain, this "Norin-10" wheat found its way to Norman Borlaug, a Rockefeller Foundation researcher and the eventual "father" of the Green Revolution.[131] The Revolution will be

---

[127] Beadle quoted in ibid., 213.

[128] See the "No-Treatment Policy" in M. Susan Lindee, *Suffering Made Real: American Science and the Survivors at Hiroshima* (Chicago: University of Chicago Press, 1994), 117–142, quote on 248.

[129] Ibid., 222–230. The Foundation's English language website is: www.rerf.jp/index_e .html.

[130] John L. Creech, "The Diplomacy of Genetic Resources: The Key Role of Plant Introduction in U.S./Japanese Relations before and after World War II," *Diversity* 15 (2000): 18–20.

[131] Ibid., 20.

recounted in Chapter 3, here it is important to note the original seeds germinated in occupied Japan, were nourished in Washington state and quickly became part of Cold War geopolitics. As the federal government authored science policies for Japan and Germany, American interest in European research and economic reconstruction grew.

Science and European Reconstruction

US officials assessed postwar European scientific proficiency to aid recovery efforts and protect American national security. The Office of Naval Research in London acted as a "window on European science," while the State Department established a science staff in London in 1947.[132] As State considered posting science attachés overseas, the Truman administration charged the department with "collection abroad for all government agencies of information in the basic sciences;" the newly established CIA would determine which countries had "potential in fields of basic and applied sciences."[133] ONR, asked to consider the "Rehabilitation of Science in Europe," produced a detailed report favoring the rehabilitation of physics with caveats about the presence of left-wing scientists in positions of power (especially in France).[134]

Fear of communism colored US postwar reconstruction assistance. The first nations aided after Truman's containment policy, Greece and Turkey, benefited from a variety of applied science projects, including hydroelectric power, improvements to agriculture, and malaria prevention through DDT spraying.[135] The European Recovery Act, or "Marshall Plan," was mutually beneficial: war-torn European economies would receive desperately needed funds for recovery, while American manufacturers would receive millions of new customers and expand their global market share. However, although industrial recovery was a critical component of the plan, support for science was almost negligible at first,

---

[132] Regarding the Department of State, see Franklin P. Huddle, "Science, Technology, U.S. Diplomacy: History and 1978 Legislation," *Technological Forecasting and Social Change* 17 (1980): 353–363. Regarding the Navy, see Harvey M. Sapolsky, *Science and the Navy: The History of the Office of Naval Research* (Princeton, NJ: Princeton University Press, 2014), 50.

[133] National Security Council Intelligence Directive No. 10 (January 18, 1949): 1–2. Available via DDRS.

[134] Krige, *American Hegemony and the Postwar Reconstruction of Science in Europe*, 32.

[135] Specialized sections on Turkey, Italy, Greece, and West Germany can be found in Barry Machado, *In Search of a Usable Past: The Marshall Plan and Postwar Reconstruction Today* (Lexington, VA: George C. Marshall Foundation, 2007), 57–112.

despite pleas from prominent scientists.[136] Eventually support for laboratories and industrial research increased as part of the effort to stabilize West Germany. The American government also turned to more indirect means of influence, encouraging the Rockefeller Foundation to increase its support for science in the Netherlands.[137] However, US administrators turned down Italian requests in physics, according to historian John Krige, because of "fears that left-wing university scientists could not be trusted."[138] Relations with France were worse.

French support for communism and internationalism worried American policy-makers. Following the war, the French appointed Frederic Joliot-Curie, well respected for his work in physics and the Resistance, as head of the CEA (the French AEC). Yet Joliot-Curie, perhaps politicized by his wartime experiences, was also a founder of the World Federation of Scientific Workers (1946), which promoted internationalism. When the United States offered the "Baruch Plan" on atomic energy, the French physicist argued for international collaboration to lessen American influence: "The smaller nations certainly would be well advised to link up with us in this domain. My point of view is that we can help England release herself from the grip of the United States."[139] Joliot-Curie went further as the Cold War heated up, signing the Stockholm appeal to ban nuclear weapons and informing the French Communist Party that progressive "scientists shall not give a jot of their science to make war against the USSR."[140] The American press condemned the statement and the CIA charged half the French CEA were "communist sympathizers" (Georges Tessier, a "fellow traveler," headed the French nuclear research center (CNRS)).[141] Such outspokenness upset domestic researchers and politicians: French physicists wanted American support for a reactor and the political right embraced anti-Communism (the French government dismissed Communist party members in 1947). The combination was powerful; both Tessier and Joliot-Curie lost their

[136] See "Science and the Marshall Plan," in Krige, *American Hegemony and the Postwar Reconstruction of Science in Europe*, 30–56.
[137] Krige, *American Hegemony and the Postwar Reconstruction of Science in Europe*, 54.
[138] Ibid., 45.
[139] Joliot-Curie quoted in Spencer R. Weart, *Scientists in Power* (Cambridge, MA: Harvard University Press, 1979), 234.
[140] Joliot-Curie quoted in Frank Close, *Half-Life: The Divided Life of Bruno Pontecorve, Physicist/or Spy* (New York: Basic Books, 2015), 157.
[141] Ibid.

positions by 1950.[142] Yet even as American aid shaped European reconstruction, the United States broke with its European allies over patenting wartime research.

Penicillin illustrated the new commercial geopolitics of medicine. Discovered and refined for medical use in Great Britain, penicillin arrived in the United States during the war. To help the British mass-produce the critical antibiotic, OSRD contracted with eleven private companies and more than thirty universities and hospitals. The American government orchestrated, in the words of historian Nicholas Rasmussen, "competition among drug firms without preventing them from patenting the fruits of their government brokered research."[143] But the British advised medical researchers to refrain from patenting their work and thus American patent applications created upheaval in the "moral economy" of medicine.[144] The State Department, for example, refused Spanish requests to tour penicillin production facilities and insisted on private production and sales.[145] Eventually, Merck and other American firms profited in the Spanish market, while the British created the National Research Development Corporation (1948) to secure future patent rights to publicly financed British research.[146] Nor was penicillin alone; a market in radioisotopes arose alongside.

Radioisotopes, a byproduct of atomic reactors, became a critical tool for researching metabolism and biological processes. Before the war, American Ernest Lawrence's cyclotrons were among the few sources for radioisotopes and the physicist provided samples to friends and colleagues. Wartime production of plutonium and uranium made radioisotopes plentiful but also increased government oversight and created a backlog: Oak Ridge laboratory had dozens of unfilled orders in 1947. However, AEC desires to promote peaceful atomic energy and capture markets overseas convinced policy-makers to increase manufacture and distribution. In the first decade of the AEC program, Oak Ridge sent out

---

[142] Krige, *American Hegemony and the Postwar Reconstruction of Science in Europe*, 61.

[143] Nicholas Rasmussen, "Of 'Small Men,' Big Science and Bigger Business: The Second World War and Biomedical Research in the United States," *Minerva* 40 (2002): 115–146, quote on 125. For the figures of participating institutions, see Robert Bud, "Upheaval in the Moral Economy of Science? Patenting, Teamwork and the World War II Experience of Penicillin," *History and Technology* 224 (2008): 173–190, 175.

[144] Bud, "Upheaval in the Moral Economy of Science."

[145] Ana Romero de Pablos, "Regulation and the Circulation of Knowledge: Penicillin Patents in Spain," *Dynamis* 31 (2011): 363–383.

[146] Bud, "Upheaval in the Moral Economy of Science?" 180.

64,000 shipments of radioactive materials which were ultimately used in more than 10,000 scientific publications.[147] Of course, the program benefited American diplomacy as well: radioisotopes were a key means of aiding Cold War diplomacy under the Marshall plan; AEC commissioner Henry Smyth observed, "the foreign distribution of isotopes has had a very good effect on our foreign relations."[148]

Radioisotopes settled into a series of stable global markets. The AEC program slowed when critics alleged shipments to Norway and Finland undermined American security, while state-subsidized reactors allowed Great Britain and Canada to undercut American prices.[149] By the early 1950s, radioisotope distribution followed geopolitical lines: the British provided most radioisotopes for European research, the Soviets did the same for Warsaw Pact nations and the United States for Latin America.[150] Both the United States and Soviet Union used radioisotopes to court allies in the developing world, one reason why the United States initiated export controls on many scientific and technical products.

American economic warfare against the Soviet Union included preventing the export of high-tech goods to the Soviet bloc. In 1949, the United States passed the Export Control Act and pressured its European allies and Japan into forming the Coordinating Committee on Multilateral Exports or COCOM. Tensions plagued the Committee from the start: The United States wanted to block all products necessary for economic growth, while many European nations wanted to block only goods with an obvious military use.[151] Additionally, France and the Netherlands threatened to leave if the organization became public, causing COCOM to operate in secret.[152] Participating American authorities included the departments of State, Commerce, Defense and the AEC.[153]

---

[147] Angela N. H. Creager, *Life Atomic: A History of Radioisotopes in Science and Medicine* (Chicago: University of Chicago Press, 2013), 5.

[148] Smyth quote in Creager, *Life Atomic*, 7.

[149] Angela N. H. Creager and María Jesús Santesmases, "Radiobiology in the Atomic Age: Changing Research Practices in Comparative Perspective," *Journal of the History of Biology* 39 (Winter 2006): 637–647. See also Alison Kraft, "Between Medicine and Industry: Medical Physics and the Rise of the Radioisotope 1945–65," *Contemporary British History* 20 (2006): 1–36, esp. 7–8.

[150] Creager, *Life Atomic*, 136.

[151] Gary K. Bertsch and Steven Elliott-Gower, eds., *Export Controls in Transition* (Durham, NC: Duke University Press, 1992).

[152] Michael Mastanduno, *Economic Containment: CoCom and the Politics of East-West Trade* (Ithaca, NY: Cornell University Press, 1992), 105.

[153] Bert Chapman, *Export Controls: A Contemporary History* (Lanham, MD: University Press of America, 2013), 189.

Ultimately, COCOM maintained the International Atomic Energy list, while the Commerce Control List and Export Administration Regulations oversaw the following categories: nuclear materials, electronics, computers, telecommunications, sensors and lasers, navigation and avionics, aerospace, and propulsion.[154] A form of American scientific and technical containment, export controls provided leverage in diplomatic negotiations, as the United States could relax or expand access to embargoed items depending on an adversary's behavior.[155] Export controls are addressed throughout the text; here is worthwhile to note a recent study which concluded COCOM contributed to Soviet "backwardness by helping to keep it two decades behind the west in avionics and computers."[156]

World War II and the immediate postwar period demonstrated the importance of science to American foreign relations. American science diplomacy was critical to maintaining alliances and atomic energy redefined national power. As such, postwar American policies sought to control military research and maintain international scientific leverage. The United States refused to cede control of atomic energy to the new United Nations and was unwilling to support a system of international laboratories. Instead, American policies began reconfiguring Western European science through grants, fellowships, and training programs; in a unique process of "co-production," transatlantic research promoted long-term prosperity and economic growth while allowing Americans to benefit from European discoveries.[157] With the establishment of NATO in 1949, the transatlantic bond firmed and enlarged, extending American export controls. Whether through cooperation or occupation, the incorporation of science into foreign policy also expanded the cast of characters and departments involved in American diplomacy: Scientists, engineers, and doctors worked alongside advisors from Commerce and Agriculture and intelligence agents. At home, the government became the primary scientific provider, giving it more influence on research agendas, institutions, and individual allegiances. Simultaneously, Cold War fears and funding deflated the early activism; even Niels Bohr recommended work on the hydrogen bomb to meet the

---

[154] Chapman, *Export Controls*, 48.

[155] Robert D. Blackwell and Jennifer M. Harris, *War by Other Means: Geoeconomics and Statecraft* (Cambridge, MA: Harvard University Press, 2016), 163–165.

[156] James K. Libbey, "CoCom, Comecon and the Economic Cold War," *Russian History* 37 (2010): 133–152, quote on 152.

[157] Krige, *American Hegemony and the Postwar Reconstruction of Science in Europe*, 1–11.

Soviet threat. By the time Congress established the National Scientific Foundation (1950), its influence was minimal and scientific internationalism was out of favor; the Foundation warned scientific organizations to avoid "using the phrase 'science is international,' and any implication of organizing science in a highly international way."[158] After the disputes of the early Cold War, many American policy-makers considered scientific internationalism, like political neutralism or non-alignment, contrary to American national interest.

[158] National Science Foundation quoted in Manzione, "Amusing," 40.

2

Science Contained

Genetics stood trial in August 1948. T. D. Lysenko, president of the Lenin All-Union Academy of Agricultural Sciences, declared the biological discipline reactionary and bourgeois. The Soviet Union banned genetics soon thereafter, ignominiously replacing it with Lysenko's own "agrobiology." For the next decade, Soviet state control over biology demonstrated how political ideology can shape a scientific field, whether by determining research priorities and funding or by enforcing ideological purity among researchers. As the Soviet Union and United States advocated different agricultural practices, the resulting controversy sparked questions about which superpower could best harness research for their people, coloring perceptions of Soviet science. Indeed, the Lysenko affair and the arms race framed superpower scientific competition for American policy-makers and the public until Sputnik.

Lysenko's swindle had a long history. A poorly trained agronomist, he rose to fame amid the struggles of collectivization, promising bountiful harvests and claiming to have converted winter wheat into spring wheat. Lysenko proposed heating and cooling seeds to change their growing season, suggesting the environment could influence heredity, a theory known as the "inheritance of acquired characteristics" and discredited by geneticists at the time.[1] When Soviet purges began in 1936, Lysenko embraced Marxist and nationalist rhetoric, casting his work as practical

---

[1] Lysenko called his seed technique "vernalization" and received attention in scientific journals at first, see Nils Roll-Hansen, *The Lysenko Effect: The Politics of Science* (Amherst, NY: Humanity Books, 2005), 17.

and home-grown, whereas genetics was theoretical and foreign.[2] The gambit worked: The USSR canceled a genetics conference in Moscow in 1937 and Lysenko and his fellow agrobiologists began removing geneticists from Soviet institutes. Experimental results and international support proved worthless before party loyalty. In 1939, for example, the famed Soviet geneticist Nikolai Vavilov fatefully pointed to the success of American hybrid corn as testament to genetics; the same year, the International Congress of Genetics elected Vavilov president, but only in absentia.[3] Prevented from attending by Soviet authorities, the geneticist died after mistreatment in Soviet jails.[4]

The politicization of genetics increased after World War II. Hopes of postwar cooperation faded as the United States and Soviet Union engaged in numerous scientific disputes between 1946 and 1948 (see Chapter 1). Biology became a flashpoint as well. In 1945, Soviet geneticist and propagandist Anton Zhebrak published an article in *Science* dubbing Lysenko "naïve" but avowing the government "never interfered" in genetic research.[5] Within months, however, Stalin and Lysenko began corresponding about improving Soviet agriculture; over the next two years, the agrobiologist kept the dictator up-to-date (and in the dark) on the progress of his new "Stalin Branched Wheat" (the crop failed).[6] As Lysenko's influence grew, Soviet geneticists appealed to outside groups for help; the Genetics Society of America (GSA) established a "Committee to Aid Geneticists Abroad."[7] But international appeals were dangerous: In 1947, the state censured Zhebrak for writing anti-patriotic articles and

[2] Lysenko was a follower of Ivan Michurin, an earlier Russian advocate of cross-fertilization. For detail on Vernalization and Michurin, see William Dejong-Lambert, *The Cold War Politics of Genetic Research: An Introduction to the Lysenko Affair* (New York: Springer, 2012), 41–72. See also David Joravsky, *The Lysenko Affair* (Chicago: University of Chicago Press, 1970), 39–62.

[3] See "Stalinist Ideology and the Lysenko Affair," in Loren R. Graham, *Science in Russia and the Soviet Union: A Short History* (Cambridge: Cambridge University Press, 1993), 121–134.

[4] Peter Pringle, *The Murder of Nikolai Vavilov: The Story of Stalin's Persecution of One of the Great Scientists of the Twentieth Century* (New York: Simon & Schuster, 2011).

[5] Anton R. Zhebrak, "Soviet Biology," *Science* 102 (October 5, 1945): 357–358. Regarding Zhebrak and propaganda, see Roll-Hansen, *The Lysenko Effect*, 266–267.

[6] Ethan Pollock, *Stalin and the Soviet Science Wars* (Princeton, NJ: Princeton University Press, 2006), 47–48.

[7] Nicholas Krementsov, "A 'Second Front' in Soviet Genetics: The International Dimension of the Lysenko Controversy, 1944–47," *Journal of the History of Biology* 29 (Summer 1996): 229–250.

removed him from positions of influence.[8] The USSR refused requests to hold an international genetics congress in early 1948 (the geneticists relocated to nearby Sweden) and convicted the field in August.

Lysenkoism played a starring role in Stalin's global propaganda campaign (see Chapter 1). The Soviet Union published transcripts of the trial and released a feature film in multiple languages (it opened in New York City as "Life in Bloom").[9] Introduced to the communist bloc as "creative Darwinism," the new biology promised to allow researchers to direct evolution, thereby empowering Soviet plans to transform nature. Lysenko, meanwhile, blamed genetics for Nazi eugenics, racism, and imperialism. Sympathetic societies formed throughout Eastern Europe; in Poland, the state branded biologists resistant to Lysenkoism "agents of British and US imperialism."[10] While such ideological control alarmed Polish geneticists, American anticommunists found it useful.

The Lysenko controversy provided the United States a perfect propaganda opportunity: The Soviet Union had violated a key scientific principle by callously altering the content of a field to fit political prejudice, allowing the United States to position itself as the defender of scientific freedom. The DOD's Research and Development Board, for example, asked its biological consultants to speak out against Lysenkoism, while the CIA assembled geneticists to lecture on the Voice of America across the globe and provided funds for the British anti-Marxist Society for Freedom in Science.[11] When German researchers split along political lines, the United States relocated geneticists from East to West.[12] In addition, the Congress for Cultural Freedom, a CIA front, sponsored conferences in Berlin and Hamburg on "Science and Freedom," using the Lysenko affair to condemn Soviet authoritarianism.[13] Funded by

---

[8] Zhebrak lost his position in the central committee as well as the Belorussian Academy of Science, see Roll-Hansen, *The Lysenko Effect*, 266–268.

[9] William deJong-Lambert and Nikolai Krementsov, "On Labels and Issues: The Lysenko Controversy and the Cold War," *Journal of the History of Biology* 45 (Fall 2012): 373–388, esp. 375–376.

[10] William deJong-Lambert, "Lysenkoism in Poland," *Journal of the History of Biology* 45 (Fall 2012): 499–524, quote on 500.

[11] See "Proletarian and Bourgeois Science,' in *Science, Politics and the Cold War*, 16–37.

[12] deJong-Lambert and Krementsov, "On Labels and Issues," 383.

[13] Richard H. Beyler and Morris F. Low, "Science Policy in Post-1945 West Germany and Japan: Between Ideology and Economics," in Mark Walker, ed., *Science and Ideology: A Comparative History* (New York: Routledge, 2003): 97–123, esp. 107.

grants from the Rockefeller Foundation and others, the Hamburg conference welcomed 900 scholars from nineteen different countries.[14] Chemist Michael Polanyi, unaware of CIA sponsorship, wrote in the *Conference Proceedings*: "we were concerned to reveal to the whole world the cruel suppression of intellectual freedom under totalitarianism."[15] Following the conference, American intelligence reported denunciation of Lysenkoism prompted the Soviets to invite the Science Council of Japan for a visit, one of many invited as part of Soviet attempts to repair their scientific standing.[16]

Japan presented a unique problem for American diplomats. Japanese biologists studied environmental factors in heredity before the war and the introduction of Lysenkoism in 1946 provoked interest.[17] Scientific freedom was not an issue: Historian Kaori Iida suggests Japanese researchers "did not identify with the West and thus did not strongly feel 'attacked' by Lysenkoists;" instead, left-wing Japanese researchers formed an association (MINKA), which equated Lysenkoism with democratization in science.[18] But American administrators in Japan, worried about Soviet influence, prohibited a pro-Lysenko institute, approving instead the Japanese National Institute of Genetics (1949) and removing faculty with communist affiliations from Japanese universities (the "Red Purge"). The occupation government outlawed the Communist Party in 1950 and many influential Japanese joined the Congress for Cultural Freedom; biologists realized Lysenkoism limited careers and MINKA disappeared within a few years.

---

[14] Francis Stonor Saunders, *The Cultural Cold War: The CIA and the World of Arts and Letters* (New York: The New Press, 1999), esp. 214. See also Peter Coleman, *The Liberal Conspiracy: The Congress for Cultural Freedom and the Struggle for the Mind of Postwar Europe* (New York: The Free Press, 1989).

[15] Anon., *Science and Freedom: Proceedings of a Conference Convened by the Congress for Cultural Freedom and Held in Hamburg on July 23rd–26th, 1953* (Boston: Beacon Press, 1955), quote on 1. Polanyi eventually resigned from the CCF when he learned of its CIA funding, see Mary Jo Nye, *Michael Polanyi and His Generation: Origins of the Social Construction of Science* (Chicago: University of Chicago Press, 2011), 211–213.

[16] William J. Morgan, "Memorandum to Col. Bryon Enyart [?]," (July 28, 1953): 1–2. Available via DDRS.

[17] Kaori Iida, "Practice and Politics in Japanese Science: Hitoshi Kihara and the Formation of a Genetics Discipline," *Journal of the History of Biology* 43 (Fall 2010): 529–570, esp. 531.

[18] Kaori Iida, "A Controversial Idea as a Cultural Resource: The Lysenko Controversy and Discussions of Genetics as a 'Democratic' Science in Postwar Japan," *Social Studies of Science* 45 (2015): 546–569, quote on 560.

The Lysenko controversy lingered for decades. Even the revelation of DNA as the molecule of inheritance did not shake the Soviet verdict. Instead, as late as 1958, Lysenko argued only "reactionaries in science" considered his work a "swindle and a deceit," convincing Khrushchev to grow corn in Siberia (the crop failed).[19] Lysenkoism became synonymous with pseudoscience and quackery, especially given the success of genetics and many of the agrobiologist's later statements (he suggested warblers give birth to cuckoos after eating caterpillars).[20] The controversy tarnished public perceptions of Soviet research; historian Nils Roll-Hansen terms the Soviet propaganda campaign "a most amazing example of a government shooting itself in the foot."[21] American representatives argued democracy and the free market, as opposed to Soviet state control, protected scientific freedom and produced results, pointing to genetics and hybrid corn as the underlying theory and proof of the Green Revolution, an anti-communist initiative aimed at transforming agriculture in the developing world (see Chapter 3). Yet even as Lysenkoism warped Soviet biology, Stalin refused to allow Marxist ideology to influence physics, aeronautics, or other fields with military applications (he was using the capitalist West's design to build bombs after all).[22] Leaders on both sides knew significant prestige and advantage would accrue to the dominant scientific power and both the Soviet Union and United States exercised considerable sway over domestic research: From Atoms for Peace to Sputnik and the space race, competition in science and technology provided the background to Cold War relations for decades after World War II.

This chapter has three sections to illustrate the impact of the early Cold War on American science and diplomacy. The first, "Co-option," considers how federal dollars shaped American research to fit national security needs, highlighting the impact on physics, ecology, seismology, and

---

[19] T. D. Lysenko, quoted in Zhores A. Medvedev, *The Rise and Fall of T. D. Lysenko* (New York: Anchor Books, 1971), 137. See also Zhores A. Medvedev, "Lysenko and Stalin: Commemorating the 50th Anniversary of the August 1948 LAAAS Conference and the 100th Anniversary of T. D. Lysenko's Birth, September 29, 1898," *Mutation Research* 462 (2000): 3–11.

[20] Loren Graham, *Lysenko's Ghost: Epigenetics and Russia* (Cambridge, MA: Harvard University Press, 2016), 1–5. Regarding warblers, see deJong-Lambert, *The Cold War Politics of Genetic Research*, 143.

[21] Roll-Hansen, *The Lysenko Effect*, 275.

[22] Stalin shut down a proposed Marxist conference on nuclear physics, see Pollock, *Stalin and the Soviet Science Wars*, 72–103. See also Holloway, *Stalin and the Bomb*.

geography. The DOD coupled civilian and military experiments, while defense funding requests blurred the line between basic and applied research to secure support. As American power increasingly relied on domestic science and technology, government concern over espionage grew. At the height of the McCarthy period, the Department of State scrutinized the political affiliation of foreign scientists, using visa and passport policies to limit the travel of those deemed suspicious.

"Cooperation" examines American influence on the primary scientific initiatives of the 1950s: the Atoms for Peace program, the European Center for Nuclear Research (CERN), and the International Geophysical Year (IGY). The United States had multiple reasons for supporting international cooperation, including the advancement of knowledge, reinforcing the image of the United States as a benevolent scientific power, advertising scientific and technical goods, and gaining intelligence about the scientific and technical capabilities of others. Atoms for Peace, for example, attempted to rebrand nuclear energy and capture the global market for civilian atomic energy. Although CERN represented European revitalization, American support was necessary and anticommunism influenced the Center's early leadership. Finally, American interests also shaped the International Geophysical Year, but the Soviet launch of Sputnik overshadowed other accomplishments and initiated the space race.

"Competition" considers the American response to Sputnik, as the government reorganized its scientific bureaucracy and redoubled diplomatic and intelligence efforts abroad. The 1950s and 1960s were the high point of superpower scientific rivalry and both the United States and USSR competed to claim the mantle of scientific leadership. American actions included attempts to engage scientists as spies, blocking Soviet efforts to acquire industrial research and maintaining export controls against the Soviet bloc. At the same time, tensions with France continued, whether over access to American computers, support for French-led scientific cooperation or French satellite programs. Whether with the Soviets or European allies, the role of science in American diplomacy expanded in the early Cold War.

CO-OPTION

It is because of this universality of common interest in science, which concerns itself very little with national borders, that scientists are by nature internationally minded ... It is important to the future of the world that individual scientists on opposite sides of national borders should remain in contact with one another even in times of political tension.

– Harvard Astronomer Bart J. Bok (1955)[23]

[It's hard] to tell whether the Massachusetts Institute of Technology is a university with many government research laboratories appended to it or a cluster of government research laboratories with a very good educational institution attached to it.

– Physicist Alvin Weinberg, Director of Oak Ridge National Laboratory (1962)[24]

The tension between the two quotes above complicated Cold War research: Bok emphasized the centrality of internationalism to science while Weinberg highlighted the complex realities of national funding for science. The Korean War confirmed American fears of communist aggression and cemented military support for R&D, leading to more government-directed research, secrecy, and classification of data.[25] By the mid-1950s, the federal government funded nearly 70 percent of American R&D, essentially coopting domestic research into the Cold War (see Figure I.1). Although there were no "hot" wars between 1953 and 1965, the US/Soviet conflict maintained the wartime dynamic between research and foreign policy, as national security demands shaped the country's scientific and technical infrastructure.

Even a minor diplomatic problem could require a variety of experts. The Soviet Union, for example, succeeded in jamming Russian-language Voice of America (VOA) broadcasts in 1948, undermining US diplomatic efforts. In response, the State Department and MIT inaugurated Project Troy, an interdisciplinary study of radio and psychological warfare.[26] Troy represented a new approach: The State Department would hire professors to work on national security needs over their summer break. To solve the immediate problem, MIT physicists worked on propagating long-distance radio waves to bypass jamming and facilitate broadcasts.

---

[23] Bart J. Bok, "Science in International Cooperation," *Science* 121 (June 17, 1955): 843–847, quote on 847.

[24] Weinberg quoted in Leslie, *The Cold War and American Science*, 14.

[25] Daniel J. Kevles, "K1S2: Korea, Science and the State," in Galison, *Big Science*, 312–333.

[26] See "Project Troy," in Allan A. Needell, *Science, Cold War and the American State: Lloyd V. Berkner and the Balance of Professional Ideals* (New York: Routledge, 2000), 155–175.

A secondary result was the creation of a cheap, durable radio receiver for global distribution. But Troy's impact went further. Studies in Russian history and culture – necessary for effective propaganda – eventually spurred the creation of Truman's Psychological Strategy Board as well as MIT's Center for International Studies (CIS). Funded in part by the CIA, CIS provided a space where classified information could be discussed by academics. Links between the government and academia grew; economist Max Millikan, for example, left Project Troy to head the CIA. Given that a single problem – like jammed radio broadcasts – could entangle a variety of fields, it is not surprising the Cold War entangled nearly all. Over the next two decades, the government had an unprecedented influence on American science.

American Science in the Early Cold War

The development of the thermonuclear bomb and civilian nuclear energy permanently tied atomic physics to the state. The federal sway over domestic physics was total: The US government maintained the large scientific infrastructure created during World War II (Los Alamos, Brookhaven and Argonne national labs, etc.) as well as a unique oversight role with all private atomic reactors and research through the Atomic Energy Commission. Domestic sources of radioactive material were under federal lock and key; even the distribution of atomic isotopes for medical and ecological research was part of federal (and foreign) policy. Meanwhile, the DOD and AEC demand for physicists seemed unlimited: almost overnight, physics funding, programs and PhDs skyrocketed as "physicist" became a well-respected career; the new journal *Physics Today* announced in 1950 the "springtime of Big Physics has arrived."[27] The level of federal funding in postwar physics leads some historians to question whether the government inappropriately determined the field's development through an attempt to shape it to national needs.[28] Others

---

[27] *Physics Today* is quoted in "The Physicists Established," in Kevles, *The Physicists*, 367–392, quote on 367.

[28] See Paul Forman, "Behind Quantum Electronics: National Security as a Basis for Physical Research in the United States," *Historical Studies of the Physical and Biological Sciences* 18 (1987): 149–229, and Daniel J. Kevles, "Cold War and Hot Physics: Science, Security and the American State," *Historical Studies of the Physical and Biological Sciences* (1990): 239–264. See also Peter Galison, Bruce Hevly, and Rebecca Lowen, "Controlling the Monster: Stanford and the Growth of Physics Research, 1933–1962," in Galison, *Big Science*.

point out the absence of a normative path for scientific development and the influence of patrons on research for centuries. But atomic energy and the bomb shaped other fields as well. The arms race required atmospheric chemistry to detect Soviet testing aboveground; underground nuclear tests required the development of seismology. Fear of fallout and radioactive leaks drove the development of ecology.

Eugene Odum, a pioneer in modern ecology, suggested his field arose via a feedback loop with atomic energy: Ecologists used concern over atomic energy to justify research, while atomic energy provided the tools – radioactive isotopes and tracers – for ecological research.[29] As early as 1946, the AEC approached Odum to study the consequences of atomic testing on Bikini Island in the South Pacific; by 1954, he and his brother had conducted the first study of the metabolism (energy intake and release) of an entire ecological system – the Eniwetok atoll. Back in the United States, the establishment of the Savannah River reactor led to an AEC/University of Georgia (UGA) partnership; soon UGA was a major center of ecological research. When the Hanford plants began to leak radioactive waste, the AEC initiated studies on atomic storage and biological magnification (the tendency of radioactive materials to accumulate at the top of a food chain). From the perspective of the AEC, historian Joel B. Hagan reminds us, "concern for the environment was never the principal rationale for supporting ecological research."[30] Nor was earthquake prediction the primary rationale for federal support for seismology.

Seismology was a small field prior to the atomic bomb. Once the military realized large underground cavities might hide Soviet nuclear tests, however, support for research rapidly materialized. Project Vela Uniform, run by the DOD's Advanced Research Projects Agency (ARPA), supported nearly every American seismologist in the late 1950s and 1960s.[31] In the words of one practitioner, "representatives of ARPA have made the statement that they bought the entire discipline of seismology when the issue of detection of underground nuclear tests came to their

---

[29] See "Ecology and the Atomic Age," in Joel B. Hagen, *An Entangled Bank: The Origins of Ecosystem Ecology* (New Brunswick, NJ: Rutgers University Press, 1992), 100–121, esp. 100.

[30] Hagen, *An Entangled Bank*, quote on 119.

[31] ARPA ran three "Vela Programs": Vela Uniform to detect underground nuclear tests; Vela Sierra for nuclear tests in atmosphere; and Vela Hotel to detect nuclear tests in space. See Sharon Weinberger, *The Imagineers of War: The Untold Story of DARPA, the Pentagon Agency That Changed the World* (New York: Alfred A. Knopf, 2017), 98.

attention and it is hard to disagree."[32] Vela Uniform eventually led to the World-Wide Standard Seismograph Network (WWSSN), which maintained more than 120 stations in more than sixty countries and along the ocean floor. Deployed from 1960 to 1967, the WWSSN cost less than $10 million and provided standardized, public measurements, fulfilling the American commitment to atomic testing restrictions. When federal funding dried up in the early 1970s, seismologists turned toward earthquake prediction and prevention; data from the WWSSN soon confirmed the theory of plate tectonics.

Earth sciences and oceanography also relied on federal funds. Supported by the Joint Research and Development Board, scientists designed the Geographic Information System (GIS) for military needs; the preeminent Lamont Geological Laboratory at Columbia received more than 90 percent of its funding from military research contracts.[33] Meanwhile, funds from the Office of Naval Research flooded oceanography, as the Office supported research into the Navy's operating environment (all basic information – whether on underwater thermoclimes, topography, or acoustics, etc. – was immediately applied by the Navy). Given the dual-use nature of oceanography, the Navy struggled with how information should be attained and what should be made public. In Project Ice Pick, for example, the Navy originally tried to recruit civilian scientists as cover for covert research off the coast of the Soviet Union. After failing to locate a ship and crew, the Navy realized public scientific records contained most of what they needed. In the words of historian Jacob Hamblin, "Ice Pick began as a lie, a plan for pretended engagement in cooperative research; but it ended as a truth, a major lesson regarding the potential value of international cooperation to the Navy."[34] A quick learner, Eisenhower declassified information on US depth soundings in 1953 to facilitate American access to foreign oceanographic data.

Distinctions between basic and applied, civilian and military, blurred even further over time. In Project Tattletale, for example, the Navy

[32] Kai-Henrik Barth, "The Politics of Seismology: Nuclear Testing, Arms Control and the Transformation of a Discipline," *Social Studies of Science* 33 (October 2003): 743–781, quote on 752.

[33] Ronald E. Doel, "Constituting the Postwar Earth Sciences: The Military's Influence on the Environmental Sciences in the USA after 1945," *Social Studies of Science* 33 (October 2003): 635–666, esp. 641.

[34] Jacob Darwin Hamblin, "The Navy's 'Sophisticated' Pursuit of Science: Undersea Warfare, the Limits of Internationalism, and the Utility of Basic Research, 1945–56," *Isis* 93 (March 2002): 1–27, quote on 18.

planned to have a satellite transmit electronic intelligence to receiving stations on the periphery of the Soviet Union. They just needed "scientific cover to deflect public attention" away from the stations.[35] So the Naval Research Laboratory contacted Herbert Friedman, a well-known astrophysicist looking to put Alpha and X-Ray detectors into space (the Solar Radiation or Solrad experiment). Friedman explained "coupling" the two needs: "We weren't destitute for opportunity. But the fact that intelligence people were happy to have a cover for what they were doing made it opportune for us to move in there" and launch sooner.[36] Operating at night and in secret, navy technicians integrated the intelligence electronics into the Solrad experiment. Once in orbit, the experiment/project was a grand success: Solrad confirmed solar flares generate x-rays while the intelligence community learned about Soviet radar and anti-ballistic missile systems, helping plot potential attack routes.

Coupling also occurred at the experiment level. For example, the DOD maintained a "two-title" policy on all research grants: one for DOD files and one for public (and scientific) consumption. Consider the following contract with the Office of Naval Research. The Defense Documentation Center title was "Weaponry – lasers for increased damage effectiveness" with the description: "Damage mechanisms allowed by laser weapons is under intense investigation."[37] The investigator, R. Pantell, a researcher at Stanford, publicly titled his work "High-power broadly tunable laser action in the ultraviolet spectrum" and argued for financial support because ultraviolet lasers were "sorely needed in the areas of medicine, long distance communication, and high energy physics."[38] The basic research was the same; the results could be applied to a variety of needs. But work on classified projects or in national security fields meant intensified scrutiny.

McCarthyism and the Global Scientific Community

The Truman administration increased the scrutiny of American researchers and worked with the British to enforce loyalty in the Cold War.

---

[35] David K. van Keuran, "Cold War Science in Black and White: US Intelligence Gathering and Its Scientific Cover at the Naval Research Laboratory, 1948–62," *Social Studies of Science* 31 (April 2001): 207–229, quote on 221.

[36] Friedman quoted in ibid., 221.

[37] Deborah Shapley, "Defense Research: The Names Are Changed to Protect the Innocent," *Science* 175 (February 25, 1972): 866–868, quotes on 866.

[38] Ibid.

Truman's executive order in 1947 required more than 60,000 researchers sign an anti-communist pledge. Revelations of atomic espionage brought renewed attention; Senator McCarthy alleged hundreds of communist sympathizers were among those listed in the American Men of Science and that a clique of fellow travelers dominated the American Association for the Advancement of Science. The arrest of physicist Klaus Fuchs in February 1950 brought to light an atomic spy ring including Harry Gold (an industrial chemist) and David Greenglass (an army machinist). Also implicated were the husband and wife team, Julius and Ethel Rosenberg, who were convicted in 1951 and executed in 1953. No one was above suspicion: in June of 1954, Robert Oppenheimer, the public face of the Manhattan Project, lost his security clearance. Fear of espionage aided state control over sensitive research. When Italian physicist Bruno Pontecorvo defected to the Soviet Union, the American and British governments allowed the press to construct a false narrative about his role as a spy. In the words of historian Simone Turchetti, the false narrative "fostered the introduction of tighter measures of control at government laboratories, including the infamous 'positive vetting.' The construction of Pontecorvo's image as an 'atom spy' therefore served various political, security, and media agendas."[39] After defection, the FBI claimed Pontecorvo had communist affiliations and should not have been allowed to travel, pushing for tighter travel security.[40]

Visa and passport regulations provided the State Department leverage over individual scientists. In the early 1950s, the Passport Office often denied passports to prominent Left-leaning American scientists like Linus Pauling and Hermann Müller. At the same time, the department frequently denied visas to foreign scientists wishing to attend conferences in the United States. An FAS committee estimated half of all foreign scientists encountered difficulty getting a visa, while more than two-thirds of French scientific visas were delayed or refused; the International Congresses of Genetics, Biochemistry, and Astronomy refused to hold conferences in the United States until policies changed.[41] As historian Jessica Wang observes, "The State Department's visa policies caused more than inconvenience and anguish for individual foreign scientists. They

---

[39] Simone Turchetti, "Atomic Secrets and Governmental Lies: Nuclear Science, Politics and Security in the Pontecorvo case," *The British Journal for the History of Science* 36 (December 2003): 389–415, quote on 392. See also Close, *Half-Life.*
[40] Turchetti, "Atomic Secrets and Governmental Lies," 409.
[41] Wang, *American Science in an Age of Anxiety.*

damaged the image of freedom that the United States wished to project abroad as a contrast to the Soviet Union, and they disrupted international scientific life."[42] Additionally, enhanced scrutiny of scientists lasted far beyond McCarthy's brief career; as late as 1961, the consular office refused visas for French physicists Michel Langevin and wife Helene (daughter of Frederic and Irene Joliot-Curie, who were themselves regularly harassed).[43]

Private foundations also participated in American containment of science abroad. In 1954, Shepard Stone became one of three in charge of international programs at the Ford Foundation – a major source of research funds. Worried about the growing Non-Aligned Movement, Stone and the Ford Foundation vigorously supported the CIA's Congress for Cultural Freedom, contributing half the budget by the early 1960s. The Rockefeller Foundation, another major scientific patron, refused to fund left-leaning researchers. However, there was a fine line between patriotism and violating an informal scientific code: when the government asked Warren Weaver, an official at the Rockefeller Foundation, to hand over foundation files on European scientists, he refused, worried they might run the "risk of having Europeans think that we have people, under our auspices, acting as scientific spies for the government."[44]

Perhaps the most ironic consequence of American anti-communism involves Qian Xuesen – the "father of Chinese rocketry."[45] A former professor of aeronautical engineering at MIT, co-founder of the Jet Propulsion Laboratory at Caltech and member of the Manhattan Project, Qian hoped to visit his ailing father after the establishment of the People's Republic of China in 1950. But the FBI argued he was a communist sympathizer and had him arrested. After five years of house arrest, during which his father died, talks began about exchanging Qian for American prisoners in China.[46] The DOD worried about letting him go, noting he would "take back with him high competence in his professional field;" after the exchange, Chinese Premier Zhou Enlai observed, "We had won back [Qian Xuesen]. That alone made the talks worthwhile."[47]

---

[42] Wang, Ibid., 279.

[43] Badash, "Science and McCarthyism," *Minerva* 38 (2000): 53–80, esp. 58.

[44] Weaver, quoted in Krige, *American Hegemony and the Postwar Reconstruction of Science in Europe*, 124.

[45] For an excellent look at this case, see Iris Chang, *Thread of the Silkworm* (New York: 1995).

[46] Lawrence Badash, "Science and McCarthyism," *Minerva* 38 (2000): 53–80.

[47] Chang, *Thread of the Silkworm*, quotes on 188 and 190.

Soon thereafter the PRC founded the Institute of Mechanics with Qian as director, beginning a thirty-year career as the head of Chinese aerospace programs.

Qian's case highlights the importance of elite scientists. The prominent role of science in national security led to increased prestige and political access for elite researchers. While the government directly and indirectly employed thousands of scientists, only a handful, often from a few federal laboratories or elite universities, occupied positions of authority. President Truman's Science Advisory Committee, established in 1951, tended to be staffed by Los Alamos or Rad Lab (MIT) personnel. Nor did this circle of advisors widen in the early Cold War; later PSAC appointments (29 of 41) were trained at seven universities (Berkeley, Caltech, Chicago, Columbia, Harvard, MIT, and Princeton).[48]

The Cold War entangled science and scientists in American domestic and foreign policies to an unprecedented and unexpected degree. Government funding shaped institutional priorities and research while government regulations shaped scientific activity by limiting scientists' movements and access. At the same time, scientists used public concerns over national security or atomic energy as leverage in negotiations for federal research dollars. In his *Farewell Address* (1961), President Eisenhower offered his assessment of the change:

[There] is a recurring temptation to feel that some spectacular and costly action could become the miraculous solution to all current difficulties. A huge increase in the newer elements of our defenses; development of unrealistic programs to cure every ill in agriculture; a dramatic expansion in basic and applied research – these and many other possibilities, each possibly promising in itself, may be suggested as the only way to the road we wish to travel. Today, the solitary inventor, tinkering in his shop, has been overshadowed by task forces of scientists in laboratories and testing fields. In the same fashion, the free university, historically the fountainhead of free ideas and scientific discovery, has experienced a revolution in the conduct of research. Partly because of the huge costs involved, a government contract becomes virtually a substitute for intellectual curiosity. For every old blackboard there are now hundreds of new electronic computers ... The prospect of domination of the nation's scholars by Federal employment, project allocations, and the power of money is ever present – and is gravely to be regarded ... Yet, in holding scientific research and discovery in respect, as we should, we must also be alert to the equal and opposite danger that public policy could itself become the captive of a scientific-technological elite.

[48] Daniel S. Greenberg, *The Politics of Pure Science* (Chicago: University of Chicago Press, 1999), 15. Originally published in 1967.

Ike's words are remarkable for the breadth of his analysis and awareness. Nearly every aspect of the American relationship to science is present, beginning with the American faith in science – the "temptation" to support "unrealistic programs" in hopes of finding "the miraculous solution." Reflecting the times, the president highlighted the shift toward large scientific enterprises – moving from the "solitary inventor" to "task forces of scientists" overly reliant on federal dollars, while decrying the impact of "the power of money" on universities and "intellectual curiosity." However, though Eisenhower worried about public policy becoming "the captive of a scientific-technological elite," government influence over science at home meant influence abroad, as the United States could leverage scientific cooperation, access, and aid as part of Cold War diplomacy.

COOPERATION

In the tragic situation which confronts humanity, we feel that scientists should assemble in conference to appraise the perils that have arisen as a result of the development of weapons of mass destruction, and to discuss a resolution in the spirit of the appended draft. We are speaking on this occasion, not as members of this or that nation, continent, or creed, but as human beings, members of the species Man, whose continued existence is in doubt. The world is full of conflicts; and, overshadowing all minor conflicts, the titanic struggle between Communism and anti-Communism.

Bertrand Russell and Albert Einstein, "Manifesto," 1955

If the United States successfully launches the first satellite, it is most important that this be done with unquestionable peaceful intent. The Soviet Union will undoubtedly attempt to attach hostile motivation to this development in order to cover her own inability to win this race. To maximize our cold war gain in prestige and to minimize the effectiveness of Soviet accusations, the satellite should be launched in an atmosphere of international good will and common scientific interest.

Secretary of State John Foster Dulles, 1956[49]

The Russell/Einstein "Manifesto" represented the internationalism championed by a few prominent scientists. Updating a classic tradition, the two claimed to represent not only themselves, but the "species Man."

---

[49] Dulles quoted in Needell, *Science, Cold War and the American State*, 340.

However, while Russell and Einstein imagined the scientific community as above politics, the reality was far murkier. Only a handful of scientists aspired to a geopolitical role and vocal support for internationalism could arouse suspicion in an era of highly charged anti-communism; support for research was a far greater concern for most. Nor was internationalism always idealistic: Instead, it often provided a useful vehicle for scientists to secure funding and for nations to garner intelligence and influence – consider the Dulles quote above about using "international good will and common scientific interest" as the background to an American satellite coup. Of course, Sputnik would surprise Dulles, as the landscape was more crowded and complex than either the Manifesto or the American Secretary of State imagined.

The European-American scientific community contained numerous actors in the early Cold War: nations looking to utilize international research for nationalist goals; individual researchers supportive of both scientific internationalism and/or increased funding; international science organizations, such as ICSU and UNESCO, looking to increase prestige and influence; and various other blocs, like the European Economic Community or the Warsaw Pact. Nuclear energy, representing the pinnacle of scientific achievement and power, was the primary preoccupation of most nations; two of the largest international scientific undertakings in the 1950s were the Atoms for Peace campaign and the creation of a European nuclear reactor. Prominent individuals played key roles in proposing international undertakings, often using personal connections and the media to pressure hesitant nations. In turn, national priorities shaped international science. The United States, for example, played the dominant role in the major scientific initiatives of the period, beginning with civilian nuclear energy.

American Nuclear Diplomacy: Atoms for Peace

The Atoms for Peace (AFP) program served multiple diplomatic ends. The atomic bomb remained a problematic symbol of American power in the early Cold War and the destruction of Hiroshima and Nagasaki tainted the greatest accomplishment of American ingenuity to date – nuclear energy. After the failure of the Baruch plan and the UNAEC, President Truman restarted disarmament talks in 1950 to minimize criticism of US testing and engage the Soviets. As the race for the thermonuclear bomb heated up a few years later, President Eisenhower initiated Operation Candor – a public-relations campaign to shape perceptions of atomic energy and ease global fears. Speaking before the UN in 1953, Eisenhower committed the United States to the peaceful exploitation of atomic energy through an "Atoms for Peace" program.[50] Ike invited the Soviets, who had limited nuclear reserves, to contribute to an energy stockpile for the "power starved" areas of the world. In response, the Soviets decried the program as a ploy to direct attention away from America's continuing weapons program and advocated complete disarmament. American motives were complex: Historian John Krige observes, "Atoms for Peace was not simply an instrument of propaganda, an attempt to promote a non-bellicose image of the United States abroad and to allay the fear of the nuclear at home. It was also intended to divert skills and resources from Moscow's military program and to restrict developing nations to purely civil activities."[51] Of course, many countries wanted access to atomic power regardless of American motives.

AFP was a diplomatic and commercial success. The scope of the campaign was unprecedented: Universal Pictures produced a short documentary called "Atomic Power for Peace" translated into forty-one languages, USIA distributed more than 3 million AFP pamphlets, and traveling AFP exhibits drew large crowds in more than 200 diplomatic posts abroad.[52] Spurred by the United States, the United Nations sponsored an International Conference on the Peaceful Uses of Atomic Energy, held in Geneva in 1955. Attended by representatives from more than seventy nations, the conference presented an opportunity for the United States to project a peaceful image while gathering intelligence on Soviet nuclear programs. It was also an excellent occasion to sell American wares; General Electric (GE), for example, installed a working reactor in Geneva (later sold to Switzerland for $180,000).[53] Subsidized by American aid programs, reactors built overseas recycled foreign aid money back into the American economy while providing influence over atomic programs abroad.

Reactors became bargaining chips in American Cold War diplomacy. Eugene Skolnikoff, who served on White House science committees during the Eisenhower and Kennedy administrations, remembers, "the United States mounted a bilateral, and in many ways competitive [to the

---

[50] Kenneth Osgood, *Total Cold War: Eisenhower's Secret Propaganda Battle at Home and Abroad* (Lawrence: University of Kansas Press, 2006), esp. 162.

[51] John Krige, "Atoms for Peace, Scientific Internationalism and Scientific Intelligence," *Osiris* 21 (2006): 161–181, quote on 163.

[52] Osgood, *Total Cold War*, esp. 169–171.   [53] John Krige, "Atoms for Peace," 175.

UN] Atoms for Peace Program in which it supplied nuclear research reactors or atomic materials to more than forty countries."[54] American nuclear diplomacy solidified foreign alliances, often securing basing rights and favorable trading conditions, whether in Argentina, Belgium, and Brazil (major suppliers of uranium) or fascist Spain and apartheid South Africa.[55] At the same time, the State Department, fearing potential criticism, worked to sideline UNESCO in assessing civilian atomic power.[56] In the words of a later congressional study, AFP was "unique as an example of international cooperation in scientific fields, in that [cooperation] came about as the result of deliberate decision of governments rather than of scientific communities."[57] The study also concluded the program "helped open markets for the infant US nuclear industry."[58] Japan and Yugoslavia provide two instructive examples.

Atomic energy was understandably controversial in Japan. The destruction of two major cities and the American occupation's unwillingness to treat atomic casualties (see Chapter 1) undermined Japanese-American agreement at first. However, the United States removed prohibitions on nuclear research in 1953; following Eisenhower's Atoms for Peace speech at the UN, Japan committed 250 million yen toward nuclear energy. After the infamous *Fukuryu Maru* incident – in which fallout from the US Bravo atomic test sickened Japanese fishermen (killing one) – the United States launched the Peace Mission for Atomic Power in Japan. Tokyo hosted the first AFP exhibit in Asia, which opened with Shinto priests and State Department employees sanctifying the exposition, followed by a message from President Eisenhower.[59] USIA produced a movie – the "Blessing of Atomic Energy" – showing Japanese atomic

---

[54] Eugene B. Skolnikoff, *Science, Technology and American Foreign Policy* (Cambridge, MA: Massachusetts Institute of Technology PressMassachusetts Institute of Technology Press, 1969), 27.

[55] Krige, "Atoms for Peace, Scientific Internationalism and Scientific Intelligence,"161–181.

[56] Jacob Darwin Hamblin, "Exorcising Ghosts in the Age of Automation: United Nations Experts and Atoms for Peace," *Technology and Culture* 47 (October 2006): 734–756.

[57] See "Case Two – Commercial Nuclear Power in Europe: The Interaction of Diplomacy with a New Technology," in House of Representatives, *Science, Technology and Diplomacy in the Age of Independence* (Washington, DC: Government Printing Office, 1976), 28–45.

[58] Ibid.

[59] Robert Trumbull, "Japan Welcomes Peace Atom Show: Shinto Purification Rites Open U.S. Exhibit in Tokyo – The President Sends Message," *NYT* (November 1, 1955): 14.

scientists working with American assistance.[60] The campaign worked: Japan set up an Atomic Energy Commission in 1956 and imported its first American reactor the following year; the "Blessing" traveled throughout Asia, affirming American-Japanese cooperation.[61]

In Yugoslavia, the United States hoped to use atomic energy to maintain the communist country's independence from the Soviet Union. The Eisenhower administration advocated for the inclusion of Yugoslavia in a new European research reactor and made the nation a priority after the construction of a Soviet reactor. Eisenhower's operations board concluded, "opportunities should be utilized for cooperation in the unclassified, peaceful uses of atomic energy, including the training in the United States of Yugoslav scientists in non-sensitive fields."[62] Reactors were part of "economic and technical assistance, both of which are aimed at helping avoid undue Yugoslav dependence on the Soviet bloc."[63] Historian Jacques Hyman suggested the country received a rare American "trifecta:" a reactor grant and research grant combined with additional funding and assistance.[64] Civilian nuclear energy furthered American diplomacy for decades, whether to keep a European nation non-aligned (Yugoslavia) or to improve relations and sales in the Asian nation (Japan) most affected by the atomic bombs and fallout. Indeed, the success of the American program stimulated alternatives to US dominance.

The UN and European Community also pursued atomic energy. The International Atomic Energy Agency (IAEA), established in 1957 with American support, resurrected earlier UN plans for an internationally supervised inspection regime. Meanwhile, the European Economic Community included atomic energy in the Treaty of Rome, creating the European Atomic Energy (EURATOM). But neither the IAEA nor EURATOM began well: The nuclear nationalism of France, the UK, and Germany undermined EURATOM, increasing each individual European states' dependence on the United States, while US bilateral agreements

---

[60] Tsuchiya Yuka, The Atoms for Peace USIS Films: Spreading the Gospel of the "Blessing" of Atomic Energy in the Early Cold War Era," *International Journal of Korean History* 19 (August 2014): 107–133.

[61] Grunden, "Wartime Nuclear Weapons Research in Germany and Japan," 128.

[62] Operations Coordinating Board, "Operations Plan for Yugoslavia (August 6, 1958)" *FRUS, Foreign Relations, 1958–1960, vol 10*: 350–357, quote on 350.

[63] Ibid., 357

[64] Jacques E. C. Hymans, "Proliferation Implications of Civil Nuclear Cooperation: Theory and a Case Study of Tito's Yugoslavia," *Security Studies* 20 (2011): 73–104.

superseded IAEA oversight. By the time of the Geneva conference on atomic energy, for example, the United States had already negotiated two dozen bilateral reactor agreements, beginning with Turkey, a recent NATO member.

In hindsight, Atoms for Peace illustrated the complexities of atomic energy in American foreign relations. The political scientist David Dickson considered AFP "successful" because "it created an institutional framework within which the United States was able to reap a maximum economic and political advantage from its position of world leadership in nuclear science and technology."[65] From his perspective, the conference in Geneva was "a science fair, where the rich nations competed against each other to sell their technological wares to the poor."[66] Historian John Krige highlighted the relationship between scientific "internationalism" and "intelligence," observing, "[Scientific internationalism] pushed back the frontiers of security restrictions and mutual distrust, enabling scientists to build together a shared body of public knowledge. [Scientific intelligence] exploited that trust to learn what others were doing, to establish the limits of what they could speak about freely, and to assess the dangers that may lurk behind what they left unsaid."[67] Finally, historians Richard Hewlett and Jack Holl add another caveat: "The problem was that international promotion and control of atomic energy were contradictory; the success of the one tended to hurt the cause of the other."[68] As the Eisenhower administration promoted civilian nuclear energy worldwide, it also promoted a shared research reactor to foster European anti-communist solidarity.

The European Center for Nuclear Research (or CERN)

The European reactor project aimed to assert European nuclear preeminence. The origins of the *Conseil European pour la Recherche Nuclaire* (CERN) are clouded by memory; many accounts tend to emphasize the role of leading scientists, rather than politics, in its creation.[69] Numerous

---

[65] Dickson, *The New Politics of Science,* 174   [66] Ibid., 194

[67] Krige, "Atoms for Peace,"167

[68] Richard G. Hewlett and Jack M. Holl, *Atoms for Peace and War, 1953–1961: Eisenhower and the Atomic Energy Commission* (Berkeley: University of California Press, 1989), quote on 307.

[69] For example, although Danish physicist Niels Bohr is often cited as an early supporter, that was not the case. See Dominique Pestre, "Commemorative Practices at CERN: Between Physicists' Memories and Historians' Narratives," *Osiris* 14 (1999): 203–216.

scientists discussed the idea of a European research reactor after World War II: The Italian physicist Edoardo Amaldi discussed a joint reactor with his British partner in 1946; French physicist Louis de Broglie made a similar proposal a few years later.[70] Finally, the American physicist Isidor Rabi proposed a joint research reactor at the UNESCO General Assembly in 1950. While Rabi hoped to "preserve the international fellowship of science" and guarantee that American physicists had "somebody to talk to," European nations had their own reasons for participating.[71]

By the mid-1950s, West Europeans agreed on a few common assumptions: first, resources such as coal and steel should be shared and no single state had the funds for a research reactor; second, the proposed reactor would not intrude on national military programs and would guarantee access to cutting-edge research; and third, the reactor would help support the revitalization of European physics.[72] Additionally, the facility would be constructed on neutral European soil – Switzerland – to avoid Cold War geopolitics. It was an excellent opportunity: by 1953, twelve governments had contributed funds.[73]

American anti-communism influenced early CERN policies. Rabi's proposal bore the hallmarks of the US State Department, which considered the reactor a means of countering Soviet influence.[74] The State Department promoted CERN as a "safe way" for West Germans to engage in atomic research, helping foster an attitude of international cooperation among West German physicists.[75] The center also excluded suspect French physicists, like the Left-leaning Joliot-Curies, at American insistence; when Stanford researcher Felix Bloch became the first director general in 1954, his dual citizenship (Swiss and US) caused French researchers to argue the reactor maintained US domination of atomic

[70] Sameen Ahmed Khan, "CERN's Early History Revisited," *Physics Today* 58 (April 2005): 87–89.

[71] Rabi quoted in Krige, *American Hegemony and the Postwar Reconstruction of Science in Europe*, quotes on 58 and 59.

[72] Dominique Pestre and John Krige, "Some Thoughts on the Early History of CERN," in Gallison, *Big Science*, 78–99.

[73] The twelve founding member states were: Belgium, Denmark, France, the Federal Republic of Germany, Greece, Italy, the Netherlands, Norway, Sweden, Switzerland, the United Kingdom, and Yugoslavia. Note that the German Democratic Republic, or East Germany, was not included. See Francois de Rose, "Paris 1951: The Birth of CERN," *Nature* 455 (September 2008): 174–175, quote on 175.

[74] Needall, *Science, Cold War and the American State*, 141–149.

[75] Richard H. Beyler and Morris F. Low, "Science Policy in Post-1945 West Germany and Japan: Between Ideology and Economics," in Walker, ed., *Science and Ideology*, 97–123.

research.[76] Finally, CERN invited Yugoslavia to participate, leading the Soviets to establish the Joint Institute of Nuclear Research for Soviet-bloc countries in 1956 (the year before, Mao and Stalin signed an agreement to cooperate on nuclear research).[77]

CERN soon discarded its Cold War trappings. Within fifteen years, the center employed more than 2,800 and maintained a budget of $50 million. A political study of CERN researchers determined most were cosmopolitan in outlook – not politically active but interested in grand, global questions.[78] The study concluded the physicists tended toward the political Left and supported the further integration of Europe, though without US participation, stressing European independence. Indeed, the study argued scientists from Britain, France, and Germany were more like each other than non-scientists from their own countries; that is, scientists at CERN tended toward "supra-nationalism" – politics beyond nationalism – and were often willing to unilaterally surrender potentially dominant nuclear forces.[79] A similar spirit of supra-nationalism also infused the largest scientific undertaking of the 1950s, the International Geophysical Year (IGY).

The International Geophysical Year

Like CERN, the IGY's origins are often romanticized. A standard narrative begins with a dinner party in Silver Springs, Maryland, hosted by American astrophysicist James Van Allen in 1950. Other notable Americans in attendance were geophysicist Merle Tuve and physicists Fred Singer and Lloyd Berkner. At the time, the refraction of radio waves in the upper atmosphere particularly intrigued Allen and others because refraction allowed the waves to circle the globe rather than travel straight out into space. After much discussion, the diners proposed adding atmospheric physics and other fields to a third international polar year. Over time, the narrative continues, these individual scientists cajoled hesitant nations into supporting the largest scientific undertaking to date.

While correct, the narrative downplays the influence of the US government and international scientific organizations. For example, both Singer

---

[76] John Krige, "Felix Bloch and the Creation of a 'Scientific Spirit' at CERN," *Historical Studies in the Physical and Biological Sciences* 32 (2001): 57–69.

[77] Holloway, *Stalin and the Bomb*, esp. 354 and 368.

[78] Daniel Lerner and Albert Teich, "Internationalism and World Politics among CERN Scientists," *Bulletin of the Atomic Scientists* (February 1970): 4–10.

[79] Ibid.

(the ONR liaison in London) and Berkner – who held positions in the DOD, CIA, and State Department – saw the initiative as aligned with American national security goals.[80] As head of Project Troy, Berkner suggested international cooperation to strengthen the anti-communist scientific community. Spurred by the physicist and advances in instrumentation, a few scientific unions proposed a third polar year only a month after Van Allen's fateful dinner.[81] The International Council of Scientific Unions (ICSU) and UNESCO considered the project an excellent scientific and institutional opportunity, welcoming the initiative and American dollars. And with Berkner installed as chairman, the IGY planning committee represented American interests alongside ICSU and the UN.

US representatives shaped the initiative throughout the planning stages. Given the unique global opportunity, the IGY committee emphasized research requiring simultaneous observations at multiple locations and in regions considered inaccessible without extraordinary effort. Special attention would be paid to the Arctic and Antarctic as well as three meridians: one running down the West coast of the Americas, one through Europe and Africa, and one through Japan and Australia. Berkner's previous work at the DOD highlighted the need for North–South observing stations; not coincidentally, these were among the meridians chosen for study.[82] Research goals and national needs often went hand-in-hand: Research in the upper atmosphere, which required developing rockets and artificial satellites to conduct experiments, would provide cover for American missile and reconnaissance programs and establish the "freedom of space."[83] Soon the program included oceanography, nuclear radiation, and other areas of national interest; by 1952, the "Third International Polar Year" had become the "International Geophysical Year." To take full advantage of the natural sun cycle, the actual "year" would begin in July 1957 and continue until December 1958.

---

[80] Lloyd Berkner was the Secretary of the DOD's Research and Development Board, worked with the Office of Scientific Intelligence at the CIA, and headed Project Troy for the State Department, where he suggested a reorganization of the foreign service to increase scientific intelligence gathering. See Needall, *Science, Cold War, and the American State.*

[81] The scientific organizations involved were URSI, IUGG, and IAU. See Frank Greenaway, *Science International: A History of the International Council of Scientific Unions* (New York: Cambridge University Press, 1996).

[82] Needall, *Science, Cold War and the American State*, 302.

[83] See "Proposing Satellites for the IGY," in Rip Bulkeley, *The Sputniks Crisis and the Early United States Space Policy: A Critique of the Historiography of Space* (Bloomington: Indiana University Press, 1991), 89–103.

The participation of communist countries, especially China, troubled American officials. Neither the Soviet Union nor the People's Republic of China was among the twenty-six original supporting countries in 1952. Worse, the United States was fighting a proxy war with both in Korea. Nonetheless, the USSR was a member of ICSU, with significant capabilities in astronautics and meteorology. After the death of Stalin and the Korean armistice, the Soviet Academy of Sciences accepted a long-standing invitation to participate. The Chinese, following the Soviet lead, soon signaled their intention to participate; the *People's Daily* called the IGY "not only a great movement in the development of geophysics, but also a new approach to international scientific collaboration."[84] However, when the Eisenhower administration pushed to include Taiwan, the PRC withdrew, arguing that the United States was using international scientific collaboration to create "two Chinas." With various Taiwan straits crises as the backdrop, geopolitical tensions with China worsened; when the Soviets refused to grant visas to Taiwanese scientists hoping to attend an IGY conference in Moscow, the US State Department urged American scientists walk out.[85] Ultimately, Chinese non-involvement weakened the status of Western-leaning scientists in China and undermined collection of data. But the United States and Soviet Union remained committed.

With both superpowers in support, the International Geophysical Year became the supreme example of international scientific collaboration during the Cold War. Coordinated by ICSU and national scientific academies, more than sixty countries participated in hundreds of research projects. Tens of thousands of scientists and volunteers staffed more than 8,000 observation stations around the world. The United States played the critical role: the US Navy handled much of the logistics (an example of military/civilian scientific collaboration), various national agencies guided people and equipment through customs and US contributions probably ran into the hundreds of millions.[86] Among the many scientific triumphs,

[84] The *People's Daily* is translated and quoted in Zuoyue Wang and Jiuchen Zhang, "China and the International Geophysical Year," in R. Launius and J. Fleming, eds., *Globalizing Polar Science: Reconsidering the International Polar and Geophysical Years* (New York: Palgrave-Macmillan, 2011), 143–153, quote on 148.

[85] Fae L. Korsmo, "The Genesis of the International Geophysical Year," *Physics Today* (July 2007): 38–43.

[86] In 1976, a Congressional study noted Congress appropriated $43 million for the IGY, although the study also concluded total U.S. contribution for travel, accommodations, and other services probably ran to $500 million. See House of Representatives, *Science, Technology, and Diplomacy*, 49.

IGY research confirmed the existence of the Van Allen radiation belts and demonstrated deep ocean trenches were inappropriate for dumping nuclear waste because currents quickly carried the waste upward. However, the data was only one consequence, as the Year also led to the still functioning Antarctic Treaty, multiple global scientific networks and the launch of Sputnik.

The Antarctic was of interest to many countries before the IGY. Immediately after World War II, a minor scramble took place, with the United States, Chile, and Argentina building bases. A decade later, the IGY provided the occasion to reassess the region, although geology was officially left out to dampen a potential gold rush of claims. The frozen continent afforded a few opportunities for geopolitical competition: the American base was "deliberately sited at the conjunction of six of the seven territorial claims," while the Soviets recorded the lowest temperature.[87] Not to be outdone, the Japanese advertised their Antarctic expedition as a renewal of national spirit: the *Asahi* newspaper proudly proclaimed Japanese researchers could "successfully perform their duties and attain results that were equal to or greater than that of the Europeans or Americans."[88] In the end, however, the continent became a symbol of international cooperation, rather than competition, with the signing of the Antarctic treaty, which banned nuclear testing, stipulated a country must be doing research to become a member and required the free exchange of information.[89] The Scientific Committee on Antarctic Research (SCAR), created simultaneously, still oversees the treaty.

The IGY spawned numerous global institutions and networks, like SCAR, which provided regular opportunities for continued US and Soviet dialogue and cooperation. As early as 1952, UNESCO and ICSU established a publications board, based in Belgium, to provide scientific abstracting in English and French. A few years later, both the United States and the Soviet Union volunteered to house World Data Centers containing all IGY data; Japan, Australia, and other European states maintained partial archives. The Data Centers pioneered microcard

---

[87] Adrian Howkins, "Science, Environment, and Sovereignty: The International Geophysical Year in the Antarctic Peninsula Region," in *Globalizing Polar Science* 245–264, quote on 255.

[88] The Asahi newspaper is quoted in Walter R. Stevenson III, "The Polar Years and Japan," in *Globalizing Polar Science* (123–142), 134.

[89] The Antarctic treaty is reproduced in Richard Fifield, *International Research in the Antarctic* (Oxford: Oxford University Press, 1987), quote on 5.

databases and served as the foundation for the CODATA system set up in Bombay a decade later (which archives quantitative scientific data down to the present). IGY success also laid the groundwork for future global collaborations, including the Upper Mantle Program, the International Geodynamics Project, and the International Biology Program (covered in Chapter 4).[90] Yet even with this impressive legacy, the IGY is most remembered for Sputnik.

The Freedom of Space and Sputnik

The Eisenhower administration considered satellites an especially sensitive area because the rockets used to launch research satellites could also serve as the basis for intercontinental missiles. Additionally, satellites had the potential to conduct reconnaissance over enemy territory or serve as the platform for the weaponization of space. American policy-makers had to prioritize whether national security or science came first in releasing information on satellite/missile/reconnaissance programs. By 1950, for example, there were three competing rocket programs within the military: the Air Force (Atlas), the Army (Redstone), and the Navy (Vanguard). At the same time, the RAND corporation and Air Force worked on high-altitude photo-reconnaissance of the Soviet Union.

The administration hoped to use IGY satellites to establish the freedom of space for Cold War reconnaissance. When the Soviets rejected Eisenhower's "Open Skies" proposal to allow for reconnaissance flights, multiple groups within the administration, including the NSC, recommended expanding satellite programs.[91] NSC 5520 stated: "Considerable prestige and psychological benefits will accrue to the nation which first is successful in launching a satellite. . . . Furthermore, a small scientific satellite will provide a test of the principle of 'Freedom of Space.'"[92] The analysis focused on the benefits of satellite reconnaissance and maintaining a separate, and secret, missile program, relegating projected scientific gains to an appendix.[93] As planned, Eisenhower announced the American

---

[90] CRS, *Science, Technology and Diplomacy,* 53–55.

[91] On "Open Skies," see David Tal, "From the Open Skies Proposal of 1955 to the Norstad Plan of 1960: A Plan Too Far," *Journal of Cold War Studies* 10 (Fall 2008): 66–93.

[92] NSC 5520 – "U.S. Scientific Satellite Program" (May 20, 1955) is archived online at the Marshall Institute at www.marshall.org/pdf/materials/805.pdf.

[93] Scientific benefits included learning about air drag from orbital decay patterns and the gravitational field of the earth, as well as determining the exact content of the ionosphere. See ibid.

satellite program within the civilian context of the IGY, helping establish the legal legitimacy of overflights: flights done for science would be permitted, flights done for military surveillance would not.

The American announcement set off waves at home and abroad. Throughout the summer of 1955, the services vied to provide the IGY launch vehicle. Ultimately, the Navy's Project Vanguard was chosen because it was not a military vehicle, although Von Braun and others at the Army's Redstone Arsenal remained heavily funded and eventually succeeded. Meanwhile, the Soviets presented their IGY research program in Brussels in 1955, but failed to mention rockets, choosing to announce their satellite program in Barcelona the following September (one year before Sputnik). Japan soon followed suit, as several countries proposed to use rocket launched instruments to experiment on the upper atmosphere. The involvement of multiple countries required additional diplomacy.

During negotiations, US representatives determined many of the new satellite regulations. Given the sensitive nature of satellite programs, the administration did not propose nations share details on launch vehicles. Instead, IGY members, led by the US representatives, established rules mandating a two-hour warning notice before a launch, data reporting, and specific frequencies for transmissions. Additionally, participating nations were expected to create a ground system to track the satellite once in orbit. But the committee had few means of enforcement and the Soviet Union disagreed with many stipulations.

The American tracking system illustrated the scientific and diplomatic benefits of participation. At the request of the US IGY committee, the Smithsonian Astrophysical Observatory and Harvard University set up a satellite tracking system with branches in Japan, Australia, India, Peru, Iran, and Ethiopia. The United States outfitted each station with a new Baker-Nunn camera telescope, specifically designed for tracking atmospheric objects. J. Allen Hynek, head of the program, believed the outposts left behind a legacy of "many semi-permanent nuclei for even greater scientific cooperation.[94] For diplomats, the Baker-Nunn stations provided geopolitical benefits as well as observations. The State Department report on the Japanese station, for example, was blunt: "We feel that in the long run scientific interchange is the best remedy for Japanese emotion and

---

[94] Hynek quoted in Teasel Muir-Harmony, "Tracking Diplomacy: The International Geophysical Year and American Scientific and Technical Exchange with East Asia," in *Globalizing Polar Science*, 279–306, quote on 285.

ignorance and we intend to push such projects. ... This is essential if we are to count upon the use of Japanese bases and other cooperation in any future conflict."[95] Of course, most participants assumed they would track an American, not Soviet, satellite first.

Three hours after launch, on October 4, 1957, Radio Moscow announced the successful orbiting of Sputnik, stunning the American public. Although only a small sphere, roughly twenty-two inches in diameter and weighing around 184 pounds, Sputnik galvanized the world's attention with its faint beeps sent from outer space. The Eisenhower administration tried to downplay the significance of the Soviet success; Ike went on a planned golf excursion while Secretary of Defense Wilson called the satellite "a nice scientific trick."[96] But Sputnik called into question American leadership in science and technology. Indeed, the State Department privately reported Soviet claims of scientific superiority were being reevaluated in capitals around the world; Khrushchev proclaimed Sputnik should convince "the people of Russia, China, India as well as Europe that our system is the best."[97] To make matters worse, the Soviets launched Sputnik II – weighing more than 1,000 pounds and carrying the dog Laika – on November 3. In less than a month, the Sputniks overturned American assumptions regarding Soviet science, made space science and technology the focus of public attention, and created a public relations nightmare for the government at home and abroad.

The Eisenhower administration tried to contain the damage. In the words of historian Rip Bulkeley, the administration quickly created a "bogus but meretricious distinction between its own peaceful, scientific, and allegedly disinterested satellite project and the somehow more sinister and less noble, if more effective, Soviet one."[98] All federal departments were to congratulate the Soviets and avoid discussing the military implications of satellites or any possibility of a space "race." Instead, rumors

---

[95] The State Department memo is quoted in ibid., 285.

[96] Wilson quoted in Robert A. Divine, *The Sputnik Challenge: Eisenhower's Response to the Soviet Satellite* (New York: Oxford University Press, 1993), quote on xv. American intelligence had been warning the administration of a possible Soviet launch, see Amy Ryan and Gary Keeley, "Sputnik and US Intelligence: The Warning Record," *Studies in Intelligence* 61 (September 2017): 1–16.

[97] Khrushchev quoted in Michael Sheehan, *The International Politics of Space* (New York: Routledge, 2007), quote on 22. Regarding the State Department, see Osgood, *Total Cold War*, esp. 338.

[98] Bulkeley, *The Sputniks Crisis*, 160.

circulated that captured Nazi scientists, rather than Soviets, were truly responsible (although America's own captured Nazi, Werner Von Braun, would later deny this to Congress). When the US Vanguard rocket managed to climb only a few hundred feet into the air before bursting into flames on December 6, the situation became a crisis. What began as the United States using international scientific cooperation – the IGY – as the backdrop for the American satellite program had become a major source of global scientific competition.

COMPETITION

Governmental support for basic research should be directly influenced by psychological considerations. Obviously if the United States were able to score "firsts" in such areas as the significant prolongation of life or controlled thermonuclear reaction, the value to national prestige would be enormous ... some basic research should be concentrated in those fields which hold the most promise for scientific discoveries which will enhance our prestige abroad.

President's Committee on Information Activities Abroad, 1958[99]

In early 1965 [President Johnson] telephoned me (in my kitchen) to complain of the preparations for the visit the next morning by the Prime Minister of Japan. He wanted a good idea to present. After all night telephone discussions across the country I presented him with a proposal while we waited for the Prime Minister to arrive. One hour later the US-Japan Medical Program was born. ... It became clear to me that being benevolent and progressive, science is a marvelous lubricant in international affairs.

President Science Advisor Donald F. Hornig, reminiscing in 1980[100]

A grim determination to beat the Soviets gripped the three administrations after Sputnik. The faith in American supremacy in science and technology, so shaken by the Soviet satellite, had to be repaired; the psychological blow demanded a grand response. Indeed, immediately after Sputnik, General Thomas D. White, the Air Force chief of staff, told

---

[99] President's Committee on Information Activities Abroad, "Conclusions and Recommendations," 38.
[100] Donald F. Hornig, "The President's Need for Science Advice: Past and Future," *Technology in Society* 2 (1980): 41–52.

Congress, "we ought to hit the moon as fast as we can."[101] The President's Committee on Information Activities Abroad suggested concentrating research in areas where the United States could score "firsts" and gain the most prestige; proposals ran from using nuclear bombs to excavate harbors and canals to global desalination and the infamous "Mohole" project to drill into the Earth's crust. But George Allen, director of the US Information Agency, advised leaders space had become "the primary symbol of world leadership in all areas of science and technology" after Sputnik.[102] Presidents Eisenhower, Kennedy, and Johnson agreed, as all three supported an evolving space program to varying degrees (Eisenhower, for example, maintained a stingy approach to the moonshot, whereas Kennedy lavished funds on the Apollo program). Over the next twelve years (1957–1969), the United States reorganized its scientific bureaucracy, renegotiated relations with allies, and landed a man on the moon.

Reorganization at Home

The federal scientific bureaucracy expanded after the Sputniks. American politicians and the public wanted reassurance the country remained the preeminent scientific nation. Scientists, meanwhile, recognized the opportunity to secure more funding, driving federal outlays for research to new heights. By 1962, Don Price, the Dean of Public Administration at Harvard, could opine, "science has become the major establishment in the American political system: the only set of institutions for which tax funds are appropriated almost on faith."[103]

Institutionalization occurred almost immediately in the executive branch. On November 7, 1957 – only four days after Sputnik II – Eisenhower introduced James Killian as the first Presidential Science Advisor and head of a new President's Science Advisory Committee (PSAC).[104] Although created in crisis, Ike emphasized long-term benefits: "In conclusion, although I am now stressing the influence of science on defense... the peaceful contributions of science – to healing, to enriching

---

[101] White quoted in Divine, *The Sputnik Challenge*, 108.
[102] Allen quoted in Sheehan, *The International Politics of Space*, 21.
[103] Greenberg, *The Politics of Pure Science*, 270–271.
[104] PSAC arose from the Office of Defense Mobilization's Science Advisory Committee. See Zuoyue Wang, *In Sputnik's Shadow: The President's Science Advisory Committee and Cold War America* (New Brunswick, NJ: Rutgers University Press, 2008). See also Gregg Herken, *Cardinal Choices*, 101–107.

life, to freeing the spirit – these are the most important products."[105]
One of PSAC's original recommendations was to coordinate federal
programs, leading to the Federal Council on Science and Technology
(FCST). Established by executive order and composed of all relevant
government agencies, FCST set up an International Committee to recommend "scientific activities compatible with our foreign policy."[106] As
science and American foreign relations intersected, PSAC's responsibilities expanded, adding an Office of Science and Technology to partner
with the legislative branch.[107]

Congress acted quickly to meet the Soviet challenge in scientific man-
power through education. The issue predated Sputnik; as early as 1955,
the scarcity of American researchers concerned the US Office of Scientific
Personnel and NATO allies.[108] NAS and NSF studies stressed the Soviet
emphasis on science and engineering. Although President Eisenhower
questioned the communist party's ability to "inspire all of their people
with science," both he and the Congress knew the United States needed to
improve.[109] As such, the National Defense Education Act (NDEA),
passed in 1958, broke a long-standing deadlock on the federal role in
education. Federal standards for state science curricula, accompanied by
federal dollars, went out to the states, introducing evolution to a generation of students. But the NDEA was minor compared to the federal
investment in space research.

The first space-related agency created in response to Sputnik was the
DOD's Advanced Research Projects Agency (ARPA, February 1958).
Since the satellite not only demonstrated Soviet ambitions in space, but
also the ability to launch intercontinental missiles, ARPA originally
focused on three key areas: space, missile defense, and nuclear-test
detection.[110] However, given ARPA's home in the Department of

[105] President Eisenhower quoted in Neal and Smith, *Beyond Sputnik*, 28.
[106] House of Representatives, *Science, Technology and Diplomacy*, 147.
[107] For an overview of PSAC-OST relations, see Skolnikoff, *Science, Technology, and American Foreign Policy*, 226–244. See also Wang, *In Sputnik's Shadow*, 196–198.
[108] John Krige, "NATO and the Strengthening of Western Science in the Post-Sputnik Era," *Minerva* 38 (2000): 81–108, esp. 84–94.
[109] Eisenhower's questioning is noted in Anon., "Memorandum of a Conference with the President, White House, Washington, October 15," *1957 FRUS: Foreign Relations, 1955–1957* 19: 607–610.
[110] Richard Van Atta, "Fifty Years of Innovation and Discovery," available on the DARPA website at: www.darpa.mil/About/History/History.aspx. The article contains a copy of the February Defense Directive creating the agency. See also Weinberger, *The Imagineers of War*.

Defense, members of the Eisenhower administration questioned whether it should be the public face of the American space program; Killian and PSAC supported a civilian institution. Over the course of 1958, ARPA slowly took over more sensitive initiatives, including satellite reconnaissance, and faded from view. Another organization would represent the United States instead.

Legislation established the National Aeronautics and Space Administration (NASA) on July 29, 1958. Following the precedent set by the American IGY program, the Act created an explicitly civilian administration and appealed for international cooperation. The agency's position against the militarization of space pleased academics; a few years earlier, the International Congress on Astronautics voted to ban its members from using research for military purposes.[111] Assembled out of NACA and Von Braun's team of missile experts from the US Army, NASA had unclear goals at the outset. One of the agency's first uses was as a civilian decoy: Project Discoverer – the mission to launch animals into space – provided cover for the CIA to retrieve satellite photographs of the Soviet Union. Soon Project Mercury – the program to place a man into orbit and return him safely – took center stage; by the April 1959 press conference introducing astronauts, NASA's role in the manned space mission guaranteed it maximum publicity and funding, as the space program became a centerpiece of American diplomacy.

American Diplomacy after Sputnik

The United States redoubled diplomatic efforts after the Soviet satellite. To improve relations, the State Department and Congress initiated exchanges, while the CIA attempted to enlist scientists to gather intelligence. The United States expanded cooperation in unclassified research with NATO allies, but disagreements over limitations surfaced, especially with France. As the space program reached orbit, the United States advertised its success throughout Asia, while European allies struggled to achieve liftoff.

The State Department reorganized after Sputnik. The secretary appointed seven new science attachés within months; a few years later there were more than twenty attachés in seventeen embassies. The department also introduced scholarly exchanges in 1958, leading Congress to

---

[111] Sheehan, *International Politics of Space*, 6.

establish the much broader Fulbright program. Widespread interest caused the program to grow rapidly: A later survey found thirty-one government agencies administering exchanges.[112] Researchers felt they were useful. Dr. Harrison Brown, a longtime foreign secretary of the NAS, argued US–Soviet scientific exchanges led to "some rather extraordinary foreign policy changes. I have seen attitudes of scientists of one country change enormously as a result of these contacts."[113] Some hoped for more than a change in attitude.

The State Department and CIA attempted to use exchanges to gather intelligence. Embassy officials suggested visits to Soviet facilities would allow for "more reliable and detailed appraisal of Soviet science," offering Polaroid cameras to those interested.[114] But George Kistiakowsky, Eisenhower's science advisor, vetoed the suggestion, later writing, "I served an ultimatum that I would challenge the State Department action in requiring 'instruction' of scientists ... these instructions on how to vote, whom not to contact, etc., were causing considerable indignation in scientific academic circles."[115] Not to be outdone, the CIA initiated Operation Lincoln to train scientists for espionage, but they too had little success. Language and expertise presented barriers and the CIA struggled with recruitment; one operative reported: "In addition, many scientists, on finding out that some of its duties are on behalf of intelligence, will have nothing to do with it. They feel the association with 'spying' may jeopardize their scientific careers."[116] The second phase of the program, 1961–1962, focused on "elicitation," which the CIA defined as "friendly conversation to the point of revealing something useful."[117] It was also more successful; the CIA received "outstanding" intelligence at an international astronautics conference.[118] As American intelligence

---

[112] Study highlighted in House of Representatives, *Science, Technology, and Diplomacy*, p136 n183.

[113] Brown quoted in ibid., 151.

[114] David E. Mark, Despatch from the Embassy in the Soviet Union to the Department of State, *FRUS, Foreign Relations, 1958–1960* vol. 10, pages 14–18, quotes on 14.

[115] George B. Kistiakowsky, *A Scientist at the White House: The Private Diary of President Eisenhower's Special Assistant for Science and Technology* (Cambridge, MA: Harvard University Press, 1976), quote on 33.

[116] Wilton Lexow, "The Science Attache Program," *Studies in Intelligence* 10 (Spring 1966): 21–27, quote on 25.

[117] Robert Vandaveer, "Operation Lincoln," *Studies in Intelligence* 7 (Winter 1963): 69–73, quote on 72.

[118] Ibid., 73.

sought to learn about science behind the Iron Curtain, American allies hoped to expand collaboration with the United States.

NATO established a Science Committee only two months after the Soviet launch. European collaboration predated Sputnik: allies founded a number of shared research centers in the early 1950s, including an advisory group for aeronautics in Brussels, an air defense center in the Hague, and an underwater research station at La Spezia in Italy.[119] By 1957, the foreign ministers of Norway, Italy, and Canada had concluded further research was a promising area of cooperation.[120] Spurred into action, the new NATO Science Committee recommended research to knit countries together, beginning with summer schools and fellowship programs.[121] The United States took the lead: The first five NATO science advisors were Americans and the United States paid half the budget.[122] Only defense-related projects were off limits because not all members had access to classified research.

The suggestion to increase collaboration exacerbated tensions within the alliance. French representatives, unhappy with the classified Anglo-American partnership, proposed a foundation for shared research. However, while the French saw potential benefits, the British and Americans did not. In response, NATO released a report on "Increasing the Effectiveness of Western Science" in 1960.[123] The study argued that the Soviet planned economy made it easier to coordinate research, supporting the French plan for an International Institute of Science and Technology (IIST).[124] Members debated for two years, with many smaller countries (Luxembourg, Italy, Turkey, and Greece) favoring the IIST (both Italy and Greece petitioned to host the institute), while the Americans, British, and West Germans worried the institute would drain national resources and challenge domestic industries and

---

[119] Skolnikoff, *Science, Technology and American Foreign Policy*, 178–183. See also W. A. Nierenberg, "NATO Science Programs: Origins and Influence," *Technology in Society* 23 (2001): 361–374, esp. 364–365.

[120] The role of the "three wise men" is discussed in F. Carvalho-Rodrigues, "NATO's Science Programs: Origins and Influence," *Technology in Society* 23 (2001): 375–381, esp. 377–378. See also Jens Erik Fenstad, "NATO and Science," *European Review* 17 (October 2009): 487–97.

[121] Anon., "NATO Progress in Science," *Science* 129 (June 12, 1959): 1598.

[122] John Krige, "NATO and the Strengthening of Western Science in the Post-Sputnik Era," *Minerva* 38 (2000): 81–108.

[123] Anon., "NATO Proposes Establishment of International Science Institute," *Science* 138 (November 30, 1962): 962.

[124] Nierenberg, "NATO science programs," esp. 369.

universities. Although the IIST plan eventually failed, the debate stimulated European partnerships like the Anglo-French project for a supersonic airliner (the Concorde).[125]

Satellites and rockets further complicated US–European relations. American offers to launch satellites for NATO allies failed to alleviate fears of European dependence, while competition – whether from Canada, Japan, or the PRC – led to calls for European cooperation.[126] Dozens of aerospace companies formed a multinational lobbying front (Eurospace, 1961), while Great Britain, France, and others created organizations to build a launch vehicle (ELDO, 1962) and conduct space research (ESRO, 1964).[127] But nationalism grounded the Europeans at first: The British and French, for example, disagreed over using American rockets.[128] Meanwhile, American departments and policies were at cross-purposes: NASA and the State Department welcomed a European launch capability, but national security memos directed the United States limit rival satellite systems (NSAM 338) and contain French military technology (NSAM 294).[129] Soviet Premier Brezhnev tried to exploit the divide during a speech in Paris, arguing the United States was domineering in space, but American influence continued to expand.[130]

American diplomats appreciated science as "a marvelous lubricant in international affairs" during the space race. The State Department initially turned to atomic energy: Less than six months after Sputnik, the United States signed a bilateral nuclear power agreement with Italy. In return, the Eisenhower administration discouraged Italy from recognizing the PRC or East Germany and expected the country "win support for US policies," including Atoms for Peace, "US Science and the Geophysical

---

[125] Krige, *American Hegemony and the Postwar Reconstruction of Science in Europe*, esp. 220–223.

[126] Sheehan, *The International Politics of Space*, 74–77.

[127] Schwarz, "European Policies on Space Science and Technology, 1960–1978," 227 and Sheehan, *The International Politics of Space*, 81.

[128] Michiel Schwarz, "European Policies on Space Science and Technology, 1960–1978," *Research Policy* 8 (1979): 204–243.

[129] John Krige, "Technology Transfer with Western Europe: NASA-ELDO Relations in the 1960s," in John Krige, Angelina Long Callahan, and Ashok Maharaj, *NASA in the World: Fifty Years of International Collaboration in Space* (New York: Palgrave Macmillan, 2013), 51–64. "(NSAM 294) U.S. Nuclear and Strategic Delivery System Assistance to France" and "(NSAM 338) Policy Concerning U.S. Assistance in the Development of Foreign Communications Satellite Capabilities," are both available online at www.fas.org.

[130] Bryce Nelson, "Hornig Committee: Beginning of a Technological Marshall Plan?" *Science* 154 (December 9, 1966): 1307–1309.

Year," and "Open Skies and Disarmament."[131] When renewal of the Japanese–US security treaty caused local tension, Ambassador Reischauer suggested research as a solution; the US–Japan Committee on Scientific Cooperation was born shortly thereafter.[132] Working with the NSF, the Committee funded research of critical interest to the Japanese, such as earthquake engineering and energy.[133] After John Glenn's successful orbit, the State Department deployed science to reinforce Asian relations, helping to establish the Korean Institute of Science and Technology, a nuclear science center in the Philippines, and offering a TVA-style program for the Mekong Delta (see Chapter 4). At the height of the Vietnam War, Americans wishing to see a national triumph could always look to the heavens, rather than Southeast Asia.

The Moon Landing

Space programs, anchored in terrestrial politics, symbolized national power. But as historian Michael Sheehan points out, "the 'power' in question is a multifaceted amalgam of different forces ranging from tangible military capability to unquantifiable degrees of prestige."[134] From 1957 to 1969, the space race epitomized Cold War scientific competition. Wins carried benefits at home and abroad. The launch of Sputnik, for example, quieted domestic criticism of Khrushchev for the 1956 Hungarian invasion. The Soviets pursued a series of firsts afterward: first man into space (Gagarin, 1961), first woman into space (Tereskova, 1963), and the first to photograph the dark side of the moon (achieving naming rights for the Sea of Moscow and Mt. Lenin). Given Soviet successes, American presidents were determined to do better.

Presidents Kennedy and Johnson invested heavily and controversially in the space program. In 1960, outgoing science advisor Killian stated: "In the long run we can weaken our science and technology and lower our international prestige by frantically indulging in unnecessary competition and prestige-motivated projects."[135] The young president initially agreed: Kennedy's inaugural address downplayed the race and he refused to increase funding for the Apollo project (the moon landing). However,

---

[131] Operations Coordinating Board, "Operation Plan for Italy (November 7, 1958)" in *FRUS, Foreign Relations 1958–1960, vol 7*, 485–496. Quote on 495.
[132] Yoshikawa and Kauffman, *Science Has No National Borders*, 100.
[133] *STAD 1986*, 37.
[134] Sheehan, *The International Politics of Space*, 8.
[135] Killian quoted in Skolnikoff, *Science, Technology, and American Foreign Policy*, 218.

Yuri Gagarin's flight jolted Kennedy into action, as he stated, "if we can get to the Moon before the Russians, then we should."[136] Kennedy pressed for the lunar landing against the wishes of PSAC; his hand-picked science advisor argued there was no scientific rationale for the spending and the two reached an agreement: "Henceforth [Kennedy] never called it science; and I never opposed it once he made the decision."[137] After Kennedy's martyrdom, Johnson took up the space mantle with characteristic fervor, stating, "In the eyes of the world, first in space means first, period. Second in space is second in everything."[138] Nor did Johnson like to be second: By the late 1960s, space research received more federal funding than all other non-defense research combined.

The moon landing remains a consummate American achievement. The Apollo mission transformed NASA from a small, multifaceted organization into a large, focused organization with a wide network of associates: the space agency grew from 8,000 employees at its founding to more than 36,000 by 1969, with an additional 350,000 contractors spread throughout universities and industry.[139] After more than a decade of work and twenty billion dollars, the United States planted the American flag on the moon, an enduring symbol of American determination and scientific and technical accomplishment. Although critics question if the Apollo project was a wise use of resources, it was an undeniable diplomatic triumph, inspiring millions around the world.[140]

Yet NASA struggled after the landing. Economic concerns and waning public and scientific interest led the United States to cancel the last three Apollo missions and the agency suffered budget cuts only months after Armstrong's famous foray. NASA administrator Thomas O. Paine sought investors abroad, visiting Canada, Japan, Australia, and Western Europe

---

[136] Kennedy quoted in Sheehan, *The International Politics of Space*, 49. Audra Wolfe argues the Bay of Pigs failure and Alan Shepard's orbital flight also influenced Kennedy, see Audra J. Wolfe, *Competing with the Soviets: Science, Technology, and the State in Cold War America* (Baltimore, MD: Johns Hopkins University Press, 2012), 94.

[137] Weisner quoted in Herken, *Cardinal Choices*, 130.

[138] Johnson quoted in Sheehan, *International Politics of Space*, 52. Regarding Johnson and space more generally, see W. Henry Lambright, *Presidential Management of Science and Technology: The Johnson Presidency* (Austin: University of Texas Press, 2012), 103–111.

[139] Alfred K. Mann, *For Better or for Worse: The Marriage of Science & Government in the United States* (New York: Columbia University Press, 2000), 107.

[140] Wolfe, *Competing with the Soviets*, 99.

to gauge foreign interest in participating in the American program.[141] Many countries, led by West Germany, considered the offer a great opportunity, although others, especially France, feared US domination (France also partnered with the USSR and India). Ultimately, NATO allies decided to participate in the American space program as well as the new European Space Agency (ESA, 1975). Manned missions faded in importance, replaced by launch services and satellite communications, while space became the location of détente with the Soviet Union (see Chapter 5).

The Legacy of Early Cold War Scientific Competition

Scientific and technical competition was a central component of foreign relations in the 1950s and 1960s. Because nuclear weapons prevented direct military engagement, the United States and Soviet Union fought to demonstrate scientific superiority, which bolstered claims of leadership and secured allies. Genetics, atomic energy and space sciences became proxies in the Cold War. The competition required both superpowers expand state influence over domestic research: in hindsight, the alphabet soup of American institutions and programs created is remarkable, including PSAC, FCST, ARPA, NASA, NDEA, the NATO Science Council, and science attachés at the State Department. A network of industrial and university contacts completed the American system. Overseas, the United States, unlike the Soviet Union, partnered with global organizations such as UNESCO, the WHO, and ICSU after World War II, validating and extending American diplomacy. Armed with considerable funding, the United States capitalized on its position to project power and influence, whether with European and Asian allies or the Soviet Union.

The American influence on science in the early Cold War was pervasive. The United States was the primary financial supporter of every major international scientific activity, especially the Geneva Conference on Atomic Energy and the International Geophysical Year. These international events reinforced American foreign policy goals: The Geneva conference, for example, complemented an American push for bilateral nuclear agreements. Regarding the IGY, an NSF member reminisced, "... the proposal for an International Geophysical Year, which was

---

[141] For an overview of this process, see chapters 4–6 of Krige, Callahan, and Maharaj, *NASA in the World*, 65–125.

essentially a very interesting and exciting scientific proposal, involved highly significant political considerations and in many ways became an important tool of US foreign policy."[142] Although Sputnik shocked the American system, the United States began offering launch services to NATO allies shortly thereafter. Even CERN, the symbol of European scientific renewal, accommodated American anti-communism.

Allied nations accepted American influence out of necessity and self-interest. In return for aid and access, West European nations allowed American policies to shape cooperative research. Indeed, at the height of the space race in the mid-1960s, allies clamored for more. The Italian prime minister proposed a new ten-year "Marshall" plan to reduce the scientific gap among NATO allies, suggesting increased collaboration in critical fields like computer technology, aeronautics, desalination, and atomic and energy research.[143] But the Johnson administration balked and the proposal, like the IIST plan before it, failed.[144] Still, the growth of applied research and development, especially in corporate laboratories, began to command international attention and national protection.

American intelligence guarded American research. The CIA monitored access to US facilities, denying Soviets the opportunity to survey American science and technology. In 1962, for example, the Soviets pressured NYU to accommodate a visiting scholar, tried to send students to a computer conference, and added a stop at IBM without getting approval first. In response, the CIA denied the exchange with NYU, blocked the students from the conference, and canceled the trip to IBM, informing the Soviet embassy, "future requests of this kind were to be addressed to the State Department, *not* directly to a US industry or research laboratory [italics in original]."[145] The Soviets often requested access to American corporate research, offering monitored access to state facilities in return: In one rare exchange, science advisor Horning toured Soviet chemical and computer industries with members from Bell Labs and IBM; although the visit did not lead to partnership, the poor state of Soviet facilities convinced Hornig American export controls worked.[146]

[142] Bulkeley, *The Sputniks Crisis*, 121.
[143] Bryce Nelson, "Hornig Committee: Beginning of a Technological Marshall Plan?" *Science* 154 (December 9, 1966): 1307–1309, esp. 1307.
[144] Ibid.
[145] James McGrath, "The Scientific and Cultural Exchange," *Studies in Intelligence* 7 (Winter 1963): 25–30, quote on 28.
[146] Wang, *In Sputnik's Shadow*, 255.

Yet the sciences also provided a useful if difficult bridge in US–Soviet relations. Beginning in the 1950s, scientific exchanges and arms-control summits became regular points of contact between the two rivals. Large cooperative research projects – like the IGY – required coordination across Cold War lines. The United States and Soviet Union also co-signed the Antarctic Treaty, acquiesced to IAEA oversight, and established procedures for rescuing astronauts and cosmonauts, even agreeing to share lunar samples. In many ways, scientific and technical agreements illustrated superpower cooperation before détente (see Chapter 5).

The early Cold War permanently changed science and international relations. The competition for prestige and influence drove funding for research to unprecedented levels. The enterprise created during World War II expanded, producing the hydrogen bomb, satellites, global positioning systems, ICBMs, and other components of the national security state. Meanwhile, collaborative research like the IGY connected scientists and non-governmental organizations in a global network, leading to impressive scientific achievements, and the United States landed on the moon. But international relations do not take place on the lunar surface. Across the globe, newly independent countries needed science and technology for development. In hindsight, the arms race, atomic espionage, and space race were the public face of the early Cold War, while medicine, the Green Revolution, and birth control were the focus of a quiet war in the developing world.

3

The Quiet War

Chinese Foreign Minister Zhou Enlai sent American General George C. Marshall a series of letters in 1946. After pleasantries, Zhou wrote: "Stalinism [is] an impractical system for China under present conditions ... the prosperity and peace of China could be promoted by the introduction of the American political system, science, and industrialization. Mao was ... ready to cooperate."[1] A second letter reemphasized the point: "In saying that we should pursue the American path, we mean to acquire US- styled democracy and science, and specifically to introduce to this country agricultural reform, industrialization, free enterprise, and development of individuality."[2] The overtures illustrate more than Zhou's disarming charm; they demonstrate the widespread desire for foreign aid, especially "science and industrialization." Of course, Cold War geopolitics quickly clouded the imagined relationship. Marshall, in China to support Chiang Kai-Shek (while trying to appear neutral), cabled President Truman, "I think it would be harmful to me here to give any publicity to these two papers."[3] For his part, Zhou carried the request into the Non-Aligned Movement and then the United Nations, where, as communist China's representative, he and other world leaders consistently asked for access to science and technology.

Faith in science and technology was universal throughout the Cold War: Capitalists, communists, and the non-aligned alike assumed the pair underlay national security, economic growth, and an improved quality of

---

[1] Zhou quoted in General Marshall to President Truman and the Secretary of State, *FRUS: Foreign Relations, 1946* 9: 149–150.
[2] Ibid.   [3] Marshall quoted in Ibid., 150.

life. Non-aligned leaders across Asia and Africa spoke often about the relationship between science and modernization:

It was science alone that could solve these problems of hunger and poverty, of insanitation and illiteracy, of superstition and the deadening custom and tradition, of vast resources running to waste, of a rich country inhabited by starving people.
Jawaharlal Nehru, speaking before Indian independence (1937)[4]

My interest largely consists in trying to make the Indian people and even the government of India conscious of scientific work and the necessity for it.
Jawaharlal Nehru, speaking after Indian independence (1951)[5]

Yes, we have learned a lot from Europe and America. ... We have been inspired by Lincoln and Lenin, by Cromwell and Garibaldi. And indeed there is still much that we must learn from these many, in many fields. But at this time, the fields that we must study are in the area of technology and science, not concepts and movements that are dictated by ideology.
Sukarno, speaking at the UN (1960)[6]

By creating a true political union of all the independent States of Africa, we can tackle hopefully every emergency, every enemy and every complexity. This is not because we are a race of superman, but because we have emerged in the age of science and technology in which poverty, ignorance and disease are no longer the masters, but the retreating foes of mankind. We have emerged in the age of socialized planning, when production and distribution are not governed by chaos, greed and self-interest, but by social needs.
Kwame Nkrumah, speaking at the First Organization of African Unity Conference (1963)[7]

Each leader sought science and technology as part of a pathway to national development and socioeconomic equality. As Prime Minister, Nehru suggested Indians embrace science as part of a broader cultural shift fusing a "scientific temper," secularism, and non-alignment.[8] He stressed the apolitical nature of research to WHO and UNESCO audiences in India and took aid from both the United States and Soviet Union.[9] The Indonesian President shared Nehru's willingness to learn from all sides and emphasized science, rather than political ideology, as

---

[4] Nehru quoted in David Arnold, "Nehruvian Science and Postcolonial India," *Isis* 104 (June 2013): 360–370, quote on 366.

[5] Ibid.

[6] Sukarno quoted in Andrew Goss, *The Floracrats: State-Sponsored Science and the Failure of the Enlightenment in Indonesia* (Madison: University of Wisconsin Press, 2011), 157.

[7] Ghanian President Kwame Nkumrah, "Inauguration speech of the First OAU Conference," Addis Ababa, Ethiopia, 1963.

[8] Bhikhu Parekh, "Nehru and the National Philosophy of India," *Economic and Political Weekly* 26 (January 5–12, 1991): 35–39 and 41–43.

[9] Arnold, "Nehruvian Science," 368.

critical to national progress. Finally, after the winds of change blew across the African continent, Ghanian President Nkrumah believed science and technology could realize the hopes of his long-suffering people.

Access to science and technology mattered in international relations. Both the Soviet Union and the United States leveraged science, technology, and arms in return for economic and political influence in newly independent nations. Policy-makers in both countries believed aid and education would speed the transition to modernity and cement Cold War alliances.[10] But the two nations were not equal: The American experience in World War II provided a unique foundation for overseas expansion, especially in agricultural and medical sciences (see Chapters 1 and 2). The Soviet Union, preoccupied with domestic rebuilding, expanded aid only after Stalin's death (the United States, in contrast, needed to maintain foreign demand for manufactured goods and military equipment). Additionally, for decades after World War II, partnerships with UN agencies allowed Americans to frame US actions and programs as arising from the international community, rather than Cold War diplomacy.

Whether Point Four, the Alliance for Progress or the Green Revolution, American diplomacy in the developing world incorporated science on multiple levels. Leaders of newly independent nations desired agricultural, industrial, and public health sciences; in response, those fields were prominent components of aid packages to anti-communist allies. Scientific and technical assistance programs generated positive public relations at home and abroad, often receiving coverage far beyond their funding and impact. Social scientists, funded by the government or private institutions, justified US diplomacy through modernization theories. Overseas, multiple federal agencies, private foundations, universities, and international organizations contributed to American assistance, creating a web of global relationships. Undertakings ranged from civilian atomic power and large engineering and agricultural projects to basic sanitation and birth control. Yet the hopeful programs began after World War II ended humbled two decades later. Military requirements and subsidies vastly overshadowed the scientific and technical component of aid packages, while insistence on private investment undermined local support. This chapter has three sections to illustrate American science diplomacy in the developing world.

[10] Michael Adas, "Modernization Theory and the American Revival of the Scientific and Technological Standards of Social Achievement and Human Worth," in David C. Engerman, et al., eds., *Staging Growth: Modernization, Development and the Global Cold War* (Boston: University of Massachusetts Press, 2003), esp. 37–39.

"Modern Commitments" examines the American pledge to development as part of anti-communism. Building on the model established by FDR, President Truman promised scientific and technical aid as a benefit of participating in the American mutual security program. Leaders in newly independent nations were hopeful but wary. At home, anti-communists assailed Point Four as a leftist program, although the CIA believed it would increase American influence and market access overseas and strengthen the UN.

"On Point" looks at American aid programs overseas from Point Four to the Alliance for Progress. Iran, one of the earliest and largest recipients, provides a case-study in American commitment as aid to the Shah continued through seven presidencies. But the geopolitics of development changed: the Eisenhower administration shifted focus from aid to trade while the participation of the Soviet Union in UNESCO complicated US/UN relations. By the 1960s, the Soviet Union operated its own technical aid programs and many Latin American nations saw the American assistance as meddling or cover for political and economic influence.

The final section, "Demographic Containment," considers the Green Revolution and contraception as aspects of American anti-communism overseas. The United States-Mexican agricultural program transformed harvests around the world, remaking the relationship between agricultural sciences and international corporations. At the same time, after American companies researched and tested contraceptives in Puerto Rico, the United States advocated for birth control and family planning in the developing world, eventually making it a requirement for aid. From agricultural science and birth control to medicine and atomic energy, science and technology were critical aspects of American diplomacy in the developing world.

MODERN COMMITMENTS

[HR 6326 – The Latin American Arms Program] is beyond the economic capacity of the Latin American countries, and if carried through it will substantially increase the armaments of Latin America ... the program envisioned under HR 6326 approaches one billion dollars ... [which] is fifty times the total funds expended by this Government in the cooperative program with Latin America in the fields of agriculture, science, civil aviation, education, et cetera, since 1940.

– US Assistant Secretary of State (1946)[11]

---

[11] Memorandum by the Assistant Secretary of State (Braden) to the Secretary of State (December 16, 1946) FRUS: *Foreign Relations, 1946* 11: 108–110.

Fourth, we must embark on a bold new program for making the benefits of our scientific advances and industrial progress available for the improvement and growth of underdeveloped areas. . . . This should be a cooperative enterprise in which all nations work together through the United Nations. . . . The old imperialism – exploitation for foreign profit – has no place in our plans. . . . Greater production is the key to prosperity and peace. And the key to greater production is a wider and more vigorous application of modern scientific and technical knowledge.

– Pres. Harry S. Truman, *Inaugural Address* (1949)

Following enactment of the Rio Pact in 1948, the United States initiated bilateral defense agreements with more than ten countries south of the border, promising significant military support but little development aid.[12] Yet the following year, President Truman offered development assistance through a "more vigorous application of modern scientific and technical knowledge." With the Cold War heating up, Truman envisioned American scientific and technical aid as a critical compliment to military assistance and "Point Four" came to represent such aid worldwide. The program fit the postwar international ethos; UNESCO established science coordination offices in Jakarta, New Delhi, Cairo, and Montevideo in 1946.[13] But Point Four (P4), unlike the UN, could build upon initiatives left over from the war.

The Background to Point Four

The United States incorporated technical aid into anti-communist diplomacy from the start. Postwar assistance to Greece and Turkey, early examples of Truman's containment policy, included technical aid within military and economic packages.[14] The Economic Cooperation Act of 1948 provided scientific and technical assistance to the Middle East: the Department of Agriculture worked in Egypt, Saudi Arabia, Lebanon, Iraq, and Afghanistan, while the Commerce Department's Civil Aeronautics Administration helped set up Turkish air traffic; others did

---

[12] The Rio Pact refers to the Inter-American Treaty of Reciprocal Assistance, signed in 1947 in Rio de Janeiro. This pact, along with many others, was part of the origins of the U.S. Military Assistance Program. See Chester J. Pach, *Arming the Free World: The Origins of the United States Military Assistance Program, 1945–50* (Charlotte: University of North Carolina Press, 1991).

[13] Walter H. C. Laves and Charles A. Thomson, *UNESCO: Purpose, Progress, Prospects* (Bloomington: Indiana University Press, 1957), 95.

[14] Patterson, "Foreign Aid under Wraps," 120–123.

extensive work in public health (detailed below).[15] Across the globe in Asia, the US Military Government in Korea (USMGIK) proposed a complete five-year plan for modernization, introducing fertilizers, hydroelectric dams, and public health clinics alongside educational reforms. Supported by the UN, USMGIK removed Confucian classics in favor of anti-communism and scientific education, ultimately supplying 30 million textbooks.[16] As American funds rebuilt Western Europe and Asia, Latin American nations asked for similar aid: in Bogota, crowds besieged the Secretary of State with requests for a southern Marshall Plan (following a minor riot, the secretary took refuge in the US embassy).

Such passionate requests stimulated many in the Truman administration. Benjamin Hardy, the speechwriter responsible for the Point Four section of Truman's inaugural address, worked for the IIAA in Brazil and advocated for technical aid programs within the administration. In a pre-speech memo entitled the "Use of Technological Resources as a Weapon in the Struggle with International Communism," Hardy suggested TVA and Marshall Plan programs for the developing world:

It would make full and affirmative use of one of the resources in which the U.S. is richest and the Soviet Union is poorest. Our overwhelming superiority in a field of constructive effort would be apparent to even the most backward and illiterate people ... The program could be developed in such a way as that the facilities of international organizations, such as FAO, ILO, WHO would be utilized in coordination with U.S. resources to the greatest possible extent ... This kind of program would be practically invulnerable to the charge of imperialism, since we would extend technical help only to those governments that ask for it.[17]

His boss was receptive. President Truman also wanted to partner with the UN, private capital, and foundations, later writing he dreamt about Point Four after implementing the Marshall Plan.[18]

Although Truman's inaugural address is best remembered for its fourth point on technical aid, the speech took pains to embed the aid within a larger Cold War context. The first and second points were support for the United Nations and global free trade, buttressed by the European Recovery Act (Marshall Plan). The third point suggested

---

[15] Dorothea Seelye Franck, "The Interchange of Government Experts," *Middle East Journal* 4 (October, 1950): 410–426.
[16] See "The Proving Ground," in Ekbladh, *The Great American Mission*, 114–152.
[17] Benjamin Hardy, "Use of Technological Resources as a Weapon in the Struggle with International Communism," reprinted in Paul William Bass, *Point Four, Touching the Dream: A Bold, New U.S. Foreign Policy* (Stillwater, OK: New Forums Press), 151–153.
[18] Ibid., 32.

collective security pacts, such as the Rio Pact, as a bulwark against international aggression (i.e., communism). Truman's fourth and last point was a program to make American science and technology available to the developing world. Even here, the president was careful to limit expectations, noting America's "limited" material resources for assistance. Nonetheless, the president claimed American science and technology were without par ("preeminent") and promised a "cooperative enterprise" through the UN, preemptively rebuffing Soviet critics by assuring, "the old imperialism – exploitation for foreign profit – has no place in our plans." Truman closed by drawing a clear line from the application of science and technology to the creation of greater industrial production, prosperity, and peace. In a single speech, the president reaffirmed the importance of science and technology to American prosperity and promised to use the pair to create similar prosperity abroad.

The Domestic and International Response

The American public responded positively. The president had invited a wide cross-section of society – including businesses, foundations, universities, churches, and individuals – to participate in a national crusade to aid the world's poor. Truman stressed the broad nature of the campaign to the State Department: "I made it clear that all existing private and governmental activities would be utilized. American business enterprises overseas and private non-profit organizations such as the Rockefeller Institute...could furnish much valuable information and assistance."[19] Clark Clifford, Truman's aide, remembered the moment with pride: "The public reception to his Inaugural Speech exceeded our hopes, but it was the overwhelming response to Point Four that most gratified and surprised me ... I am convinced that Point Four and its offspring remain one the noblest commitments our nation has made in this century."[20] Jonathan Bingham, a later administrator of P4, also recollected, "public interest in the program was enormous. Church groups everywhere rallied behind it, recognizing in it an expansion of their own long cherished missionary ideals. The task of furthering international development became a favorite topic at meetings of all kinds, including labor groups,

[19] Truman, quoted in Minutes of Meeting (UM-1), Department of State, February 3, 1949, n2. This is included in "The Genesis of the Point Four Program" in *FRUS, Foreign Relations, 1949*, 1, 757–787, quote on 763.
[20] Clifford quoted in Bass, *Point Four*, 37.

women's clubs, businessmen's luncheons and student debates."[21] But the program ran into political headwinds in Congress.

Legislation for Point Four ran afoul of congressional anti-communists. Senator McCarthy (R-WI) accused Haldore E. Hanson, the State Department official assigned to Point Four, of being on a "mission to communize the world."[22] Others assailed the program as a "scheme" for worldwide economic aid; Senator Jenner (R-IN) deemed Point Four a Soviet program.[23] Senator Taft (R-OH) went further, warning: "I do not know whether Mr. Haldore Hanson is a card-carrying Communist or not, but I do know that as head of the State Department's technical staff on Point Four he is helping to draft this basic program which the Communists in the Politburo in Moscow are counting on as essential in their expanding conquest of the Western world."[24] Esther Caukin, the staffer assigned to UNESCO, was asked whether "the Communists or the anti-communists will receive aid under the Point Four program with Hanson in charge?"[25] Privately, the president wrote to the Speaker of the House, Representative Rayburn (D-TX): "In countries where the choice between Communist totalitarianism and the free way of life is in the balance, action through such program as the Point Four can tip the scales toward the way of freedom."[26] As Americans debated the geopolitics of aid, Truman's proposal inspired a hopeful wariness overseas.

Foreign leaders and intellectuals lauded the idea of Point Four while acknowledging larger geopolitical realities. Soedjatmoko, a member of the Indonesian UN delegation, opened an essay with "Point Four is an assertion of faith in a world based on the recognition of the interdependency, both economic and political, of all nations."[27] However, he turned quickly to American support for the French in Indochina: "America's failure to make a sufficiently early stand on the issues of colonialism, and the impression thus created of her acquiescence in continued colonial warfare by the metropolitan powers, left very serious doubts in the

[21] Jonathan B. Bingham, *Shirt-Sleeve Diplomacy: Point 4 in Action* (New York: John Day, 1954), 13.

[22] McCarthy quoted in William S. White, "McCarthy Accuses Point Four Official," *New York Times* (March 14, 1950): 1 and 4, quote on 1.

[23] C. P. Trussell, "Foreign Aid Fate Up to Senate Today," *New York Times* (May 25, 1950): 14.

[24] Taft quoted in ibid.   [25] White, "McCarthy Accuses Point Four Official," 4.

[26] Truman quoted in Tarun C. Bose, "The Point Four Programme: A Critical Study," *International Studies* 7 (1965): 66–97, 81.

[27] Soedjatmoko, "Point Four and Southeast Asia," *Annals of the American Academy of Political and Social Science* 270 (July 1950): 74–82, quote on 74.

minds of many Asians as to America's true intentions." Georges Hakim, offering a "Middle Eastern" perspective, asked "What guarantees are there that private American capital will not try to maximize its profits at the expense of the peoples of the underdeveloped countries?"[28] Americans struggled to be sensitive to such concerns. Alan Valentine, a US advisor to the Netherlands and India, asked perhaps the key question: "Is it primarily a device for containing Russia, humanitarian, or a way to save American private enterprise?"[29] He continued, "We must supply enough aid to compete successfully with the powerful appeal which communism makes to ignorant and hungry people," suggesting the government require efforts to bring populations "under control" before sending aid.[30] But Americans weren't the only ones appealing to the "ignorant and hungry people."

The Soviet response to Point Four was swift. On January 22 – only two days after Truman's address – Soviet newspapers tarred the program as an attempt "to provide maximum opportunities for penetration of American capital into backward countries" and as part of the "formation of a world-wide empire."[31] *Pravda* criticized it as "an attempt to replace British, Dutch, and French colonial empires with a world-wide American monopoly."[32] Within weeks the Soviets established the Council for Economic Mutual Assistance (CEMA) to coordinate trade between Soviet-bloc countries. Soon Commissions for Scientific and Technical Cooperation linked Soviet satellites; the CIA later reported technical aid flowing between "joint-stock" companies and free access to patents and technology for Poland (a key Soviet ally).[33] In November 1949, when the UN General Assembly approved a plan for technical assistance (backed by American money, personnel, and equipment), the Soviet delegate accused Truman of using P4, like the Marshall plan, to divide the world in two.[34]

---

[28] George Hakim, "Point Four and the Middle East: A Middle East View," *Middle East Journal* 4 (April 1950): 183–195, quote on 192.

[29] Alan Valentine, "Variant Concepts of Point Four," *Annals of the American Academy of Political and Social Science* 270 (July 1950): 59–67, quote on 59.

[30] Ibid., quotes on 60 and 66.

[31] Moscow newspapers quoted in Sergius Yakobsen, "Soviet Concepts of Point Four," *Annals of the American Academy of Political and Social Science* 268 (March 1950): 129–139, quotes on 129.

[32] *Pravda* quoted in Bose, "The Point Four Programme," 85.

[33] CIA Office of Research and Reports, "Soviet Economic Assistance to the Sino-Soviet Bloc Countries," (June 13, 1955). CIA Doc No/ESDN: CIA-RDP79T01003A001000190002-8.

[34] Yakobsen, "Soviet Concepts of Point Four," 129–139.

CIA analysts remained nonplussed. An agency study observed, "Soviet capabilities of impeding the program are relatively limited. The USSR will probably prohibit Satellite participation."[35] Instead, analysts worried about needy countries trying to "turn the emphasis of the program from technical to financial assistance on the order of the European Recovery Program."[36] The agency affirmed use of UN machinery strengthened the global organization while minimizing "US liabilities to charges of imperialism;" further noting the US- maintained control because it was "the only country capable of supplying most of the required funds." Finally, American analysts agreed with their Soviet counterparts, concluding Point Four would "inevitably result in the spread of US machinery and methods to the countries aided, thus forming closer ties with the United States economically, socially, and politically."[37] Though the CIA considered Point Four good business abroad and a clear anti-communist coup, the program struggled to become law. In 1950, after withstanding Republican opposition, Truman signed the Act for International Development.[38]

Point Four

Backed by the UN, Point Four was both an expression of international goodwill and an instrument of American diplomacy. After debate about whether to expand the IIAA, Congress provided an initial $34.5 million to a new Technical Cooperation Administration (TCA) on the assumption Point Four would be a self-help program supplemented by private capital, philanthropy, universities, and international organizations. At the same time, the United States announced it would pay more than half the cost of UN technical programs (contributing $45 million out of a total $85 million – UN members promised the rest).[39] But the difference between American assistance to European allies and the developing world was striking: the Marshall Plan ran into the billions (around

[35] Central Intelligence Agency [directorate and office unspecified], "Difficulties in the Implementation Abroad of Point Four (ORE 54–49)," (June 13, 1949): 1–10, quote on 2. DNSA Soviet Estimate Collection, record # SE00068.
[36] Ibid., 3.   [37] Ibid.
[38] The best source for internal conversations about the structure of the program, especially the role of the UN and the IIAA, including funding and sample contracts, is found in "The Point Four Program," *FRUS, Foreign Relations, 1950, Vol. 1*, 846–874. See also Sergei Y. Shenin, *The United States and the Third World: The Origins of Postwar Relations and the Point Four Program* (Huntington, NY: Nova Science Publishers, 2000).
[39] Office of Public Affairs, Department of State, *The Point Four Program* (Washington, DC: Government Printing Office, 1949).

$13 billion), while the TCA's original budget was in the millions ($34–$45 million). Such lower funding, according to the *American Journal of Public Health*, demonstrated Point Four was altruistic and divorced from the Cold War.[40] Support from the WHO, FAO, and a new UN technical agency provided further evidence of the program's neutrality. Brochures on the US/UN partnership advertised four goals: "Strengthening the UN, Building Political Democracy, Expanding International Trade, and Raising Standards of Living (in that order)."[41] Communism went unmentioned, although the brochures acknowledged prioritizing global trade over living standards.

Point Four contrasted American prosperity with foreign poverty. Statistics easily quantified need: The State Department compared developing nations to the American standard in caloric intake, energy production, rates of disease, number of physicians and research infrastructure, etc. Aid fell "roughly into two categories: technical, to improve food production, health, education, and productive skills; and financial, to develop transportation, communication, water control and power, and productive industries."[42] To illustrate previous success, the State Department pointed out IIAA Health and Sanitation programs reduced malaria infection across Latin America, while a single horticulturist doubled coffee yields after only three years of experimental work. Participating nations would also receive research and laboratory facilities along with teacher and student exchanges. Funding proved the US commitment: $18 million for agriculture and forestry, $17 million for health, and $9 million for education, with less than a million for geological surveys, weather, and fisheries.

Point Four grew quickly. Its UN counterpart received pledges for $20 million from member nations even before congressional approval; the *New York Times* enthused: "If any loss was sustained through the Soviet-and-satellite boycott it was not apparent. Indeed, there may even have been some gain, since there was a noticeable absence of recrimination, suspicion, and propaganda."[43] Truman used the program to criticize the Soviets, telling a delegation from thirty-four interested countries: "If the Soviet Republic would to its neighbors as we act to ours, I don't

[40] Editorial Board, "Point Four," *American Journal of Public Health* 40 (June 1950): 740–743.
[41] Department of State, *The Point Four Program.*   [42] Ibid.
[43] Anon., "Point Four at Havana," *New York Times* (May 30, 1949): 12. On the $20 million in pledges, see Anon., "Rally for Point Four," *New York Times* (June 16, 1950): 24.

think there would be any chance for a third world war."[44] Ceylon, Iran, Libya, Nicaragua, Paraguay, and Ethiopia, among others, signed agreements the first year. By the beginning of 1952, there were 216 Point Four projects in 34 countries in Asia, Africa, and Latin America.[45] Confident of success, Congress appropriated $240 million for technical cooperation, with initiatives already underway in artificial fertilizers, irrigation and ecology, health and malaria control, and strengthening research and educational curricula.[46] The need for agricultural expertise globalized US colleges: for example, Arizona State in Iraq, Oklahoma State in Ethiopia, the University of Arkansas in Panama, and Michigan State in Colombia and Vietnam.[47] Harry Bennett, the first Point Four Administrator, praised the agricultural program at a UN FAO conference:

A great group of scientists and technicians in laboratories and experiment stations around the world is continually increasing our knowledge of food production, food preservation, nutrition, and distribution of food. These advances in science and technology are not the monopoly of any nation or group of nations, but are made available for the use of mankind ... We have borrowed grasses from Africa; melons from Africa and the Middle East; horticulture products from all over this world; our citrus fruit from the Middle East, from Spain, Italy and Brazil...We have borrowed scientific data. We have borrowed and used in the Western world and now we come with technical assistance. It is a two-way street. We are cooperatively sharing the advances that have been made in science, technology, genetics and nutrition."[48]

Nor was the program limited to agricultural sciences. In 1952, the AEC Commissioner suggested nuclear power plants "may have an important place in any future Point Four programs."[49] Atoms for Peace (see Chapter 2) began the following year; soon, atomic reactors, like agricultural sciences, were integral to US aid packages to the developing world.[50]

[44] Truman quoted in Anon., "Truman Tells Point Four Delegates Russia Could Ease Fears by Being a Good Neighbor," *New York Times* (September 27, 1952): 3.

[45] Technical Cooperation Administration, Department of State, *Point Four Projects: July 1, 1950 through December 31, 1951* (Washington, DC.: Government Printing Office, 1952), ii.

[46] Samuel P. Hayes, "An Appraisal of Point Four," *Proceedings of the Academy of Political Science* 25 (May, 1953): 31–46, figure on 33.

[47] Bose, "The Point Four Programme," 87.

[48] Bennett quoted in Bass, *Point Four*, quotes on 142–144.

[49] Glennan quoted in Anon., "Atomic Power Plants in Point Four Program," *Science Newsletter* 62 (July 5, 1952): 7.

[50] See "Appendix VI" the list of Agreements for Cooperation in the Civil and Military Uses of Atomic Energy in Hewett and Holl, *Atoms for War and Peace*, 581. The list tracks with the recipients of other types of American aid.

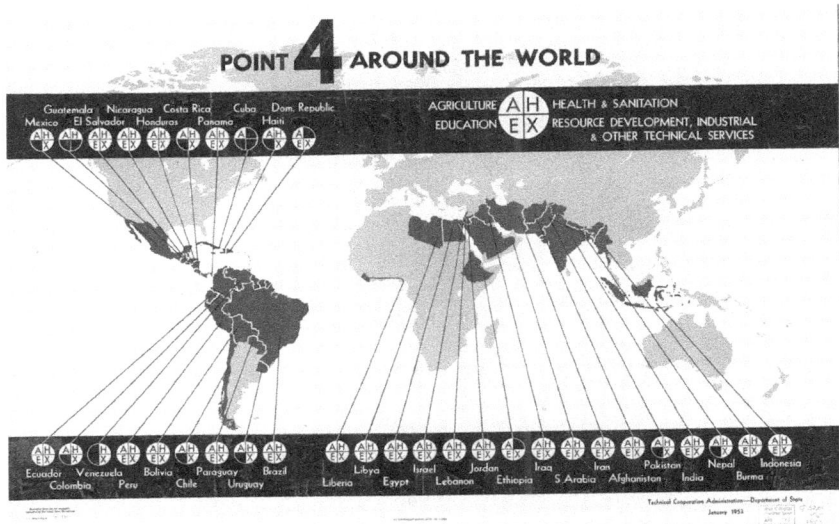

FIGURE 3.1 Map of Point Four activities around the world (1953)
This map from the Technical Cooperation Administration of the Department of
State details American assistance around the world, highlighting programs in
agriculture, health and sanitation, education, resource development, and
industrial and technical services. Tensions with the United States during the Peron
presidency led to Argentina's exclusion from American programs in Latin
America, while the map does not show the agricultural and technical specialists
working in French Indochina at the time.

Within three years of Truman's inaugural address the United States
constructed a network of scientific and technical aid programs around the
world. Although aid programs were never consolidated, and frequently
changed administration, "Point Four" became a recognized term for
scientific and technical assistance. Jonathan Bingham, a P4 administrator,
observed: "In Latin America it is *Punto Quarto*. In India the English
words are used, spelled out in the Hindu [sic] alphabet. Not long ago
the director of the program in Jordan received a letter addressed in
English (obviously translated from Arabic) to the 'Master of the Fourth
Spot'."[51] William Warne, director of the Iranian program, added: "Point
4 is understood around the world. In the most remote places the words
become a passport, a warranty of friendship, a whole language when no

---

[51] Jonathan B. Bingham, "The Road Ahead for Point Four," *New York Times* (May 10,
1953): SM12 & 67, quote on 12.

other words have meaning."[52] Indeed, scientific and technical programs provided cover from Soviet attacks of imperialism, although questions remained about relations with military and international organizations. As early as 1951, UN agencies felt "outbid" by their American partners, who were often considered too aggressive and willing to work with corrupt governments.[53] Supporters suggested the program just needed to get in the field and win local hearts and minds.

ON POINT

From the village level upward, all that we do relates to American Science and to American Science teaching ... Is it better wheat, better corn, better fruit, better vegetables? American science points the way. Is it better livestock, better animal care, elimination of livestock disease, the uses of inoculation? American science points the way. Is it reforestation or broad experimentation in range management? American science points the way. Is it the elimination of malaria, the battle against trachoma, the problems of sanitation, the provision of pure water? American science and engineering point the way.

Cedric Seager, Chief of the Iran Division of the Technical Cooperation Administration (1953)[54]

A traditional society is one whose structure is developed within limited production functions, based on pre-Newtonian science and technology, and on pre-Newtonian attitudes toward the physical world ... The preconditions for take-off were initially developed, in a clearly marked way, in Western Europe of the late seventeenth and early eighteenth centuries as the insights of modern science began to be translated into new production functions in both agriculture and industry, in a setting given dynamism by the lateral expansion of world markets and the international competition for them.

Walt W. Rostow, *The Stages of Economic Growth:
A Non-Communist Manifesto* (1961)[55]

[52] William E. Warne, *Mission for Peace: Point 4 in Iran* (Bethesda, MD: IBEX Publishers, 1999), quote on 12. Originally published in 1956.
[53] Michael T. Hoffman, "U.S. Outbidding U.N. in Technical Help," *New York Times* (March 25, 1951): 15.
[54] Cedric Seager, "Point Four and American Science Teaching," *School Science and Mathematics* 53 (March 1953): 201–206.
[55] W. W. Rostow, *The Stages of Economic Growth: A Non-Communist Manifesto* (London: Cambridge University Press, 1961), 4 and 6.

Cedric Seager and Walt W. Rostow represent American modernization programs from different perspectives: one, on the ground trying to teach science; the other, theorizing about the role of science in international development. Both accepted science and technology could transform pre-modern societies and both believed in what they were trying to accomplish. And yet both seem naive in hindsight: Seager failed to acknowledge the geopolitical complications of America's role in Iran, while Rostow assumed a causal link between "Newtonian science and technology" and the creation of modern industry and society, ignoring the role of unique cultures and contexts to development. Nonetheless, Rostow and other modernization theorists shaped American aid packages for decades, although the focus on science and technology rarely materialized. Instead, whether as part of Point Four or the Alliance for Progress (AFP), American assistance remained merely an attractive military adjunct. Consider Iran.

Three Decades of Assistance to Iran

American interests in Iran arose during World War II. After the shah offered oil to Nazi Germany, British and Soviet forces invaded Iran in August of 1941, eventually setting the shah's twenty-two-year-old son, Muhammad Reza Pahlavi, on the peacock throne. The following March, after America's entrance into the war, Iran began receiving Lend-Lease aid. Burgeoning oil interests in Saudi Arabia soon required weakening non-American influences in Iran. Secretary of State Cordell Hull stated the region's economic and strategic importance to FDR: "it is in our interest that no great power be established on the Persian Gulf opposite the important American petroleum development in Saudi Arabia."[56] By 1943, the United States operated six missions in Iran, including intelligence services and collaboration with the gendarme (GENMISH) and army (ARMISH).[57] After V-E Day, the new shah appealed to the new president for aid, reminding Truman of Iranian contributions and sacrifices in wartime. The following year, Morrison-Knudson (an American engineering and construction firm) signed a contract to study economic

---

[56] Hull quoted in James A. Bill, *The Eagle and the Lion: The Tragedy of American-Iranian Relations* (New Haven: Yale University Press, 1989), 19.
[57] John Prados, *Safe for Democracy: The Secret Wars of the CIA* (Chicago: Ivan R. Dee, 2006), 97–107.

development in Iran; ARMISH helped retake Northern provinces after denying Soviet oil concessions in the region.

Iranian wariness of foreigners limited American influence at first. During the war, the outspoken parliamentarian Muhammad Mossadegh railed against foreign concessions and advisors; after the war, US economic surveys aroused suspicion.[58] When the United States pledged $10 million in surplus military equipment in 1948, Mihdi Davudi, an influential Iranian writer, asked "Why should a poor nation such as ours that has gone through years of poverty be armed to defend the selfish interests of the millionaires of American and England? This is the story of the wolf and the lamb. Why doesn't the United States give us aid to help us improve our education, agriculture and health...This is a $10 million baited trap that we must jump away from."[59] But the Pahlavis couldn't refuse equipment worth millions and so the American presence expanded the following year. So did the opposition. After surviving an attempt on his life, the shah toured the United States in late 1949, receiving a twenty-one-gun salute at West Point and visiting GM, Lockheed, and the Hoover Dam. Months later, Iran signed one of the first Point Four agreements.

US–Iranian agreements prioritized military aid. Cyrus Waynick, acting Point Four administrator, argued Iran was "a nation under the ramparts of the Soviet Union."[60] As such, agreements provided $65 million for military aid and half a million dollars for education, agriculture, and health (the three areas desired by Davudi and the focus of Iranian Point Four).[61] Mossadegh warned against corruption, worried about the influx of American dollars, but American anti-communism presented a more pressing problem.[62]

The geopolitics of aid created tension. In late 1951, the Mutual Security Act absorbed Point Four, explicitly tying funds to support for American anti-communism. In response, the Syrians rejected P4 aid and Mossadegh restated Iranian opposition to a Western defense pact.[63] Meanwhile, Iranian nationalization of oil worried investors; the *New*

[58] James F. Goode, *The United States and Iran, 1946–51: The Diplomacy of Neglect* (New York: St. Martin's Press, 1989), 43.

[59] Davudi quoted in Bill, *The Eagle and the Lion*, 50.

[60] Waynick quoted in Anon., "Iran Gets First Point Four Grant in a $500,000 Rural-Aid Program," *New York Times* (October 20, 1950): 1 & 14.

[61] Ibid.   [62] Mossadegh quoted in Warne, *Mission for Peace*, 29.

[63] Albion Ross, "Iran Will Accept $23,000,000 U.S. Aid," *New York Times* (January 6, 1952): 15. Regarding the Syrians, see Bingham, *Shirt-Sleeve Diplomacy*, 213. See also Shenin, *The United States and the Third World*, 153.

York Times expected Mossadegh's policies to be a "deterrent" to banks, foundations, and businesses looking to invest, further depressing aid funds.[64] Desperate to counter charges of American imperialism, Point Four administrator Bennett flew to Tehran to convince the Iranians to accept an increased aid package, unfortunately perishing in a plane crash.[65] Ultimately, the Mossadegh government agreed to an increased $23 million out of necessity: on a monthly basis, P4 funds alone were equivalent to one-fifth of government income (which barely covered the army, police, and civil servants).[66] Iran needed help; William Warne, the new local administrator, quickly proposed more than sixty new projects.

The expanded program drew criticism. "Suspicion of 'foreign agents' is real and deep," chronicled *New York Times* reporter Albion Ross.[67] The Marxist Tudeh party organized against the expansion. By April, crowds had destroyed four P4 offices, causing the wives and children of the newly arrived scientists and technicians to leave. The following year more than 500 Iranians stormed a warehouse and private houses in Shiraz.[68] Tudeh argued Point Four meant "more guns and tanks will be sent to Iran," greeting aid workers with "Yankee Go Home" signs.[69] When P4 imported larger chickens and donkeys to improve Iranian livestock, the Marxists jeered the initiative: "Ah, the Great United States comes to help Iran and what does Iran get? A few jackasses! The people are hungry, but what do the promises of the wealthy Imperialists amount to? Jackasses!"[70] The Soviets echoed the party line: Radio Moscow labeled Warne an "imperialist warmonger" determined to establish "a new colonial regime in Iran."[71] But many rural Iranians wanted jackasses; administrator Bingham happily reported Soviet propaganda advertised the program and boosted requests for livestock.

Point Four staff fought back, enlisting local governors for support and creating media to showcase the American commitment to helping Iran. P4

---

[64] Anne McCormick, "The Iran Crisis as It Touches Point Four," *New York Times* (May 23, 1951): 34.

[65] Albion Ross, "Bennett Planned Aid Plea to Iran," *New York Times* (December 25, 1951): 22.

[66] Albion Ross, "Point 4 Looms Big in Iran's Economy," *New York Times* (July 16, 1952): 4.

[67] Albion Ross, "U.S. Point 4 Aides in Iran Disturbed," *New York Times* (January 15, 1952): 7.

[68] Anon., "Iran Point 4 Staff Stays Despite Raid," *New York Times* (April 19, 1953): 25.

[69] The quote comes from a Tudeh party newspaper quoted in Warne, *Mission for Peace*, 117.

[70] Ibid., 49.   [71] Ibid., 207.

printed a magazine in Farsi ("Land and People") and the Farsi service became the first postwar VOA program in the Middle East (Truman delivered the inaugural broadcast).[72] Point Four administrator Cyrus Waynich also took to the airwaves, telling the Farsi service, "This country's Point IV program ... offers people of underdeveloped areas a chance to fight their way out of the vicious circle of poverty and disease ... [Point IV] has to do with the stuff of life ... with birth and death, hunger and food, sickness and health."[73] Appropriately, the radio lineup included "Science Roundup" – a broadcast promoting American scientific achievements, including artificial rain-making to improve crop growth, medical advances in fighting deafness and heart disease, profiles of famous scientists, and highlights of the US space program. Moscow tried to jam broadcasts.

The Iranian Point Four mission reflected both humanitarian and Cold War diplomacy. The United States spent nearly $40 million on Iranian aid between 1950 and 1953. Such sums funded an effective anti-malarial campaign via widespread DDT spraying (14,000 villages in a single year), the introduction of new crops and livestock, the establishment of sanitary water and sewage systems, milk pasteurization plants, mobile health units with x-rays, and increased medical education, among others. Administrator Warne proudly boasted, "there were few areas of Iranian life into which Point 4 did not extend."[74] America paid 60 percent of the United Nations technical aid program in Iran. Yet the United States also conspired with the British to overthrow the democratically elected Prime Minister Mossadegh and eliminate opposition to the shah, thereby cementing the US/Pahlavi bond and guaranteeing more scientific and technical aid.

Within a few years, Iran had the largest Point Four program in the world, with hundreds of Americans employed.[75] The United States and Iran signed an agreement on civilian atomic energy in 1957, initiating a two-decades-long program in nuclear assistance, including providing reactors and training Iranian scientists and engineers.[76] At the same time,

---

[72] Deborah Kisatsky, "Voice of America and Iran, 1949–1953: US Liberal Developmentalism, Propaganda and the Cold War," *Intelligence and National Security* 14 (Autumn 1999): 160–185, esp. 166.

[73] Ibid., 168.   [74] Warne, *Mission for Peace*, 183.

[75] Anon., "Iran Aid Report Held Misleading," *New York Times* (April 17, 1956): 3. See also Sam Pope Brewer, "Iran Taking over Point Four Work," *New York Times* (April 21, 1956): 3.

[76] Mustafa Kibaroglu, "Good for the Shah, Banned for the Mullahs: The West and Iran's Quest for Nuclear Power," *Middle East Journal* 60 (Spring 2006): 207–232.

David Lilienthal – the first chair of the AEC – and his private Development and Resources Corporation began a hydroelectric project in oil-rich Kuzestan.[77] After years of the wrangling over financing, the Iranians opened the Dez Canyon Dam in 1963. Atomic and hydroelectric power, modern agriculture and medicine, education; America and Point Four seemed to be re-making Iran.

Not everything went smoothly. American agricultural and scientific equipment often went unused (even unopened). US participants struggled with the complexities of Iranian life.[78] Point Four staff supported modernizing agriculture, for example, but analysts thought the shah's program to redistribute royal lands would take forty-five years to create independent landowners (by one estimate, more than 90 percent of Iranian villages existed in a tenancy system described as serfdom).[79] Unfortunately, according to administrator Bingham, the State Department was "reluctant to speak out for any principles of social justice."[80] Point Four attempts to expand basic education met resistance as well; William Warne wrote: "Under gentle pressure from Point 4 little girls are beginning to be admitted to some village schools, though not in the same number, as boys."[81] Nor were tensions confined to domestic issues. Administrator Bingham believed Point Four would succeed "only so long as its longrange, non-political character is maintained."[82] But American aid programs, including Point Four, were always political.

Indeed, the politics of aid assured continued assistance. In 1967, for example, Secretary of State Dean Rusk praised Iran for becoming the second "non-European" country to "graduate" from Point Four.[83] Among the statistics he cited at the time: Iran had received $605 million in technical assistance, $900 million in military aid, and $122 million in food aid.[84] This translated, in part, to two squadrons of F4 phantoms,

---

[77] Christopher T. Fisher, "'Moral Purpose Is the Important Thing:' David Lilienthal, Iran, and the Meaning of Development in the US, 1956–63," *The International History Review* 33 (October 2011): 431–451.

[78] Regarding the struggles in importing American agricultural practices, see Jessie Embry, "Point Four, Utah State University Technicians, and Rural Development in Iran, 1950–64," *Rural History* 14 (April 2003): 99–113.

[79] Albion Ross, "Point Four Aiding Land Reform," *New York Times* (August 4, 1952): 4.

[80] Bingham, *Shirt-Sleeve Diplomacy*, 216. [81] Warne, *Mission for Peace*, 174.

[82] Ibid., 217.

[83] The first country was Taiwan; remember that the MSA officially absorbed Point Four in 1951, sixteen years earlier. Rusk quoted in Anon., "U.S. and Iran Celebrate the End of Aid Program," *New York Times* (November 30, 1967): 13.

[84] Ibid.

schools for more than 100,000 Iranians, a 95 percent immunization rate against smallpox, and protection from starvation. But had Iran really graduated? A decade later, the United States agreed to provide eight new nuclear reactors, MIT signed a contract to train Iranian scientists, and President Carter signed an atomic agreement with Iran in July 1978.[85] Even as the shah's government teetered in December, the US embassy in Iran highlighted the expanding nuclear program and other joint initiatives, citing ongoing projects by the NAS, NRC, NSF, USGS, NOAA, and NASA (these included work in solar energy, population and family planning, environmental sciences, meteorology and landsat applications, among others).[86] The embassy, expecting political calm to return, proposed placating the shah's critics with an Iranian Institute of Science and Technology.[87] Clearly, scientific and technical assistance, regardless of the name, was a key part of US–Iranian relations to the end.

The Iranian Point Four program stretched through seven presidencies – Truman to Carter – demonstrating how aid programs bolstered long-term relationships. At the same time, military obligations overburdened these high-profile but comparatively inexpensive initiatives. In 1965, critic Tarun Bose argued, "tying technical assistance, which is more humanitarian, to military and security raises fears in scientists and foreign countries, that assistance is becoming secondary."[88] Although the Iranian case is unique in its length, complexity, and outcome, the fusion of military and scientific and technical assistance was common.

An Evolving Mix of Programs, Goals, and Participants

Since American and UN aid programs adopted a humanitarian rhetoric, their use in American diplomacy created friction, especially after the mutual security program absorbed Point Four in 1951. The requirement to participate in America's anti-communist crusade concerned developing nations and the emphasis on commerce, rather than aid, undermined belief in the program's altruism. Meanwhile, American funding of UN technical assistance meant the United States could direct international aid to Cold War allies.

---

[85] Mustafa Kibaroglu, "Good for the Shah, Banned for the Mullahs: The West and Iran's Quest for Nuclear Power," *Middle East Journal* 60 (Spring 2006): 207–232.

[86] United States Embassy of Iran to the Secretary of State, "Science and Technology in Our Foreign Relations," (December 6, 1978) – accessed via the Digital National Security Archive, Iranian Collection, record #IR01875.

[87] Ibid.   [88] Bose, "The Point Four Programme," 96.

US representatives appreciated that UN programs "closely coordinated" with Point Four.[89] The UN Technical Assistance Administration, advised by the World Bank and IMF, operated through seven scientific and technical agencies (UNESCO, ILO, FAO, WMO, ITU, WHO, and ICAO) and sent nearly 800 specialists from sixty-one nationalities into dozens of countries within its first two years.[90] Yet the demand could not be met. Though technical assistance was the foremost UN service project, its budget hovered around $20 million, compared to almost $300 million for Point Four.[91] Others calculated a 10 to 1 disparity in the funding for American and UN operations, not including military assistance.[92] Thus Point Four eclipsed its UN counterpart, while American dollars shaped UN priorities: most UN aid went to pro-American regimes in Cold War hot spots such as Iran, Greece, Philippines, Thailand, Brazil, Pakistan, Taiwan, Turkey, and Lebanon. In Indochina, for example, US and UN programs split duties – the WHO provided nursing education and child healthcare, while the United States provided supplies.

Southeast Asia revealed the confusion of military and humanitarian missions. In Vietnam, for example, Point Four arrived more than a decade before the marines. As the French fought to maintain their colony in 1952, an official American pamphlet stated: "a platoon of American irregulars – 53 men and women – are fighting the quiet war in Indochina. A small mobile team of agricultural specialists, doctors, engineers, financial experts, and educators are battling against the ignorance, disease, and poverty which threaten Indochina's existence."[93] For the next two decades, fleets of sound trucks proclaimed America's peaceful intentions as an escalating economic and military program invented South Vietnam

[89] Bart J. Bok, "The United Nations Expanded Program for Technical Assistance," *Science* 117 (January 23, 1953): 67–70, quote on 67. Bok was the U.S. representative to UNESCO and the UN technical administration.

[90] David Webster, "Development advisors in a time of Cold War and decolonization: the United Nations Technical Assistance Administration, 1950–59," *Journal of Global History* 6 (July 2011): 249–272, esp. 254. In addition to agencies already discussed, the UNTAA included the International Labor Organization (ILO), International Telecommunications Union (ITU) and the International Civil Aviation Organization (ICAO).

[91] Bok, "The United Nations Expanded Program for Technical Assistance," 67. See also Laves, *UNESCO*, 53. Finally, see Shenin, *The United States and the Third World*, 149–152.

[92] Walter R. Sharp, "The Institutional Framework for Technical Assistance," *International Organization* 7 (August 1953): 342–379, quote on 344.

[93] MSA pamphlet quoted in Stephen Macekura, "The Point Four Program and U.S. International Development Policy," *Political Science Quarterly* 128 (2013): 127–160, quote on 150–151.

(see Chapter 4).[94] In nearby Thailand, *Science* voiced concerns about US actions, observing, the "unrestrained tampering with this way of life through the importation and distribution of *untested* fertilizers, mechanical equipment, seeds, and ideas of farming practices is inviting possible outcomes we have not planned [italics in original]."[95] Tempers flared between scientific advisors and the military. After the director of his program quit, one agricultural trainer blamed Department of Defense attitudes, quoting a military official, "The trouble with you technical people is that you're too slow. There's a war going on out here, and we can't wait for you."[96] Nor was the incident unique; disputes over US diplomatic goals led to frequent reorganization.[97]

The Eisenhower administration focused American assistance on anti-communism and trade. Unwilling to defend Truman's initiative, President Eisenhower merged the TCA into the new Foreign Operations Administration (FOA) in 1953, leading five senators to argue the president was trying to smother Point Four.[98] Opponents charged the military commitments destroyed the "goodwill and morale" of participating countries.[99] The president reorganized again two years later, subsuming all assistance into the International Cooperation Administration (ICA). Aid remained primarily martial: Around 75 percent of total funding went to military assistance and defense.[100] To offload costs, the Eisenhower

---

[94] James M. Carter, *Inventing Vietnam: The United States and State Building, 1954–1968* (New York: Cambridge University Press, 2008).

[95] J. P. King, "Science versus Administration in Certain U.S. Foreign Aid Efforts," *Science* 117 (May 22, 1953): 568–569, quote on 568.

[96] Ibid., 569.

[97] Paul P. Kennedy, "U.S. Shifts System of Technical Help" *New York Times* (January 5, 1954): 52. For an overview of the five organizational arrangements and seven different administrators under Truman and Eisenhower, see W. Haven North and Jeanne Foote North, "Transformations in U.S. Foreign Economic Assistance," in Louis A. Picard, et al., eds., *Foreign Aid and Foreign Policy: Lessons for the Next Half-Century* (Armonk, NY: M.E. Sharpe, 2008): 263–301.

[98] Kennedy, "U.S. Shifts System of Technical Help," 52.

[99] Phelps Phelps, "Undermining Point Four," *New York Times* (October 7, 1953): 28.

[100] The administration did not clarify the differences between "military assistance" and "defense support." The numbers presented were: $2,703 million in total assistance, with military assistance getting 38 percent ($1,022 million) and defense support 37 percent ($999 million), development assistance got 6 percent ($162 million), technical coop 5 percent ($153 million), Asian development fund 4 percent, other 6 percent. See Stokes, "The International Cooperation Administration," 35–37. An outside analyst from the Bulletin of Atomic Scientists suggested the figure was closer to 85 percent. See Peter G. Franck, "Point Four – Five Years After," *Bulletin of the Atomic Scientists* 10 (June 1954): 218–239.

administration sought corporate participation. Joseph Stokes, an ICA official, was candid: "Our motives are not entirely philanthropic. It is plain business sense to help raise the standard of living of other countries, because only as these standards are raised can we increase our markets and expand our commerce and influence."[101] Stokes announced an expansion of Americans abroad: "At one end of the spectrum in the field of technical assistance there are programs in the peaceful uses of atomic energy, such as the Asian Nuclear Research Center . . . At the other end of the spectrum are programs in improving rice planting methods for the farmer who holds half an acre of land."[102] American assistance served a variety of goals, from scientific and technical training to expanding production and assessing foreign resources.

Geology programs, for example, reflected a mix of commercial and national security interests. Beginning in 1951, Point Four recruited geologists from the Department of the Interior to conduct surveys of strategic materials abroad. In partnership with private mining companies, Department geologists sought out deposits of tungsten, manganese, copper, and other minerals. US officials hoped to stimulate commercial development of underused resources, often pressing for liberalization of local economies. "Interior's geologic diplomats," according to historian Megan Black, "used their local influence to alter mining codes to launch further mineral exploration and foreign investment."[103] Nor was geology alone; many assistance programs aimed to increase production of raw materials or open local markets.

Nations weighed the scientific and technical benefits of Point Four against the geopolitical and economic costs. Military security requirements caused the Syrians and Burmese to withdraw from the program, while the acceptance of aid caused the Indonesian cabinet to resign.[104] Fear of economic displacement led Portugal to block American specialists from their colonies of Mozambique and Angola; Belgium refused P4 entry into the Congo. Egyptian president Gamal Nasser originally welcomed

[101] Joseph M. Stokes, "The International Cooperation Administration," *World Affairs* 119 (Summer 1956): 35–37, quote on 35. Stokes was the Deputy Director for Technical Services of the ICA.
[102] Stokes, "The International Cooperation Administration," 36.
[103] Megan Black, "Interior's Exterior: The State, Mining Companies, and Resource Ideologies in the Point Four Program," *Diplomatic History* 40 (2016): 81–110, quote on 107.
[104] Walter R. Sharp, "The Institutional Framework for Technical Assistance," *International Organization* 7 (August 1953): 342–379, esp. 347.

Point Four funds and dismissed charges the program had a "colonial aim."[105] But the obligation to support American anti-communism caused him to suspend aid, fearing it would be used to influence Suez Canal negotiations.[106] Nor was Nasser alone; the technical assistance program in Iraq, begun in 1951 and funded by $2.5 million in oil revenues, eventually collapsed in an anti-Western revolution, ending more than a dozen projects in agriculture and preventive medicine.[107] By then, the Soviets were involved.

The Soviet Challenge in the Developing World

The entrance of the Soviet Union changed UN technical assistance. Stalin had little interest in assistance outside the communist bloc, preferring to shepherd limited resources and personnel while forbidding UN participation.[108] Until his death in 1953, no Soviet bloc country participated in UN aid programs. The new leadership, however, offered money and expertise to the technical assistance program. The Soviet Union joined UNESCO in the spring of 1954, eventually pledging more than $5 million annually and raising the technical aid budget to around $37 million by 1957 (the United States contributed nearly half).[109] One of the most ambitious projects was an Institute of Technology outside Bombay – a Soviet-funded Indian/UNESCO project, which would also provide more than a thousand scholarships for students at Soviet institutions. Of course, the United States already provided similar educational opportunities at American universities as well as an allied-only school of nuclear science and engineering at Argonne national laboratory.[110]

---

[105] Nasser quoted in Samir Ahmed, Press Attaché, Embassy of Egypt, "Point Four Program in Egypt," *New York Times* (May 3, 1954): 24.

[106] Peter G. Franck, "Point Four – Five Years After," *Bulletin of the Atomic Scientists* 10 (June 1954): 218–239.

[107] Henry Wiens, "The United States Mission in Iraq," *Annals of the American Academy of Political and Social Science* 323 (May 1959): 140–149.

[108] Thomas G. Patterson, "Foreign Aid under Wraps: The Point Four Program," *The Wisconsin Magazine of History* 56 (Winter 1972–73): 119–126, esp. 126.

[109] Laves and Thomson, *UNESCO*, 334. See also Elizabeth M. Thompson, "What Next for UN Technical Assistance?" *World Affairs* 120 (Summer 1957): 49–53, esp. 51.

[110] Peter J. Westwick, *The National Labs: Science in an American System, 1947–1974* (Cambridge, MA: Harvard University Press, 2003), 164–66. For examples of Pakistani and Indian training programs, see Matthew Fuhrman, *Atomic Assistance: How 'Atoms for Peace' Programs cause Nuclear Insecurity* (Ithaca, NY: Cornell University Press, 2012), 162 and 193–194.

Cold War competition stimulated assistance. Khrushchev offered $250 million in hydroelectric and mining aid to Afghanistan, resulting in the cancellation of American and European engineering contracts.[111] Soviet "Point Four" programs with Argentina and India soon followed.[112] Like capitalists, the communists used technical assistance to further their home economy: Soviet aid came in non-controvertible rubles, requiring recipients purchase Soviet products and services.[113] The CIA closely tracked Soviet assistance, reporting the 1954 Sino–Soviet agreement required each signatory "send specialists to give technical assistance and acquaint each other with their achievements in the fields of science and technology."[114] Nonetheless, intelligence analysts believed Soviet global commitments remained a fraction of US spending, estimating the United States lent fourteen times more at the start of Point Four (1950) than the Soviets five years later.[115] But Soviet aid continued to expand: In 1956, Khrushchev traveled to India, Afghanistan, and Burma with offers of additional assistance; after the Suez Canal crisis, the Soviet premier committed to helping Nasser complete his Nile dream (the Aswan dam).[116]

The Soviet Union aimed to exploit its post-Sputnik prestige in the developing world. In 1958, Chile gave the Soviets their first opportunity to make observations in the Southern hemisphere; Soviet astronomers worked at the University of Chile and cosmonaut Alexei Leonov later toured the country.[117] In 1960, Khrushchev opened the People's Friendship University with promises to not "force our ideas, our ideology on any student."[118] The CIA estimated around 2,500 students attended – many Egyptian or Indian – and nearly half majored in engineering, medicine,

---

[111] John P. Callahan, "Soviet Said to Offer Afghans 'Point Four,'" *New York Times* (July 30, 1954): 1–2.
[112] Harry Schwartz, "Kremlin Offers Asians Its Own Point Four Plan," *New York Times* (November 7, 1954): E9.
[113] Charles R. Dannehl, *Politics, Trade and Development: Soviet Economic Aid to the Non-Communist Third World, 1955–89* (Brookfield, VT: Dartmouth Publishing Company, 1995), 4.
[114] CIA Office of Research and Reports, "Soviet Economic Assistance to the Sino-Soviet Bloc Countries," 64.
[115] Ibid.
[116] Carol Lancaster, *Foreign Aid: Diplomacy, Development, Domestic Politics* (Chicago: University of Chicago Press, 2007), 65.
[117] CIA Directorate in Intelligence, "Latin America Looks to Eastern Europe," (March 29, 1968): 1–13. CIA Doc No/ESDN: CIA-RDP79–00927A006300080005–2.
[118] Khrushchev quoted in CIA Directorate of Intelligence, "Soviet Academic and Technical Programs for Students from the Less Developed Countries of the Free World," (May 1965): 1–38, quote on 3. CIA Doc No/ESDN: 0000309819.

physics, or mathematics.[119] Nonetheless, Soviet influence remained slight. Funding remained small and Soviet attitudes toward the developing world interfered: The CIA observed, "Racial discrimination has been a major source of discontent and complaints from African students have been profuse."[120] Of course, Sputnik's orbit included Little Rock, Arkansas, in its daily itinerary, reminding global audiences of America's integration crisis.[121]

The Soviet satellite convinced American politicians to increase assistance to the developing world. Eisenhower's science advisor encouraged an expansion of aid, observing, "The striving to emulate American scientific and technological progress has become an ambitious and urgent goal for countless millions of people."[122] The Senate Foreign Relations Committee, including Senator John F. Kennedy, suggested the United States take the lead in development efforts, proposing an International Development Year patterned after the IGY.[123] As a freshman in Congress, Kennedy considered Point Four wasteful and voted against funding. However, a trip to Southeast Asia in 1951 changed his mind; within a few years Senator Kennedy supported aid for the region (he was an enthusiastic backer of Ngo Diem in South Vietnam).[124] After Sputnik, Kennedy spoke often about science and geopolitics, with the Soviet satellite providing a frequent line of attack against the Eisenhower/Nixon administration.[125] Once in office, the young president chose Walt Rostow, a prominent modernization theorist, as his foreign policy expert. Concern for US prestige underlay Kennedy's support for the moon landing (see Chapter 2) and his passion for science and technology resurfaced in policies toward Latin America – a region where anti-Americanism was on the rise.

---

[119] Ibid., 4–5.   [120] Ibid., 13.

[121] Thomas Borstelmann, *The Cold War and the Color Line: American Race Relations in the Global Arena* (Cambridge, MA: Harvard University Press, 2003), 104.

[122] G. B. Kistiakowsky, "Science and Foreign Affairs," *Science* 131 (April 8, 1960): 1019–1024, quote on 1021.

[123] John W. Finney, "Senate Unit Calls Science Vital Foreign Policy Tool," *New York Times* (September 20, 1959): 1 and 26.

[124] Michael E. Latham, *Modernization as Ideology: American Social Science and "Nation Building" in the Kennedy Era* (Chapel Hill: University of North Carolina Press, 2000). See also Jeffrey F. Taffet, *Foreign Aid as Foreign Policy: The Alliance for Progress in Latin America* (New York: Routledge, 2007), esp. 24–25.

[125] See "The Soviet Challenge to American Education" and "The Years the Locusts Have Eaten," in John F. Kennedy, *The Strategy of Peace* (New York: Harper & Brothers, 1960), 206–212 and 235–241.

The Alliance for Progress

The Alliance for Progress was President Kennedy's signature initiative in the developing world. His inaugural address emphasized the potential benefits of science and offered assistance south of the border:

Let both sides seek to invoke the wonders of science instead of its terrors. Together let us explore the stars, conquer the deserts, eradicate disease, tap the ocean depths and encourage the arts and commerce ... To our sister republics south of our border, we offer a special pledge – to convert our good words into good deeds – in a new alliance for progress – to assist free men and free governments in casting off the chains of poverty.

Pres. John F. Kennedy, *Inaugural Address* (1961)

The president moved quickly to implement his vision. Established in August of 1961, the Alliance for Progress stressed economic development and improvements in agriculture, health, and science education. A few months later, Kennedy established the United States Agency for International Development (USAID), creating a federal clearinghouse for American aid overseas. But the new agency fell prey to the problems of its predecessors.

Commercial and military requirements undermined the Alliance. Although Kennedy committed significant funds for development (a first – more than $1 billion the first year), the Alliance earmarked most dollars for purchases of American industrial, agricultural, or military equipment; critics felt the program subsidized American manufacturers by underwriting foreign purchase of their wares. Tension between development aims and military heavy-handedness plagued the administration from the start; three weeks after announcing the Alliance for Progress, Kennedy authorized the Bay of Pigs invasion. Presidential directives emphasized relying on local armed forces; David E. Bell, USAID administrator, argued, "military forces have an essential role as a stabilizing force in these countries."[126] Determined to secure anti-communist allies more than to

---

[126] Regarding Kennedy, see John F. Kennedy, National Security Action Memorandum 88, "Training for Latin American Armed Forces," September 5, 1961. The memo suggests setting up an "FBI academy" in the United States to train foreign officers. Archived online at www.jfklibrary.org. See also John F. Kennedy, National Security Action Memorandum 119, "Civic Action," December 18, 1961. Archived online at www.fas.org/irp/offdocs/nsam-jfk/nsam119.jpg. David E. Bell quoted in Willard F. Barber, "Can the Alliance for Progress Succeed?" *Annals of the Academy of Political and Social Science* 351 (January 1964): 81–91, quote on 82.

promote democratic reforms, Alliance funds soon supported a host of juntas.[127] Historian Walter Lafeber pointed out US military aid doubled yearly during the Kennedy years, USAID provided gas and helicopters to anti-riot police, and during the height of the program (1961–1966), militaries overthrew nine Latin American governments.[128] Yet even as the Alliance supported authoritarian regimes, the State Department promoted science education to counter "anti-democratic intellectual activity" (i.e. communism).[129]

Although Secretary of State Rusk spoke often about the importance of science to development, the Alliance spent little to foster research in Latin America.[130] Instead, the program focused on science education to change cultural habits.[131] The United States pledged $38 million for universities and technical institutes, including $4 million apiece for strengthening science curricula and English language training.[132] Assistant Secretary of State Martin argued, "Quantitative precision, exactness in work and production schedules, careful calculation of cost advantages, all must become second nature to have a modern society. Scientific or rational habits of thought, as opposed to the traditional or emotional approach to problems is equally imperative, as are the somewhat anti-romantic values of neatness and cleanliness and order."[133] Lawrence Harrison, USAID Director for Latin America in the 1960s and 1970s, later suggested the "values and attitudes" of Hispanic culture were "the principal obstacles

---

[127] Thomas C. Field, Jr., "Ideology as Strategy: Military-Led Modernization and the Origins of the Alliance for Progress in Bolivia," *Diplomatic History* 36 (January 2012): 147–183, esp. 148. See also Stephen M. Streeter, "Nation-Building in the Land of Eternal Counter-Insurgency: Guatemala and the Contradictions of the Alliance for Progress," *Third World Quarterly* 27 (2006): 57–68.

[128] Walter Lafeber is especially critical of the Alliance for Progress. See "Updating the System" in Walter Lafeber, *Inevitable Revolutions: The United States and Central America,* 2nd edn. (New York: W. W. Norton & Company,1993), 147–177, esp. 153–154.

[129] Department of State, "An Alliance for Progress Illustrative Program on Education, Science and Culture," (March 1, 1962): 1–14. Available through DDRS.

[130] Dean Rusk, excerpts of speech, "Dean Rusk Analyses Alliance Science Progress," *Science News Letter* 86 (October 24, 1964): 258 and 270. Rusk was speaking at a "Conference on Science and Development."

[131] Department of State, "An Alliance for Progress Illustrative Program on Education, Science and Culture," (March 1, 1962): 1–14. Available through DDRS.

[132] Agency for International Development, Department of State, *The Alliance for Progress . . . an American Partnership* (Washington, DC: Government Printing Office, 1965).

[133] Edwin Martin quoted in Latham, *Modernization as Ideology,* 95.

to progress in Latin America."[134] Samuel Butterfield, another long-time USAID veteran, added fatalism and a poor work ethic.[135] But the emphasis on commerce and anti-communism, more than local culture, undermined American assistance.

Latin American leaders tired of the Alliance. Eduardo Frei, the Chilean presidential candidate favored by the United States, described the initiative as "inoperative" after the United States spent $136 million in his country.[136] In 1967, Frei refused USAID funds, arguing the program undermined local development by over-emphasizing private capital and the military, allowing for the consolidation of unjust regimes.[137] Instead, Frei joined other Latin American presidents to support the *Punta Del Este Declaration*. Signed by the leaders of five nations (Colombia, Chile, Venezuela, Peru, and Ecuador), the *Declaration* made clear the Latin American desire for science and development: "We will harness science and technology for the service of our people. Latin America will share in the benefits of current science and technological progress so as to reduce the widening gap between it and the highly industrialized nations."[138] Only Argentina and Brazil abstained, leaving the Alliance invested in authoritarian regimes in the late 1960s.[139]

Brazil and Argentina illustrated the tensions within American assistance. The establishment of military governments in Brazil (1964) and Argentina (1966) led local scientists to request that their American counterparts condemn authoritarianism. Brazilian and Argentinian researchers, according to the National Academy of Sciences, "expressed the view

---

[134] Lawrence Harrison quoted in Samuel Hale Butterfield, *U.S. Development Aid – An Historic First: Achievements and Failures in the Twentieth Century* (Westport, CT: Praeger, 2004), quote on 72.

[135] Ibid.

[136] Willard F. Barber, "Can the Alliance for Progress Succeed?" *Annals of the Academy of Political and Social Science* 351 (January 1964): 81–91.

[137] Eduardo Frei Montalva, "The Alliance That Lost Its Way," *Foreign Affairs* 45 (April 1967): 437–448. For a contemporary analysis, see Albert L. Michaels, "The Alliance for Progress and Chile's 'Revolution in Liberty,' 1964–1970," *Journal of Interamerican Studies and World Affairs* 18 (February 1976): 74–99.

[138] *Declaration of the Presidents of America*, Punta del Este, Uruguay, April 14, 1967. Archived online at http://avalon.law.yale.edu/20th_century/intam19.asp.

[139] Of course, the United States spent $5 million to manipulate the Brazilian political scene in favor of the military. See Stephen G. Rabe, *The Killing Zone: The United States Wages Cold War in Latin America* (New York: Oxford University Press, 2012), 107. An interesting introduction to AFP in Brazil is Andrew J. Kirkendall, "Kennedy Men and the Fate of the Alliance for Progress in LBJ Era Brazil and Chile," *Diplomacy & Statecraft* 18 (2007): 745–772.

that NAS should sever all relations in these countries, thereby denouncing the regimes."[140] But NAS policy was to "avoid either endorsing or censuring governments."[141] Believing engagement would be more successful, the NAS set up a telex network to connect research centers in Argentina with the library at the University of Chicago (a major technical achievement at the time).[142] In Brazil, the NAS initiated a training program in chemistry with the National Research Council, offering workshops for hundreds of scientists and engineers. President Nixon also included the country in his 1969 "Atoms in Action" plan, helping establish a nuclear science demonstration center in Sao Paulo.[143] Within a few years, the NAS considered the Brazilian initiative a "very effective program" because of increased government funding for R&D, acceptance of American patent laws, and the creation of a pharmaceutical industry.[144]

In hindsight, however, most scholars consider the Alliance for Progress a failure. USAID administrator Samuel Butterfield calculated the United States and other donors spent more than $20 billion between 1961 and 1970 to prop up governments without insisting on political or economic reforms in return.[145] AFP funds and tax breaks frequently went to US-owned firms and local elites, worsening pre-existing economic imbalances. Some contemporary critics felt the Alliance appealed to Americans because "it provides a technological answer to the immense psychological and sociological problems of underdevelopment," but technical budgets were minimal relative to security costs.[146] The size and complexity of the Alliance proved impossible to manage. Historian Jeffrey F. Taffet argued Alliance funds were "sent in a haphazard way using a variety of theoretical approaches," adding "there was no shared or simple definition of what the Alliance for Progress meant, how it would be implemented, or

[140] Harrison Brown and Theresa Tellez, eds., *National Academy of Sciences: International Development Programs of the Office of the Foreign Secretary, Summary and Analysis of Activities, 1961–1971* (Washington, DC.: National Academy of Sciences, 1971), 7.
[141] Ibid., 7.   [142] Ibid., 25.
[143] Richard Nixon, "Opening of the 'Atoms in Action' Nuclear Science Demonstration Center in Brazil," (September 4, 1969). Available at the Digital National Security Archive, Nuclear Non-Proliferation Collection, Record # NP01257. A nearly verbatim transcript can be found for the Philippines center, opened February 11, 1969; see Digital National Security Archive, Nuclear Non-Proliferation Collection, Record # NP01243.
[144] Brown and Tellez, eds., *National Academy of Sciences*, 41.
[145] Butterfield, *U.S. Development Aid*, 69.
[146] Robert M. Smetherman and Bobbie B. Smetherman, "High Visibility Aid: The Alliance For Progress," *The Western Political Quarterly* 24 (March 1971): 52–54, quote on 53.

even who was in charge."[147] In the end, the failure of the Alliance weakened American support for modernization abroad, ending many scientific and technical programs, although those supplemented by private foundations and international agencies remained successful; indeed, perhaps the longest-lasting impact of American assistance came in public health and agricultural science.

DEMOGRAPHIC CONTAINMENT

Nor would we be any closer to the kind of world we want to live in if the effect of economic development was merely a rapid increase of population in already overpopulated countries – that is, more people surviving to struggle along at about the same old subsistence levels. Yet this is what may happen in large parts of southern and eastern Asia, in Egypt, and nearer home in Puerto Rico, unless birth control is adopted by these peoples [alongside] the importation of modern "death control," as represented by more efficient food production, sanitation, and modern medicine.

Stanford economist and modernization theorist Eugene Staley (1952)[148]

The [Senate] Committee on Foreign Relations in particular has given deep attention to America's international health goals through ICA. The committee has pointed out America's deep compassion and its desire to reduce human pain and suffering. The committee has pointed out our enlightened self-interest, as well, for we are soundly helping new nations to become stable and strong members of the world community. Thereby, they are better able to resist the totalitarian aggression which preceded and has marked the cold war – aggression which thrives on conditions of want and privation in disadvantaged nations.

Office of Public Health, International Cooperation
Administration (1961)[149]

Afraid of instability, US foreign policy-makers appealed to various sciences to shape developing world demographics. Agricultural sciences would increase food production, modern medicine and public health programs would reduce mortality rates, and contraception and family planning would limit population growth and the demand on scarce

---

[147] Taffet, *Foreign Aid as Foreign Policy*, 10.

[148] Eugene Staley, "Technical and Economic Assistance under Point Four," *Proceedings of the Academy of Political Science* 25 (May 1952): 23–32, quote on 29.

[149] Office of Public Health, International Cooperation Administration, *Technical Cooperation in Health* (Washington: International Cooperation Administration, 1961), 417.

resources. Foreign direct investment and a strong anti-communist military completed the modernization strategy. Historian John Perkins named this the "Population National Security Thesis" and proposed a causal chain: Overpopulation (leads to) Resource Exhaustion (leads to) Hunger ... Political Instability ... Communist Insurrection ... Danger to US Interests ... War.[150] Perkins's focus was the "Green Revolution" – the broad name for the American program to reduce hunger and gain allies by increasing the commercial production of new strains of wheat, corn, and rice. Of course, the revolution, complemented by medical and contraceptive aid, was part of the constellation of American scientific and technical initiatives outlined previously. But agricultural and medical programs were particularly influential: A new scientific approach helped sell the programs to locals while World War II left behind a suite of international institutions devoted to ending hunger and improving health.

Global Health and the Malaria Eradication Program

American global health programs predate the World Health Organization (WHO, 1948). During World War II, the IIAA initiated health programs with all eighteen Latin American allies. After the war, the Philippine Rehabilitation Act of 1946 provided millions for public health in the islands. Within a few years, American anti-communism stimulated further bilateral health initiatives in the Middle East (including Turkey, Egypt, Jordan, Iran, and Saudi Arabia) as well as Southeast Asia (including Formosa, Japan, Indonesia, Burma, Thailand, Indochina, the Philippines, and Korea). In return for battling communism, for example, Greece received funding for six hospitals, five tuberculosis sanitariums, three nursing schools, and improved water supplies in 375 communities.[151] The existence of such programs allowed American policy-makers to link UN and WHO programs to pre-existing American initiatives, leading the Soviet Union to withdraw.

The Cold War intruded on global health coordination. Early triumphs include the WHO response to cholera in Egypt, inoculation against tuberculosis, and the distribution of penicillin for syphilis.[152] But debates

---

[150] Perkins, *Geopolitics and the Green Revolution*, 118–120.
[151] Office of Public Health, *Technical Cooperation in Health*, 429–432.
[152] Amy L. S. Staples, *The Birth of Development: How the World Bank, Food and Agriculture Organization, and World Health Organization Changed the World, 1945–1965* (Kent, OH: Kent State University Press, 2006), 140.

over the socioeconomic basis for disease led the United States to withhold funding from the newly created WHO until the organization renounced socialized healthcare (see Chapter 1). The communist bloc then walked out between 1949 and 1950, returning only in 1957. Still, the presence of blocs frustrated WHO officials and researchers who considered geopolitics inappropriate and antithetical to the organization's mission.[153] Of course, blocs could further US interests; a later congressional study concluded communist withdrawal strengthened the American position.[154] It was difficult to disentangle politics and healthcare at the height of the Cold War; a 1957 American Medical Association editorial suggested using the "universal language of medicine, not only as a means of exchanging scientific knowledge and promoting medical progress but, more so, as an easily understood basis to explain the free and democratic way of life."[155] The American government agreed.

American global health initiatives bolstered anti-communist allies in newly independent countries. ICA brochures declared, "Technical cooperation in health has become an essential part of U.S. foreign policy," proudly noting more than twenty African and Asian nations established since World War II received aid.[156] Believing "Fast Action makes Friends Abroad," the ICA responded to smallpox and cholera outbreaks in East Pakistan and Thailand as well as hepatitis in Ceylon. Although the ICA only cooperated "in those cases where new research is essentially needed for the solution of specific problems," this led to the establishment of the American Foundation for Tropical Medicine in Liberia and a research lab in Dacca, East Pakistan (part of the SEATO Cholera project).[157] Health programs benefited both countries – most of the cost was borne by the host country, which received better health care, while the United States received positive international press, access, and anti-communist allies.

Allies welcomed healthcare initiatives. The Iranian parliament mandated oil revenues pay for public health programs; the shah posed while opening the Shiraz Medical Center, water treatment plants in Teheran, and a public health laboratory in Ishafan. To sell an increased US military presence, South Vietnamese President Ngo Diem told his national assembly in October 1959: "with regard to [American] health programs, the

---

[153] See "The Development of Blocs in the W.H.O." in Siddiqi, *World Health and World Politics*, 77–86.
[154] House of Representatives, *Science, Technology and Diplomacy*, 113–115.
[155] AMA editorial reproduced in Office of Public Health, *Technical Cooperation in Health*, 422.
[156] Ibid., 423.   [157] Ibid., 496.

equipment of our hospitals has been increased and modernized. The struggle against epidemics, and such diseases as tuberculosis, leprosy, and cancer, has yielded notable results."[158] The ICA reproduced similar statements from leaders in Thailand, Colombia, Israel, and Indonesia. American initiatives also collaborated with international organizations, making it difficult to receive aid without the United States; in 1958, for example, the Soviet Union asked the World Health Organization to begin a crash program to eradicate smallpox, listing seventeen socialist countries for WHO, but not American, aid.[159]

The Malaria Eradication Program (MEP) illustrated the coordination of US and UN programs. After federal entomologists recommended DDT as a delousing powder and anti-malarial spray during World War II, the US War Production Board mass-produced the pesticide, eventually turning out three million pounds per month by 1945.[160] America committed to DDT during the war: the Public Health Service sprayed all southern military establishments, while the malaria control project in Atlanta evolved into the Centers for Disease Control after the war. By the time the WHO began spraying, the United States already supported anti-malarial activities in twenty countries; the WHO merely extended the program to an estimated 179 million people worldwide.[161] One of the largest eradication programs to date, the MEP relied on American agencies and chemicals, leading to UN approval of DDT; by 1965 more than 100 countries sprayed the pesticide.[162] Allies praised the joint effort: Ethiopian leader Haile Selassie, for example, enthused, "The successful pursuit during the past year of the anti-malarial campaign stands as a tribute to the devotion and cooperation of experts from the Ethiopian government, the World Health Organization, UNICEF and the US

---

[158] Diem quoted in ibid., 419–420.

[159] Anon., "Russia Asks War on Smallpox," *The Science News Letter* 73 (June 7, 1958): 366.

[160] Edmund Russell, *War and Nature: Fighting Humans and Insects with Chemicals from World War I to* Silent Spring (New York: Cambridge University Press, 2001), 161.

[161] George W. Pearce, et al., "Specifications of the International Cooperation Administration for DDT Water-Dispersible Powder for Use in Malaria Control Programmes," *Bulletin of the World Health Organization* 20 (1959): 913–920. An excellent overview of the role of DDT in the Green Revolution and the MEP is David Kinkela, *DDT & the American Century: Global Health, Environmental Politics and the Pesticide that Changed the World* (Charlotte: University of North Carolina Press, 2011).

[162] Other global eradication campaigns include tuberculosis (1947–1951), yaws (1955–1970) and smallpox (1967–1980). See Nitsan Chorev, *The World Health Organization between North and South* (Ithaca, NY: Cornell University Press, 2012), 60.

operations mission to Ethiopia."[163] Ngo Diem agreed, stating, "I would like to stress the anti-malaria campaign. The campaign against this disease extends over seven years, from 1957 to 1963, with the collaboration of the U.S. Operations Mission."[164] Until the 1970s, most observers considered the MEP a successful complement to American agricultural programs, which also had roots in World War II and Latin America.

The Mexican Agricultural Program, the FAO, and PL-480 Food Aid

Agricultural collaboration brought together scientists, governments, and foundations south of the border. As FDR established the IIAA and Lend-Lease aid to Mexico, his administration encouraged the Rockefeller Foundation (RF) to fund agricultural research. The timing was perfect. The RF International Health Division thought it would complement their public health programs, while others at the foundation wanted to focus on modernizing agriculture through genetics.[165] Helpfully, Mexico's newly elected president welcomed foreign investment and reduced tariffs on American chemical products. Established in 1943, the Mexican Agricultural Program (MAP) originally focused on the control of wheat rust, the improvement of corn-breeding production and studies related to the management of soils and livestock disease.[166] Over the next two decades, the MAP helped start the Green Revolution, transforming US food aid and agriculture worldwide.

Mexico became a test-case in commercial agriculture. Within a few years, MAP researchers began focusing on expensive high-yield hybrid seeds that reacted well to chemical fertilizers and pesticides; i.e. seeds designed for large-scale industrial cultivation.[167] Carl Sauer, a Rockefeller Foundation researcher, warned: "A good aggressive bunch of American agronomists and plant breeders could ruin the native resources for good and all by pushing their American stock ... Mexican agriculture cannot

[163] Selassie quoted in Office of Public Health, *Technical Cooperation in Health*, 419–420.
[164] Diem quoted in Ibid., 421.
[165] Marcos Cueto, "Introduction," in Marcos Cueto, ed., *Missionaries of Science: The Rockefeller Foundation & Latin America* (Bloomington; Indiana University Press, 1994), ix–xvii. See also Jonathan Harwood, "Peasant Friendly Plant Breeding and the Early Years of the Green Revolution in Mexico," *Agricultural History* 83 (Summer 2009): 384–410.
[166] Ana Barahona and Francisco J. Ayala, "The Emergence and Development of Genetics in Mexico," *Nature* 6 (November 2005): 860–866.
[167] Jack Ralph Kloppenburg, Jr., *First the Seed: The Political Economy of Plant Biotechnology*, 2nd edn. (Madison: University of Wisconsin Press, 2004).

be pointed toward standardization on a few commercial types without upsetting native economy and culture hopelessly."[168] Undeterred, Norman Borlaug, chief RF plant geneticist, believed low yields resulted from a lack of nitrogen and poor disease resistance. In 1949, his first rust-resistant, nitrogen-fertilized wheat crop prospered in the Yaqui valley in Sonora, leading the Mexican government to subsidize hydroelectric dams for large-scale irrigation and fertilizer production.[169] By the mid-1950s, hybrid fields produced remarkable yields, dominating the new commercial wheat industry in Mexico. Supporters hailed the MAP as a resounding success; a similar program was soon underway for corn. With Mexico the model of modernization, American diplomacy promoted large-scale commercial agriculture around the world.

American food policy evolved after World War II. President Roosevelt helped establish the first UN agency – the Food & Agricultural Organization (FAO) – in 1943. However, when the FAO suggested a World Food Board to buy and hold agricultural products to control price and distribution, the Truman administration vetoed the idea.[170] John Boyd Orr, the Nobel Peace Prize-winning Scottish biologist and first FAO chief, became so frustrated with American (and British) attitudes he demanded the organization relocate its headquarters from D.C. to Rome and promptly resigned. Orr later recalled upon leaving the United States, he "took out [his] handkerchief ... wiped the dust of America from the soles of my shoes with it and threw it into the harbor."[171] But the FAO could not walk away.

The FAO, like other UN technical organizations, closely coordinated with the United States. By 1954, more than seventy nations participated in the organization, which maintained regional offices in Washington, Bangkok, and Cairo. Ralph Phillips, the American representative, praised the global teamwork required to bring the benefits of agricultural science

[168] Sauer quoted in Harwood, "Peasant Friendly Plant Breeding and the Early Years of the Green Revolution in Mexico," 390.

[169] See "Mexico's Way Out" in Nick Cullather, *The Hungry World: America's Cold War Battle Against Poverty in Asia* (Cambridge, MA: Harvard University Press, 2011), 43–71. A more positive take on the legacy of the Green Revolution in the Yaqui valley is Pamela A. Matson, ed., *Seeds of Sustainability: Lessons from the Birthplace of the Green Revolution in Agriculture* (Washington: Island Press, 2012).

[170] Staples, *The Birth of Development*, 92–99.

[171] Orr quoted in Matthew Connelly, *Fatal Misconception: The Struggle to Control World Population* (Cambridge, MA: Harvard University Press, 2008), 127.

to farmers around the world.[172] Funded by the United States, the FAO supplied seeds, researched fertilizers, and sprayed for locusts. In Italy, for example, the organization conducted a hybrid corn school; today, almost half of Italian corn consists of two lines developed in American experiment stations.[173] Similar hybridization programs operated in a dozen Latin American countries by the early 1950s.[174] American bilateral aid dwarfed UN and philanthropic assistance. In 1952, for example, India received $1.2 million from the Ford Foundation and around $50 million from the United States, while the entire FAO budget was around $5 million.[175] Yet partnerships were critical: Foundations paid for agricultural research and education, the UN provided geopolitical cover and coordination, and the United States subsidized both the chemical companies offering fertilizers and pesticides and the financial institutions providing the loans (India borrowed/contributed $86 million). Fear that India might turn toward communism stimulated the passage of Public Law 480 – the Agricultural Trade Development and Assistance Act of 1954.

PL-480 became the basis of American international agricultural policy. The law allowed the United States to maintain domestic agricultural prices by selling surplus grains below market cost to foreign nations who could not afford to purchase the grains at market value. Additionally, because the grains (or other agricultural goods) were paid for in local currencies, the United States acquired large foreign reserves. According to Congress, PL-480 allowed the United States "to make maximum efficient use of surplus agricultural commodities in furtherance of the foreign policy of the United States."[176] Between 1954 and 1967, exports of agricultural products amounted to $17.2 billion with recipients in 116 countries with half the world's population.[177] However, excess grains were not just a benefit of American agriculture, they were also integral to American science diplomacy.

The PL-480 program began funding American scientific and technical assistance in the late 1950s. In the words of historian Jacqueline McGlade, "the focus of PL-480 technical assistance shifted from

---

[172] Ralph W. Phillips, "International Cooperation to Improve World Agriculture," *The Scientific Monthly* 79 (September, 1954): 154–164.

[173] Kloppenburg, *First the Seed*, 122.

[174] Countries included Mexico, Guatemala, El Salvador, Venezuela, Brazil, Uruguay, Argentina, Costa Rica, Cuba, Colombia, Peru, and Chile.

[175] Perkins, *Geopolitics and the Green Revolution*, 154–155.

[176] House of Representatives, *Science, Technology and Diplomacy*, 125.   [177] Ibid.

commercial development to agronomic education as hundreds of scientists, agronomists, nutritionists, and health and sanitation experts instituted programs in soil chemistry, agronomy, crop sciences, seed production, soil, water and forestry conservation, land usage, reclamation and irrigation in recipient nations."[178] Later known as the "Food for Peace" program, American agricultural policies found fertile ground at the UN. Eisenhower administration members, fertilizer company representatives, and FAO officials met in December 1959 to "introduce the scientific method and modern technology into traditional agriculture."[179] Surplus grains, PL-480 funds, and scientific assistance provided the United States with a powerful diplomatic carrot (or stick) around the world. Consider hybrid wheat – its success in Mexico led to adoption in India and elsewhere.

The Green Revolution in India, the Philippines, and Vietnam

The Indian soil had been well tended by the time hybrid wheat was ready for transplant. India, like Mexico, suffered from the interrelated problems of hunger, population growth, and poverty. But it was officially neutral in the Cold War. As noted above, the Indian government received Point Four funds starting in 1951, but remained staunchly non-aligned and limited official American influence. Instead, Prime Minister Nehru asked the Ford Foundation for help with agricultural research. The Eisenhower administration cheered the relationship; CIA director Allen Dulles believed Ford's overseas programs were "a 'great asset' to the U.S. in its international relations" and encouraged the foundation "work in the difficult areas where the U.S. government technical assistance would be relatively ineffective because of the suspicions of the indigenous governments."[180] By the mid-1950s, the Ford Foundation trained agricultural extension staff, the Rockefeller Foundation conducted agricultural research, and the United States paid for more than 2,000 Indian trainees to travel to American agricultural stations. In 1958, foreign advisors reorganized the Indian Agricultural Research Institute (IARI, originally established in 1905) and Ralph Cummings, field director of the

---

[178] Jacqueline McGlade, "More a Plowshare Than a Sword: The Legacy of US Cold War Agricultural Diplomacy," *Agricultural History* 83 (Winter 2009): 79–102, quote on 93.

[179] House of Representatives, *Science, Technology and Diplomacy*, 113.

[180] Dulles quoted in Corinna R. Unger, "Towards Global Equilibrium: American Foundations and Indian Modernization, 1950s to 1970s," *Journal of Global History* 6 (March 2011): 121–142, quotes on 132.

Rockefeller Foundation, became dean.[181] A Ford Foundation report laid out a technocratic path to modernization relying on chemical fertilizers, hybrid seeds, and tube-well irrigation.[182]

The United States required India accept foreign investment to receive agricultural assistance. Following Foundation directives, IARI began working with hybrid wheat in 1961, even though rice was the staple of the Indian diet. Additionally, a broad American-led coalition – including USAID, the State Department, Bechtel, the Ford and Rockefeller Foundations, the FAO, and the World Bank – pressed for liberalization of the Indian economy as a requirement for agricultural aid, a sticking-point until the prime minister's passing in 1964. The following year, at the onset of drought in the Bihar region, the United States forecast an Indian famine and placed the country on a monthly tether, limiting PL-480 food aid until India devalued the rupee and accepted foreign investment.[183] In 1966, Indira Gandhi, Nehru's daughter and India's new prime minister, reversed her father's positions, devaluing the currency and opening state industries, including fertilizers, to American investment. President Johnson and Gandhi celebrated the new bilateral partnership. A few years later, Lester Brown, former head of the USDA's International Agricultural Development Service, confessed, "the U.S. and the World Bank put a great deal of pressure…especially on the Indian government, to encourage multinational corporations to invest in local production capacity."[184]

The introduction of hybrid wheat transformed Indian agriculture. Fields planted with hybrid seeds received state subsidies for irrigation, fertilizers, and roads. Farmers able to take advantage of the program frequently saw quantum leaps in yields.[185] Led by large-scale commercial producers, India escaped famine. Monkombu Swaminathan, Borlaug's counterpart in India for nearly half a century and a member of parliament, expressed a common view: "Borlaug brought a revolution in ideas, a revolution in thinking, a revolution in technology. It's a totality. It was a

[181] Vandana Shiva, *The Violence of the Green Revolution: Third World Agriculture, Ecology and Politics* (London: Zed Books, 1991), 28–29.

[182] Unger, "Towards Global Equilibrium,"133.

[183] See "The Meaning of Famine" in Cullather, *The Hungry World*, 218–231; "Wheat Breeding and the Consolidation of Indian Autonomy," in Perkins, *Geopolitics and the Green Revolution*, 157–186; Vandana Shiva, *The Violence of the Green Revolution*, 30–33.

[184] Brown quoted in W. D. Posgate, "Fertilizers for India's Green Revolution: The Shaping of Government Policy," *Asian Survey* 14 (August 1974): 733–750, quote on 742.

[185] Raju J. Das, "Geographical Unevenness of India's Green Revolution," *Journal of Contemporary Asia* 29 (1999): 167–186.

great social change. It's quite likely that a billion people have been saved in India, Pakistan and Bangladesh."[186] Agriculture, from his perspective, was critical to the shift to the "scientific temperament" so desired by Jawaharlal Nehru. Later critics argued it also shifted Indian agriculture toward foreign products such as hybrid rice, as the United States began introducing seeds from the Philippines. Soon agricultural production linked American anti-communist allies.

American rice research arose in the Philippines. Supported by the United States, the Rockefeller and Ford Foundations established the International Rice Research Institute (IRRI) in 1962. To command more influence, the Foundations pledged $7 million each to build a high-tech research facility in the Philippines. The new labs would conduct project-focused research, using the wartime OSRD as their model.[187] Additionally, the IRRI brought together scientists from eight countries, requiring cooperation among Asian national institutions and linking agricultural training and research throughout the Southern hemisphere (centers were also built in Colombia and Nigeria).[188] Within a year, IRRI developed IR-8, a rice that responded well to fertilizers. The impact was immediate: Bountiful harvests, combined with USAID funds and an American advertising firm, helped propel Ferdinand Marcos to victory in 1966. His slogan? "Progress is a Grain of Rice."[189]

Hybrid rice advertised the benefits of an American alliance throughout Southeast Asia. In 1966, USAID funded Project Spread, large-scale trials of IR-8 combined with distribution of Atlas and Esso farm chemicals. President Johnson toured the Philippines and South Vietnam, promising audiences the war on hunger was "the only war in which we seek escalation."[190] Agricultural experiment stations became showcases of modernization in the war against communism. IR-8 was essential to American allies campaigning on developmental platforms: Marcos advertised "Rice, Roads and Schools," Suharto called himself the "Father of Development," while Indira Gandhi and Ayub Khan highlighted infrastructure ("Us Build" and "Together let us build"

---

[186] Swaminathan quoted in Pallava Bagla, "A Guru of the Green Revolution Reflects on Borlaug's Legacy," *Science* 326 (October 16, 2009): 361.

[187] See Latham, *The Right Kind of Revolution*. Nick Cullather, "Miracles of Modernization: The Green Revolution and the Apotheosis of Technology," *Diplomatic History* 28 (April 2004): 227–254, esp. 232.

[188] Sterling Wortman, "Extending the Green Revolution," *World Development* 1 (December 1973): 45–51.

[189] Cullather, "Miracles of Modernization,"246.   [190] Johnson quoted in ibid., 244.

respectively). Historian Nicholas Cullather concluded, "Development rhetoric and the promise of scientific solutions to poverty allowed these regimes to neutralize class antagonisms within their coalitions, appeal to supposedly universal demands for increased consumption and national strength, and identify the opponents of progress."[191] The process was especially ironic in Vietnam.

The United States military introduced hybrid rice for the benefit of the South Vietnamese, but the crop eventually fed populations on both sides of the DMZ. Herbert Komer, head of American pacification programs, originally celebrated the hybridization program and adopted the slogan "rice is as important as bullets."[192] For demonstration, Americans planted fifty tons of IR-8 outside Saigon (US workers tended 12 hectares, Vietnamese 778 hectares). Operation Takeoff, additional plantings of IR-8, succeeded in altering local perceptions. At first, the National Liberation Front (the US foe) discouraged adoption of the foreign rice, warning it spread leprosy and impotence. However, by 1968, the insurgency welcomed IR-8. The following year, troops encountered fields of hybrid rice deep in enemy territory; by 1972, the CIA confirmed IR-8 flourished in a fifth of North Vietnam's paddies. Congress worried the rice would spread to China and Cuba. As agricultural aid reduced hunger across Vietnam, American officials focused on the remaining concern: overpopulation.

Population Controls

Organized American efforts to engineer populations arose after World War I. The American Eugenics Society formed in 1923, reflecting both the enthusiasm of early genetic research as well as concern over immigration. Within years birth control could be forced on the "unfit." The *Buck v Bell* ruling (1927) legalized involuntary sterilization, leading to thousands of unwanted procedures. At the same time, population studies boomed among modernization theorists who hoped to quantify the caloric needs and different fertility rates of entire racial groups. By 1935, researchers at the Princeton Office of Population Research, led by economist Frank W. Notestein, suggested contraceptives for global birth control. Although revelations of Nazi practices undermined the American eugenics movement, population control gained support after the World War II.

---

[191] Cullather, *The Hungry World*, 172.
[192] Komer quoted in Cullather, "Miracles of Modernization," 248.

The United Nations and international NGOs took the lead. In 1946, the UN established a Population Division (headed by Frank W. Notestein). After the failure of the World Food Board, UN administrators supported birth control as an evolutionary necessity and an example of good global governance (the World Bank, for example, required family planning programs before loans). At a population conference in England in 1948, Joseph Needham, head of UNESCO's science department, concluded, "Conscious world control of population in relation to natural resources is not an impossible dream, but a certain development having all the authority of social and biological evolution behind it."[193] Others were even more direct. Famed contraceptive advocate Margaret Sanger wrote in 1950: "I consider the world and almost all our civilization for the next 25 years is going to depend upon a simple, cheap, safe contraceptive to be used in poverty-stricken slums and jungles, and among the most ignorant people."[194] Always active, she raised $150,000 for biologist Gregory Pincus to start work on an oral contraceptive and helped establish the International Planned Parenthood Foundation in 1952. The same year, John D. Rockefeller III established the Population Council, an influential NGO composed of social scientists, pharmaceutical industry reps, and other birth control advocates. Finally, in 1954, Hugh Moore – the wealthy inventor of the Dixie cup – began mass-mailing warnings of a coming global "population bomb." As the Rockefeller and Ford Foundations funded studies on population and development in American universities, pharmaceutical companies researched contraceptives overseas.

Puerto Rico became the test case for American birth control experiments in the early Cold War. During the Depression, overpopulation concerned federal administrators of the territory. Governor Berkeley wrote to Margaret Sanger in 1933: "some method of restricting the birth rate ... is the only salvation for the island. The tragedy of the situation is that the more intelligent classes voluntarily restrict their birth rate, while the most vicious, most ignorant, and most helpless and hopeless part of the population multiples with tremendous rapidity."[195] Twenty years

---

[193] Needham quoted in Connelly, *Fatal Misconception*, 132.
[194] Sanger quoted in Betsy Hartmann, *Reproductive Rights and Wrongs: The Global Politics of Population Control* (Boston: South End Press, 1995), 174.
[195] Berkeley quoted in Laura Briggs, *Reproducing Empire: Race, Sex, Science and U.S. Imperialism in Puerto Rico* (Berkeley: University of California Press, 2002), 87. For additional works considering the testing of birth control in Puerto Rico, see Latham, *The Right Kind of Revolution*, and Connelly, *Fatal Misconception*, 175.

later economist J. K. Galbraith thought the time was ripe, writing, "At precisely the moment when the United States proclaimed its interest in Point Four, the people of Puerto Rico constituted themselves as a kind of pilot plant to demonstrate the process."[196] While it remains unclear if the Puerto Ricans saw themselves as a "kind of pilot plant," it is clear pharmaceutical companies did (including Proctor & Gamble, G. D. Searle and Company, Youngs Rubber, Ortho Pharmaceutical, Eaton Labs, and Lanteen Medical Products).[197] By 1956, Gregory Pincus, working for Searle and with Puerto Rican government approval, announced an oral contraceptive based on synthetic steroids. The pill, along with the IUD and sterilization, were ready for their global debut.

American administrations were hesitant to support birth control even as international interest grew. During the US occupation, for example, the Japanese passed a Eugenic Protection Law to provide contraceptives and abortion services; in response, MacArthur's military government downplayed the legislation while permitting the practices. By the early 1950s, the debate over birth control was global and research centers opened in both India and Chile. Fear that overpopulation led to insecurity and communist influence gained traction in the Eisenhower administration, leading the president to order General William Draper to study the issue and find an "effective two-cent contraceptive."[198] However, when the Draper Report concluded population growth was a national security problem requiring action, Ike refused to move. Instead, leaks of the report caused controversy, as the Roman Catholic Bishops of the United States labeled birth control for "economically underdeveloped countries" a "disastrous approach."[199] Alonzo Smith, press officer for the ICA, vehemently rejected the suggestion: "Not one penny of foreign aid funds ever has been used for dissemination of birth control information and there are no plans to do so."[200] Within the administration, George Kistiakowsky, Ike's science advisor, considered it inappropriate or "quite amusing" to even bring up contraceptives in administration meetings.[201] But he could not escape the topic overseas; at an agricultural fair in New Delhi, Kistiakowsky engaged with his Indian counterpart in a "long discussion on birth control in India and the need to encourage it, as otherwise a

---

[196] Galbraith quoted in Briggs, *Reproducing Empire*, 111.

[197] Briggs, *Reproducing Empire*, 123–124.

[198] Eisenhower quoted in Connelly, *Fatal Misconception*, 185.

[199] Anon., "U.S. Denies Aiding on Birth Control," *New York Times Special* (November 29, 1959): 43.

[200] Ibid.   [201] Kistiakowsky, *A Scientist in the White House*, 185.

hopeless situation will result."[202] Such global concern soon bubbled to the surface. In 1960, groups from nineteen countries and thirty-nine Nobel laureates requested the UN work on population control. In the United States, the pill became legal.

The Kennedy administration, the first Catholic presidency, warmed to global population control slowly. In his first year, the AMA ruled sterilization a safe procedure and Alan Guttmacher promoted the IUD as an alternative contraceptive. By 1963, the Population Council, World Bank, Ford Foundation, and USAID helped send a million IUDs to India.[203] After Kennedy's assassination, his brother spoke out about birth control efforts in Latin America: "We should now accelerate our own research into population control device and techniques. …We cannot attempt to compel Latin Americans to practice birth control; this would only inflame suspicions that we seek to keep them underpopulated and weak. But we should stand ready to help; we should encourage any efforts they undertake to make themselves more aware of their problem."[204] Kennedy's successor was even more expressive in private; exasperated by Indian requests for aid, President Johnson fumed, "I'm not going to piss away foreign aid in nations where they refuse to deal with their own population problems."[205]

Birth control and family planning became American policy during the Johnson administration. In his first State of the Union Address after being elected President, LBJ promised, "I will seek new ways to use our knowledge to help deal with the explosion in world population and the growing scarcity in world resources."[206] The Foreign Assistance Act of 1968 tied PL-480 funds to birth control, requiring "not less than 5% of the total sales proceeds received each year shall be used for programs to control population growth."[207] USAID quickly established birth control studies and centers overseas.[208] For domestic critics, this meant the United States promoted abortion services abroad before the same procedures were legal

---

[202] Ibid., 13.
[203] See "Birth Control for a Nation: The IUD as Technoscientific Biopower," in Chikako Takeshita, *The Global Biopolitics of the IUD: How Science Constructs Contraceptive Users and Women's Bodies* (Cambridge, MA: Massachusetts Institute of Technology Press, 2012), 33–72.
[204] Kennedy, "The Alliance for Progress," 33.
[205] Johnson quoted in Connelly, *Fatal Misconception*, 221.
[206] Johnson SOTU speech available online at: www.presidency.ucsb.edu/ws/?pid=26907.
[207] House of Representatives, *Science, Technology and Diplomacy*, 125.
[208] Latham, *The Right Kind of Revolution*, 103–105.

at home. Promotion abroad also meant the US government, through USAID and NIH, became the major funder of contraceptive research by the end of the decade, surpassing the pharmaceutical industry. Led by Dr. Reimert Ravenholt, USAID had eighty staff people assigned to population issues by 1969, in addition to funding UN and IPPF programs. The following year, twenty-seven countries committed to cutting birth rates; including US allies Ferdinand Marcos, Ayub Khan, Indira Gandhi, and Park Chung-hee (non-aligned supporters included Joseph Broz Tito, Saddam Hussein, and Gamal Nasser – communists such as Mao Zedong and Nikita Khrushchev already supported birth controls).[209] Contraception, like the Green Revolution, linked allies and foes alike in the Cold War.

Legacies of Cold War Scientific and Technical Assistance

In theory, science was critical to American development assistance. Modernization theorists spoke frequently about the role of science in development, supporting Point Four-style programs as a bulwark against Soviet communism.[210] The enthusiasm went all the way to the top: Presidents from Truman to Nixon shared their faith in science and technology to overcome communism and advance American interests, guaranteeing the expansion of American aid programs into the 1970s. But modernization schemes often foundered, perhaps because scientific and technical aid rarely amounted to more than a fraction of total assistance. Unable to address local socioeconomic conditions, American assistance labored under geopolitical burdens, leaving behind a mixed legacy.

Scientific and technical aid programs incentivized participation in American anti-communism and benefited American allies. Although it is impossible to separate development programs from their Cold War context, milk pasteurization plants, university laboratories, and basic sanitation improved allied quality of life. The United States spent more on development than the Soviet Union, but it was also richer and neither superpower emphasized such commitments. Nevertheless, Point Four and later initiatives often provided the only funding for education and scientific equipment; exchange programs meant opportunity for locals and the chance to establish personal connections with the United States, creating a

---

[209] Connelly, *Fatal Misconception*, 281.
[210] In addition to texts cited previously, see Nils Gilman, *Mandarins of the Future: Modernization Theory in Cold War America* (Baltimore: Johns Hopkins University Press, 2003).

supportive local population with American experience and training. By the 1960s, exchanges were so successful the developing world worried about a "brain drain."

Scientific and technical aid also quieted criticism of US Cold War policies. Though funding never approached military levels, Point Four and similar programs allowed Americans to see themselves and American diplomacy as benevolent. Milk pasteurization plants, university laboratories, and basic sanitation, rather than tanks and bombers, made the front page and provided clear evidence of American goodwill. Scientists and engineers, even if unwitting, made ideal political ambassadors, capable of being neutral and humanitarian while acting in support of Cold War diplomacy. Partnerships with UN agencies and NGOs further confirmed American altruism, often winning global support for anti-communism. Of course, science diplomacy in the developing world also provided access and opportunity for American corporations abroad.

Midway through the 1960s, American administrators could feel good about the place of science and technology in American foreign relations. The space race trended in American favor, the "Green Revolution" – named by USAID administrator William Gaud in 1968 – raised global agricultural yields and the United States supported public health, family planning, and industrial development through technical aid programs around the world. A remarkable variety of scientific and technical initiatives, legitimated by international organizations, supported American commercial interests and anti-communism overseas. A wide network of people, institutions and programs linked the globe as never before. Yet rumblings beneath the surface threatened a potential earthquake.

4

The Crossing Point

The Alkali Inspectorate began monitoring industrial air pollution in Scotland in 1863. A decade earlier, Scottish chemist Robert Angus Smith coined the term "acid rain" to describe the impact of industrial smoke on precipitation, which led to passage of the Alkali Act and Smith's appointment as the official inspector of the air. Other scientists continued the study of pollutants; in the 1960s, Swedish chemist Svante Odén argued European air pollution, such as sulfur dioxide and nitrogen oxide emissions from cars, increased acidity in Scandinavian lakes and rivers, thereby waging an "insidious chemical war" against fish and forests.[1] Determined to bring the problem to international attention, the Swedes proposed a global environmental conference, arguing the causes of national environmental problems transcended national borders. So in 1972, more than a century after the Scots began monitoring air pollution, delegates from 113 nations met in Stockholm at the first United Nations Conference on the Human Environment. Although the conference motto was "Only one earth," each nation on earth had its own priorities.

For American representatives, the Stockholm conference provided an opportunity to repair the nation's standing amid the Vietnam conflict. Russell Train, founder of the World Wildlife Federation and future EPA chief, and Christian Herter, a former Secretary of State, co-chaired the American delegation. Before the meeting, Train tried to convince

---

[1] Odén quoted in Marc A. Levy, "International Co-operation to Combat Acid Rain," in Helge Ole Bergesen, et al., eds., *Green Globe Yearbook of International Co-operation on Environment and Development 1995* (Oxford: Oxford University Press), 59–68, quote on 59.

National Security Advisor Henry Kissinger participation would bolster the American brand, writing: "This positive image of the United States [as a leading environmental nation] is to be contrasted with a large number of divisive issues in which we are involved abroad."[2] American goals before the meeting included establishing a UN environmental agency and finalizing the convention on ocean dumping. Reflecting the novelty of global environmentalism, domestic industrial and environmental groups remained peripheral.

The conference reflected both change and continuity in Cold War geopolitics. The previous year, the United Nations recognized the People's Republic of China (PRC); as such, Stockholm became the first major international conference attended by a communist Chinese delegation. European tensions proved more intractable: The exclusion of East Germany and inclusion of West Germany caused the Soviet bloc to boycott the conference.[3] Environmentalists pointed out Baltic Sea programs needed cooperation among blocs to no avail. To reduce further conflict, conference organizers focused on a limited number of themes: human settlements, resource management, pollution, education, and development.[4] But it was impossible to isolate environmental policies from the larger geopolitics.

Expanded UN membership, including the Chinese and recently independent nations, meant the United States faced more vocal criticism. American plans to downplay the war failed from the start: the Swedish Prime Minister criticized US policies in Vietnam during his opening address, saying, "The immense destruction brought about by indiscriminate bombing, by the large-scale use of bulldozers and herbicides is an outrage sometimes described as ecocide, which requires urgent international attention."[5] The Americans almost walked out. Tang Ke, chairman of the Chinese delegation, attacked US policy in Vietnam in a later session, arguing the United States was responsible for "massive killings of innocent old people, women and children, as well as unprecedented and

---

[2] Train quoted in Stephen Hopgood, *American Foreign Environmental Policy and the Power of the State* (Oxford: Oxford University Press, 1998), 76.

[3] Anon., "Stockholm Conference in Danger," *New York Times* (January 14, 1972): 32.

[4] Philip A. Abelson, "After the Stockholm Conference," *Science* 175 (February 11, 1972): 585.

[5] Prime Minister quoted in Hopgood, *American Foreign Environmental Policy and the Power of the State*, 101.

serious destruction of the human environment."[6] Nor was criticism limited to diplomats; an American observer wrote to *Science*: "Thousands of visitors to Stockholm, in addition to the Swedish people themselves, viewed it as bald-faced hypocrisy that the United States could profess concern for the human environment during the very days that we were (and are) ravaging a subcontinent and rendering it unfit for human habitation."[7] But US policy in Vietnam was only a short-term dispute.

Developing nations, led by China, opposed many American positions. The Chinese, for example, wanted the conference declaration to blame global capitalism for environmental degradation and encourage the free transfer of science and technology to the developing world; the American delegation demanded removal of the Chinese language.[8] However, Principle 20 of the final declaration leaned toward PRC proposal, stating:

> Scientific research and development in the context of environmental problems, both national and multinational, must be promoted in all countries, especially the developing countries. In this connection, the free flow of up-to-date scientific information and transfer of experience must be supported and assisted, to facilitate the solution of environmental problems; environmental technologies should be made available to developing countries on terms which would encourage their wide dissemination without constituting an economic burden on the developing countries.[9]

Industrialization provided another area of disagreement: Developed countries, like the United States, worried about environmental damage while developing countries emphasized modernization.[10] Tang Ke, for example, downplayed American fears, observing, "One does not not eat for fear of choking;" a Brazilian official suggested his country wanted more pollution if it meant industrialization.[11] The location of a proposed

---

[6] Tang Ke quoted in Nigel Hawkes, "Stockholm: Politicking, Confusion, but Some Agreements Reached," *Science* 176 (June 23, 1972): 1308–1310, quote on 1309.

[7] Evelyn A. Mauss, "Stockholm Conference," *Science* 177 (September 1, 1972): 746.

[8] Bao Maohong, "The Evolution of Environmental Problems and Environmental Policy in China: The Interaction of Internal and External Forces," in J. R. McNeil, ed., *Environmental Histories of the Cold War* (Cambridge: Cambridge University Press, 2010), 323–342, esp. 333–336.

[9] *Declaration of the United Nations Conference on the Human Environment* (1972).

[10] Paul G. Harris, "International Environmental Affairs and U.S. Foreign Policy," in Paul G. Harris, ed., *The Environment, International Relations and U.S. Foreign Policy* (Washington, DC: Georgetown University Press, 2001), 3–44, esp. 7.

[11] Tang Ke quoted in Hawkes, "Stockholm," 1309. Brazilian official quoted in Shawn William Miller, *An Environmental History of Latin America* (Cambridge: Cambridge University Press, 2007), 206–207.

UN environmental headquarters sparked additional friction: Developing countries felt it should be in the Southern hemisphere, while the United States preferred New York City or Geneva. After petitions from both Jomo Kenyatta and Indira Gandhi, the UN voted to locate the HQ in Kenya (the United States cast the only negative vote).[12] Remarkably, amid these disputes, the conference gave birth to a new mindset and organization.

Encouraged by the United States, participants at Stockholm reimagined global governance, basing it for the first time on environmental science. Although the recommendations of the Stockholm Action Plan were little more than "paper resolutions," the new United Nations Environmental Program (UNEP, 1972) soon set up a Global Environmental Monitoring System.[13] Backed by teams of researchers and data, the UNEP hoped to shame countries into compliance: *Science* observed, "Maybe the Brazilians will continue to hack down the Amazon rain forest, the French and Chinese to test their nuclear weapons in the atmosphere, and the Japanese to hunt the great whales, but at least they will no longer be doing so with tacit international approval."[14] The UNEP also tackled Stockholm's unresolved issues: The concept of "sustainable development" eventually reconciled concern for environmental damage with demands for development, while more than thirty nations signed the Convention on Long-Range Transboundary Air Pollution (such as acid rain) in 1979.[15] In hindsight, the Stockholm conference and subsequent actions raised the question of how much influence environmental science could have on national policies through international organizations like the UN. However, while the Conference expanded the role of science at the UN, the funding and politics of science were undergoing a transformation in the United States.

As noted in the introduction, the 1960s and 1970s were the "crossingpoint" in American R&D funding: after World War II the government funded nearly 70 percent of American R&D; by 2000, private industry funded nearly 70 percent; in the 1970s, funding was relatively equal (see Figure I.1). But the shift in funding was only one change: The period also

[12] Hopgood, *American Foreign Environmental Policy and the Power of the State*, 100–108.
[13] SSM, "What's Happened since Stockholm?" *Environment Science & Technology* 8 (March 1974): 214–216.
[14] Hawkes, "Stockholm," 1308.
[15] Adil Najam and Cutler J. Cleveland, "Energy and Sustainable Development at Global Environmental Summits: An Evolving Agenda," *Environment, Development and Sustainability* 5 (2003): 117–138, esp. 125.

witnessed the end of the public consensus on science and technology, leading to criticism of industrial development as well as American initiatives overseas. At the same time, the growth of R&D in the private sector transformed the international political economy of science, as demands for access from the developing world ushered in the formalization of global intellectual protections and patent rights.

Two sections consider this critical era. The first, "Eruption," details the impact of environmentalism and the Vietnam War on science in American foreign relations. As ecologists linked industry to pollution and environmental degradation, domestic industrial producers questioned and politicized research. Environmentalists and activists criticized American policies in Southeast Asia, altering dynamics familiar since World War II. The Vietnam War, for example, did not lead to a closer partnership between the government and academic research; instead, criticism of defense research reduced the DOD presence at American universities. Meanwhile, environmentalism reframed science diplomacy: The Green Revolution and Malaria Eradication Program, previously considered fruitful initiatives, were recast as harmful by activists the following decade. Finally, the contested domestic politics hindered American diplomacy for science, limiting US participation in the International Biology Program. "Market Biology," the second section, considers American diplomacy for genetic engineering, whether establishing global safety standards or patent rights. A growing biotech industry led the United States to advocate against a UN center for genetic engineering, while the legalization of biotechnology sparked discord between the country and its NATO allies and the UN.

ERUPTION

Science should be employed to constructively transform the conditions of life throughout the United States and the world. Yet ... One-half of all research and development in America is directly devoted to military purposes. ... Further, science and scholarship should be seen less as an apparatus of conflicting power blocs, but as a bridge toward supranational community: the International Geophysical Year is a model for continuous further cooperation between the science communities of all nations.

Students for a Democratic Society, *The Port Huron Statement* (1962)

The unprecedented advances in science and technology have created a new dimension of international life. The global community faces a series of urgent problems and opportunities which transcend all geographic and ideological borders. It is the distinguishing characteristic of these issues that their solution requires international cooperation on the broadest scale.

President Richard Nixon, *Third Annual Report to the Congress on United States Foreign Policy* (1972)[16]

With the IGY modeling ideal international relations, the Port Huron authors dreamt of a world working cooperatively to improve the human condition. In the students' imagination, science was a "bridge toward a supranational community" rather than an "apparatus of conflicting of power blocs." A decade later, President Nixon suggested environmental hazards arising from "unprecedented advances in science and technology" required the type of cooperation the students envisioned. Pollution ignored national borders and reframed threats to national security and quality of life, while critics at home and abroad protested the use of defoliants, incendiaries and other chemicals in Vietnam. Environmentalism linked anti-industrial and anti-war criticisms, undermining the postwar consensus on research.

The Environmental Movement, Vietnam War, and Collapse of Consensus

The environmental movement forced ecology into the political realm. Following upon revelations of fallout from atomic testing, Rachel Carson's *Silent Spring* (1962) ignited a firestorm by alleging the chemicals praised for the conveniences of modern life poisoned the environment. A bestseller, the work singled out DDT and suggested the USDA ignored warnings the compound accumulated in fatty tissues, creating severe health risks. The resulting public debate politicized ecology; the official response from the Ecological Society of America acknowledged: "Rachel Carson's book *Silent Spring* created a tide of opinion which will never again allow professional ecologists to remain comfortably aloof from public responsibility."[17] As ecology entered the political lexicon, the chemical industry fought back, with Dow Chemical and

---

[16] Richard Nixon, *Third Annual Report to the Congress on United States Foreign Policy* (February 9 1972).

[17] ESA statement reproduced in Dorothy Nelkin, "Scientists and Professional Responsibility: The Experience of American Ecologists," *Social Studies of Science* 7 (February 1977): 75–95, quote on 79.

Monsanto demeaning Carson's credentials and conclusions, trying to create doubt about research critical of industrial production (at the same time, the tobacco industry deployed a similar strategy to combat the anti-smoking campaign of the American Cancer Society).[18] As questions of industrial research increased, the war in Vietnam raised questions about defense research.

American policy-makers and the military believed science and technology could provide a critical advantage in Southeast Asia. The Pentagon approached a group of promising civilian physicists in 1957; a few years later, they became known as "JASON" or the "Jasons" after the mythological Greek explorer.[19] Invitation-only and funded through ARPA, the Jasons suggested motion detectors, listening devices and urine-sniffers to help locate guerrillas. The group also became known for promoting several far-fetched ideas, including Operations Popeye (artificial rain-making to render the Ho Chi Minh trail impassable) and Igloo White (an electronic barrier, combined with air targeting, to stop infiltration into South Vietnam).[20] Both operations received millions in funding but failed to achieve their objectives. Meanwhile, ARPA hired US Forest Service personnel to set fire to groundcover, but Vietnam's tropical climate foiled the plan.[21] Although chemicals proved more effective, their use came at a political cost.

The use of defoliants abroad triggered controversy at home. Operation Ranch Hand – the US defoliant program in Vietnam – began the same year Carson published *Silent Spring*. For the next decade, Ranch Hand planes sprayed Agent Orange and other chemicals to reduce foliage and underbrush (their motto: "Only we can prevent forests"). The program was a qualified success: While it cleared jungles, it also created ecological devastation, increasing domestic criticism. Researchers questioned government transparency: The leader of the AAAS herbicidal assessment team complained, "we can't get the most basic information we need – a list of areas sprayed, and when, and with what ... we were told the

[18] Naomi Oreskes and Erik M. Conway, *Merchants of Doubt: How a Handful of Scientists Obscured the Truth from Tobacco Smoke to Global Warming* (New York: Bloomsbury, 2011), 10–32.
[19] See Ann Finkbeiner, *The Jasons: The Secret History of Science's Postwar Elite* (Viking, 2006).
[20] Herken, *Cardinal Choices*, 152–159. See also "Preemption," in Lambright, *Presidential Management of Science and Technology*, 68–74.
[21] Weinberger, *The Imagineers of War*, 165.

information is classified ... though the enemy certainly knows where we sprayed."[22] The Federation of American Scientists opposed the program and Johnson's science advisor, chemist Donald Hornig, delivered a petition against defoliants signed by more than 5,000 scientists. Pressure escalated alongside the war: The AAAS and NIH issued a scathing report on Agent Orange, eventually forcing President Nixon to review American chemical weapon protocols.[23]

The Vietnam War politicized many American scientists. President Johnson originally enjoyed support from the scientific community: In a rare example of activism, more than 50,000 scientists, concerned Barry Goldwater should not be given nuclear responsibility, campaigned for Johnson in the 1964 election. But the war undermined support. In 1966, scientists in Boston organized against the use of chemical weapons, eventually forming two groups: Science for the People and the Union of Concerned Scientists.[24] In one of their largest actions, the Union led a "March 4th" movement emphasizing the social responsibility of scientists at more than thirty campuses. More than 40 professors at MIT (few in war-related fields) signed a petition decrying the militarization of science. Their platform echoed the earlier Port Huron students, suggesting the need to turn "research applications away from the present emphasis on military technology toward the solution of pressing environmental and social problems."[25] Others took more direct action.

Protestors disrupted military research and politically active scientists challenged the status quo. The Institute for Defense Analyses (IDA) at Columbia University, funded by ARPA in 1958, was one of the first non-profit university-based government research centers. The IDA and similar centers aroused little criticism before "Americanization" of the war. A few years later, SDS published a pamphlet on *MIT and the Warfare State* (1968) and students staged protests at both Columbia (which

[22] Meselson quoted in John Lewallen, *Ecology of Devastation: Indochina* (Baltimore: Penguin Books, 1971), 60. See also Barry Weisberg, *Ecocide in Indochina: The Ecology of War* (San Francisco: Canfield Press, 1970).

[23] David Zierler, "Against Protocol: Ecocide, Détente and the Question of Chemical Warfare in Vietnam, 1969–1975," in McNeil, *Environmental Histories of the Cold War*, 257–278.

[24] See Kelly Moore, *Disrupting Science: Social Movements, American Scientists, and the Politics of the Military, 1945–1975* (Princeton, NJ: Princeton University Press, 2008), esp. chapters 5 and 6.

[25] Stuart W. Leslie, *The Cold War and American Science: The Military-Industrial-Academic Complex at MIT and Stanford* (New York: Columbia University Press, 1993), quote on 233.

disaffiliated the IDA in response) and the Liquid Crystals Institute at Kent State (which produced motion detectors for combat). Activists also bombed Sterling Hall, home of the Army Mathematics Research Center at the University of Wisconsin. Dow Chemical – the primary manufacturer of napalm – became a favorite target of anti-war protests. At the same time, Lawrence Livermore researcher John Gofman founded the Committee for Nuclear Responsibility after being terminated for suggesting allowable radiation doses were too high (the Union of Concerned Scientists provided corroborating evidence).[26] Scientists also helped end illegal intelligence campaigns: after the Federation of American Scientists wrote the post office about mail tampering, the CIA shut down their operation for fear of bad publicity.[27] Tensions reached a breaking-point during the Nixon administration.

By the early 1970s, environmentalism and the Vietnam War complicated public support for government research. Disagreement reached into the executive branch: Disputes with his science advisors over the war and supersonic transport (SST) incensed President Nixon. When the SST proposal failed in congress, Nixon aide Douglass Hallett suggested a speech at Cal Tech, writing:

In light of the SST vote, I am wondering whether this might not be a good time for the president to deliver a good speech on the present assault on science and technology ... defending the importance of continual research and development. Out of the present concern with the environment, the failure of technological warfare in Southeast Asia, suspicion by students of academia's involvement in government research, and the over-all fiscal situation and the pressure on the taxpayer, there has arisen a potentially dangerous feeling that scientific progress is necessarily counter to human and ecological concerns.[28]

Although Nixon spoke in favor of research, tensions with his science advisors continued: When PSAC disagreed with him on the deployment of a limited anti-ballistic missile system, the President disbanded his scientific advisory council and abolished the post of science advisor. A writer in *Science* remarked, "The sensation of a fall from grace has

---

[26] David J. Hess, *Alternative Pathways in Science and Industry: Activism, Innovation and the Environment in an Era of Globalization* (Cambridge, MA: The MIT Press, 2007), 95.

[27] Robert Gillette, "The CIA's Mail Cover: FAS Nearly Uncovered It," *Science* 188 (June 27, 1975): 1282–1284.

[28] Douglas Hallett to Ray Price, "Memorandum for Ray Price: Possible Presidential Speech on Present Attack on Science and Technology," March 25, 1971. Available via DDRS, CK 31005950055.

grown familiar to the scientific community."[29] Meanwhile, economic concerns imposed tighter constraints on federal budgets, ending the post-World War II "honeymoon" between the government and academic science.[30]

The climate of distrust influenced defense research on campus. Members of congress first demanded more oversight of federal research after revelations of prescription-related fetal deformities in the early 1960s, eventually reorganizing the FDA. As the economy struggled later in the decade, disenchantment with the return on scientific investment grew – even the space program saw its budgets slashed after the moon landing (see Chapter 2). By the early 1970s, Vietnam War protests convinced Congress to limit DOD funding to research with clear military applications, reducing defense funding to less than a tenth of university research.[31] The Joint Chiefs felt the impact was minimal: Project Hindsight – a DOD report – concluded much of the money spent on research had been wasted or ineffective.[32] Protests ended with the war, although MIT and Stanford remained at the forefront of military research a decade after the war's conclusion.[33] Instead, criticism shifted to American scientific initiatives overseas, whether the Green Revolution or Malaria Eradication Program, and to large global initiatives like the International Biology Program.

The International Biological Program and Global Governance

Questions regarding the Green Revolution surfaced two years after Gaud coined the term. AAAS herbicide teams emphasized the war's impact: Matthew Meselson, a researcher from Harvard University, opined: "Vietnam, once a major exporter of rice, now imports American rice. What has happened is that in Indochina the Green Revolution has been overwhelmed by what can be called the Green Devolution, caused by American war."[34] The criticism broadened when *Foreign Affairs* observed that while production of wheat improved, the majority of farmers were

---

[29] John Walsh, C. H. and B. J. C., "1973: A Crisis Atmosphere: Science Policy: Détente, LDC's Add New Dimension," *Science* 182 (December 28, 1973): 1326–1328.

[30] See "End of the Honeymoon: 1965–1975," in Mann, *For Better or Worse*, 131–150. See also "The End of Consensus," in Wolfe, *Competing with the Soviets*, 105–120.

[31] Wolfe, *Competing with the Soviets*, 114.   [32] Ibid., 120.

[33] Leslie, *The Cold War and American Science*, 251.

[34] Meselson quoted in John Lewallen, *Ecology of Devastation: Indochina* (Baltimore, MD: Penguin Books, 1971), 61.

"institutionally prevented from taking advantage of the new agricultural trends," thus increasing polarization of rich and poor.[35] Others later argued the World Bank and agribusiness used the revolution to penetrate markets; opponents labeled IR-8 "seeds of imperialism" in the Philippines.[36] Large commercial monoculture agriculture, based on widespread use of fertilizers and pesticides, came to symbolize "big" science gone wrong for critics. Debate continues today, with proponents emphasizing high yields, freedom from famine, and rural electrification as clear benefits, along with the creation of global agricultural research stations.[37] In the early 1970s, however, the immediate issue was the widespread application of pesticides required.

Scientific revelations and domestic regulations ended the Malaria Eradication Program. The US/UN campaign struggled to keep spraying after *Silent Spring*, especially after the discovery of resistance among mosquitos, which necessitated heavier doses. When the United States banned the domestic use of DDT, but not its manufacture or foreign sale, the WHO ended the program. Many refused to condemn spraying: Norman Borlaug denounced the international effort to ban the chemical, telling reporters, "If most DDT uses are cancelled, I have wasted my life's work."[38] Between 1957 and 1965, the WHO spent more than $50 million and the United States another $100 million (host countries spent more than five billion).[39] With the Green Revolution under attack and MEP canceled, uncertainty over the benefits of scientific cooperation hindered US participation in the International Biology Program.

---

[35] Wolf Ladejinsky, "Ironies of India's Green Revolution," *Foreign Affairs* 48 (July 1970): 758–768, quote on 763.

[36] See, for example, Vandana Shiva, *The Violence of the Green Revolution: Third World Agriculture, Ecology and Politics* (London: Zed Books, 1991).

[37] For a quick introduction to the debate in one location, see Kathleen Baker and Sarah Jewitt, "Evaluating 35 Years of Green Revolution Technology in Villages of Bulandshahr District, Western UP, North India," *The Journal of Development Studies* 43 (February 2007): 312–339. Note the authors argue the revolution did not widen the gap between rich and poor. Others argue famine remains a concern and that rural poverty increased, see Jonathan Harwood, "Peasant Friendly Plant Breeding and the Early Years of the Green Revolution in Mexico," *Agricultural History* 83 (Summer 2009): 384–410. Still others note it saved crop land but did not solve malnourishment because diet diversity decreased, see Prabhu l. Pingali, "Green Revolution: Impacts, Limits and the Path Ahead," *Proceedings of the National Academy of Sciences* 109 (July 31, 2012): 12302–12308.

[38] Borlaug quoted in Cullather, *The Hungry World*, 246.

[39] See "Exercising International Authority: The Malaria Eradication Program," in Staples, *The Birth of Development*, 161–179.

For once, an international scientific program coalesced with only limited American support. In a repeat of history, Lloyd Berkner and others discussed a biological IGY over dinner in 1959.[40] However, the International Biological Program (IBP) struggled to congeal; scientists and nations disagreed about organizing themes. Soviet representatives wanted the "Biological Basis of Man's Welfare" to stress socioeconomic conditions; American representatives refused.[41] Nonetheless, by 1962, fifty nations framed the IBP around several themes, including human genetics, population growth, and man and ecology. As the Vietnam War escalated, however, some US officials realized the IBP offered an excellent opportunity to showcase American benevolence abroad.

Yet the State Department had to overcome skepticism from the scientific community and congress to secure US participation. The plan was simple: American ecologists would partner with colleagues in the developing world to solve local problems.[42] The State Department recommended participation and the National Research Council praised the initiative for helping "developing countries in their fight against starvation and disease" and fostering more "efficient use of the world's resources."[43] But some ecologists worried about "big" influences on the field: the president of the Ecological Society of America told his association, "I am troubled that the future of ecology seems to be increasingly determined by the councils of the National Science Foundation, the International Biology Program and possibly the Institute of Ecology."[44] The first studies announced included the global spread of insects, pollen, and germ spores as well as Inuit lifestyles and exotic ecosystems, such as Hawaii.[45] Congressional reaction was lukewarm and finding funding proved difficult; the first two bills died in committee and money only

---

[40] Elena Aronova, Karen S. Baker, and Naomi Oreskes, "Big Science and Big Data: From the International Geophysical Year through the International Biology Program to the Long-Term Ecological Research (LTER) Network, 1957–Present," *Historical Studies in the Natural Sciences* 40 (2010): 183–224.

[41] See "International Biological Programme," in Greenaway, *Science International*, 172–182.

[42] Frederick E. Smith, "The International Biology Program and the Science of Ecology," *Proceedings of the National Academy of Sciences* 60 (May 1968): 5–11.

[43] NRC quoted in Harold M. Schmeck, Jr., "U.S. Lists Research Plans for a Biology Project," *New York Times* (September 21, 1967): 19.

[44] ESA president quoted in Nelkin, "Scientists and Professional Responsibility," 86.

[45] Ibid.

surfaced in 1970 (the United States eventually contributed $40–60 million).[46] Only a decade after the IGY, American support for international science was no longer guaranteed.

The IBP failed to live up to IGY standards. While the IGY was an exercise in simultaneous data collection requiring multinational collaboration, each nation carried out its own IBP program and the resulting piles of "Big Data" were never centralized, organized, and published.[47] The IBP also lacked participation from other fields, perhaps because of the absence of a core program. The political climate split the ecological community: The ESA, concerned corporations and government agencies did not share their values, established a code of ethics in 1971 limiting use of their field to "the benefit of man and the environment."[48] However, some members argued the code was inappropriately political and the organization removed the language. Yet even as the IBP faltered, the Stockholm conference introduced a new approach to science in American diplomacy.

The United States advocated for multiple international environmental accords after Stockholm. Although geopolitics tested the Stockholm conference, the desire to appear "green" united the delegations; by one account, the Stockholm conference spurred the creation of national "EPAs" throughout Latin America (the Chileans even ruled pollution unconstitutional).[49] The conference also encouraged cooperation on hazardous chemicals across UN agencies: The UNEP and WHO, for example, established criteria for assessing environmental and human health risks while the WHO and FAO set maximum pesticide residues in food.[50] More than one hundred nations signed the convention to protect endangered species as well as multiple conventions to prevent

---

[46] For an excellent overview of the US role in IBP, see "Big Ecology," in Hagen, *An Entangled Bank*, 164–188.

[47] Ibid. See also Elena Aronova, Karen S. Baker, and Naomi Oreskes, "Big Science and Big Data: From the International Geophysical Year through the International Biology Program to the Long-Term Ecological Research (LTER) Network, 1957-Present," *Historical Studies in the Natural Sciences* 40 (2010): 183–224.

[48] The original code is reproduced in Nelkin, "Scientists and Professional Responsibility," 86. The newest ESA Code of Ethics, revised May 2013, can be found online at: www.esa.org/esa/?page_id=857.

[49] Ibid.

[50] Henrik Selin, *Global Governance of Hazardous Chemicals: Challenges of Multilevel Management* (Cambridge, MA: Massachusetts Institute of Technology Press, 2010), 55.

marine pollution (London, Oslo, and Paris).[51] Finally, environmental protection provided common ground for the Basic Treaty between East and West Germany (the Federal Environmental Agency became the first West German agency headquartered in Berlin).[52] However, while environmental sciences provided one impetus toward global governance, the biotechnology industry argued for a different type of global governance, one focused on guaranteeing the intellectual protection of private research.

MARKET BIOLOGY

Closely allied to basic research is our national policy on technology exchange. The United States' preeminence ... faces us with a policy question as to how far we should share the fruits of our research and technology ... One strategy for serving this security interest is the"Maginot Line" concept which attempts to restrict the transfer of expertise to other nations. It is based on the view that technological preeminence is a national asset to be guarded jealously from others. Another approach is to view our preeminence as an asset to be invested in building effective partnerships with other nations to create a world pattern of free sharing of scientific and technological knowledge.

National Security Council internal paper (1970)[53]

It's the same with any technology. The only difference is that, for the first time, it's the biologist who has come up with something that has commercial potential.

Molecular biologist and Genex co-founder Leslie Glick (1980)[54]

Biotechnology arose through the intersection of academic biology and industry. The government's role was initially unclear. But the innovative

---

[51] The full title is "Convention on International Trade in Endangered Species." The 1972 London Convention was the first general convention on preventing marine pollution and was followed by conventions on polluting from an aircraft (Oslo, 1972) and polluting from land-based sources (Paris, 1974). Ibid., 56–57.

[52] Kai Hünemörder, "Environmental Crisis and Soft Politics: Détente and the Global Environment, 1968–1975," in McNeil and Unger, eds., *Environmental Histories of the Cold War*, 257–278.

[53] National Security Council, internal paper, NSSM 102 related, "Science and Technology [Attribution to National Security Council Based on Format]," (December 1970): 18–24. DNSA Presidential Directives (Part II) Collection, Record # PR00639.

[54] Glick quoted in Anthony J. Parisi, "Industry of Life: The Birth of the Gene Machine," *New York Times* (June 29, 1980): F1.

university/industry partnerships promised a quick turnaround from experiment to application, leading politicians to encourage the ventures through financial incentives, legal protections, and limited regulation. Faced with competition overseas, the "commercial potential" of Genex outweighed establishing a "world pattern of free sharing of scientific and technical information," leading the US government to stress private intellectual property rights and fend off attempts to include biotechnology in United Nations development programs. Yet tensions remained because the genetic disputes raised unanticipated questions regarding the ownership of science and life itself.

The Legal and Scientific Background to Biotechnology

Universities rarely consulted patent lawyers before the advent of biotechnology. One prominent exception dates to 1925, when alumni established the Wisconsin Alumni Research Foundation (WARF) to patent a process using ultraviolet radiation to add vitamin D to milk (the foundation's most famous product is the rat poison and anticoagulant Warfarin). MIT and Columbia began similar programs before World War II; a few others worked with the Research Corporation, a firm specializing in securing university rights.[55] After World War II, questions arose about the legality of patenting research conducted with federal funding. A key case involved a University of Wisconsin professor who discovered an anti-cancer drug with funding from both Bristol-Myers and the federal government; when Bristol-Myers did not get exclusive license, companies began to consider research "contaminated" by federal money.[56] Nevertheless, Stanford University, hoping to capitalize on campus research, established a licensing office in 1970.

Genetic engineering bridged academic biology and industry. In 1972, biologists Stanley Cohen at Stanford and Herbert Boyer at the University of California discovered how to cut genetic material from one organism and paste it into another – a technique known as recombinant DNA engineering. The technique had clear commercial potential: Biologists could insert genes into bacteria to mass-produce rare animal and human proteins cheaply. Stanford sought patent protection, though the idea of patenting a scientific technique surprised Cohen, and private industry

---

[55] Henry Etzkowitz, et al., eds., *Capitalizing Knowledge: New Intersections of Industry and Academia* (Albany: State University of New York Press, 1998), 223.

[56] Ibid., 11.

rushed to fund the field.[57] An agreement between Harvard and Monsanto to develop anti-cancer drugs – Monsanto got its pick of patentable products – ushered in dozens of similar partnerships. Success with somatostatin (growth hormone) stimulated start-ups: biologist Leslie Glick and entrepreneur Robert F. Johnston founded Genex in Maryland in 1977; European businessmen founded Biogen the same year.[58] Yet biological patent rights remained unclear.

Biotechnology pushed the boundaries of patent law. Publication of the technique voided future European patent rights – the EEC required patent approval before publication – but Stanford sought the patent to secure commercial rights in the United States and reassure investors. Nor were Cohen and his university alone: Microbiologist Anand Chakrabarty and General Electric sued for patent approval on a bacterium bred to eat oil spills. The US Patent and Trademark Office, represented by Commissioner Sidney Diamond, promised to fight GE's patent application until the Supreme Court. If the Cohen-Boyer patent represented an audacious attempt to patent a scientific technique, the *Chakrabarty* vs. *Diamond* case represented an even more audacious attempt to patent a living organism. The patentability of federally funded research provided a third concern. Even as the courts struggled to get a handle on the new field, the nascent biotech industry rode a wave of investment. Eli Lily and Genentech – a company founded by Boyer and fellow faculty at the University of California – announced the synthesis of human insulin in 1978. Interferon was the next big hope.

Interferon, the Geopolitics of Overinvestment, and American Diplomacy

Interferon exemplified the biotech craze of the 1970s. Originally discovered in 1957 by virologists at the British National Institute for Medical Research (NIMR), "interferon" caught the medical community's attention because it interfered with viral growth; later research indicated the protein might also inhibit the growth of cancer tumors.[59] The War on Cancer spotlighted interferon, leading lobbyists to ask Congress for an

---

[57] Hughes, "Making Dollars Out of DNA," 548–552.

[58] Susan Wright, *Molecular Politics: Developing American and British Regulatory Policy for Genetic Engineering, 1972–1982* (Chicago: University of Chicago Press, 1994), 87.

[59] John A. Osmundson, "Interferon Gets New Cancer Role," *New York Times* (September 15, 1963): 120. See also Toine Peters, *Interferon: The Science and Selling of a Miracle Drug* (New York: Routledge, 2005), 96–99.

initiative comparable to the Apollo program. Representative Claude Pepper (D-FL) spoke in favor of increased funding during hearings: "If there were an enemy at our borders killing a thousand of our citizens a day, any member of Congress who would say that we couldn't afford to spend more to defeat that enemy would be hooted down."[60] In response, the National Cancer Institute received an additional $13 million earmarked for interferon research and NIH distributed millions more to laboratories across the country.

Acquiring the "miracle drug" took a minor miracle: The protein could only be harvested from white blood cells previously exposed to a virus and the Finnish Red Cross had the only significant supply.[61] Undeterred, the American Cancer Society (ACS) spent $2 million to buy the world's supply of interferon from Finland in 1978 (research centers in the United States competed for samples).[62] The drug became the most expensive ever used in American clinical trials as the ACS undertook its most expensive cancer research to date.[63] With trials underway and the market cornered, geneticists around the globe raced to synthesize the prized protein.[64]

Americans failed to win the race: Japanese, Belgian, and Swiss labs all synthesized interferon before their American competitors. Nor was there a miracle; the ACS trial results were poor and four patients died during French trials a few years later.[65] Only in the 1990s, after more than a decade of testing, did the class of proteins known as interferons become an accepted part of medical practice. On one hand, the interferon story illustrates how an overeager public and press can create unattainable

---

[60] Pepper quoted in Peters, *Interferon*, 148.

[61] A short overview of the "miracle drug" media is found in Dorothy Nelkin, *Selling Science: How the Press Covers Science and Technology* (New York: W. H. Freeman and Company, 1995), 2–6. Finnish doctor Kari Cantell and the Finnish Red Cross Transfusion Service, which had access to the unique white blood cell collection of the Finnish Blood Bank, operated the only large-scale production of interferon at the time. See Kari Cantell, *The Story of Interferon: The Ups and Downs in the Life of a Scientist* (New York: World Scientific Publishing Company, 1998).

[62] David Leff, "Biotechnology Transfer: A Two-Way Street," in Diane B. Bendahmane and David William McClintock, eds., *Science, Technology and Foreign Affairs: Volume II, Climate, Scientific Dialogue and Health* (Washington, DC: Department of State, 1985), 12–16. This work is a product of the Center for the Study of Foreign Affairs at the Foreign Service Institute of the US Department of State.

[63] Peters, *Interferon*, 140.

[64] See "The Wedding with Genetics," in Robert Bud, *The Uses of Life: A History of Biotechnology* (New York: Cambridge University Press, 1993), 163–188.

[65] Four of fourteen ACS patients suffered "substantial regression," see Harold M. Schmeck, Jr., "Interferon: Studies Put Cancer Use in Doubt," *New York Times* (May 27, 1980): C1.

expectations; but what other lessons can be learned from the miracle protein of the 1970s?

The race to synthesize interferon demonstrated the viability of global biotech competition, shaping American regulations, intelligence gathering, and diplomacy. While the United States was not behind – American biotech firms succeeded shortly after the others – the field exemplified a new era in which domestic corporations competed in research-driven global industries. Congress shifted policies in response: medical research received more funding, while legislation aimed to decrease the time between experiment and application, often by encouraging university/industry partnerships. Over the next two years, bills to regulate biotech died in Congress and NIH relaxed regulations.[66] Historian Susan Wright concluded: "the structure of international industrial competition entered as a major factor in shifting congressional interest from regulation of genetic engineering to its promotion."[67] As the competition intensified, genetic engineering became a concern of American intelligence.

The CIA monitored the growth of recombinant DNA research in rivals overseas, often highlighting the impact of regulations and lack of secrecy.[68] Analysts believed regulations caused geneticists to relocate, reporting, "Four foreign companies have moved recombinant DNA research projects in France. Some British scientists have relocated to West Germany … Some British researchers have transferred experiments involving a cancer-causing virus to Switzerland."[69] A detailed overview on Soviet genetics stressed facilities lagged "state-of-the-art research by several years," but warned researchers were "taking advantage of continuing access to virtually all advanced Western recombinant research which is neither classified nor proprietary."[70] Intelligence officers bemoaned the openness of the field: "Even the details of a procedure recently submitted for US patent rights have been published openly. Two of the coauthors of

[66] Sally Smith Hughes, "Making Dollars Out of DNA: The First Major Patent in Biotechnology and the Commercialization of Molecular Biology, 1974–1980," *Isis* 92 (September 2001): 541–575, esp. 568.

[67] Ibid., 449.

[68] CIA, "National Intelligence Daily Cable," (April 11, 1977): 1–20, quote on 18. CIA Doc No/ESDN: CIA-RDP79T00975A030000010018-8.

[69] CIA, "National Intelligence Daily Cable," (June 3, 1978): 1–15, quote on 11. CIA Doc No/ESDN: CIA-RDP79T00975A030700010038-9.

[70] CIA, National Foreign Assessment Center, "Soviet Recombinant DNA Research: Status and Trends," (September 1978): 1–10, i. CIA Doc No/ESDN: 0000969741.

this patented research project are now conducting research in West European laboratories."[71] Worse, NATO allies and the Japanese provided the USSR with recombinant DNA resources, while Soviet geneticists had been sent on "fact-finding" missions to international scientific meetings and proposed adding recombinant DNA to the détente exchanges. But the CIA discouraged genetic assistance to the Soviet Union, even as the Carter administration offered recombinant DNA assistance to the Chinese as a benefit of normalization (see Chapter 5).

US representatives worked to convince allies to adopt American regulations and standards. Unionized British researchers, for example, focused on regulating genetic laboratories as part of workplace safety; American negotiators downplayed such concerns. William Gartland, an NIH director and American representative, later wrote about his role: "Through participation in the [European Science Foundation] liaison committee, the United States was able to play a significant role in promoting comparable safety guidelines for recombinant DNA research."[72] OES director John Negroponte worked to ensure European biotechnology safety standards met "all U.S. research, regulatory and trade interests."[73] By 1978, the CIA reported most European countries (along with Japan and fifteen other nations) adopted NIH-based guidelines.[74] As American negotiators secured European agreement on biotech standards, developing nations attempted to access the field through the UN.

The G-77, Genetics, and the ISTC

Since the Kennedy administration, American officials worried about the growing demands for access to science and technology in the developing

[71] Ibid., 6.

[72] William J. Gartland, Jr., "Cooperative Efforts in Development of Safety Guidelines for Recombinant DNA Research," in Mitchell B. Wallerstein, ed., *Scientific and Technological Cooperation among Industrialized Countries* (Washington, DC: National Academy Press, 1984), 171–181, 177. Gartland was Director of the NIH Office of Recombinant DNA activities.

[73] Negroponte quoted in US Congress, House of Representatives, Committees on Foreign Affairs and Science and Technology, *Overview of International Science and Technology Policy: The Federal Organization, Joint Hearing before the Committees on Foreign Affairs and Science and Technology,* 99th Cong., 2nd sess. (May 20, 1986): 17.

[74] CIA, "National Intelligence Daily Cable," (June 3, 1978): 1–15. CIA Doc No/ESDN: CIA-RDP79T00975A030700010038-9. See also, Herbert Gottweis, "Transnationalizing recombinant DNA regulation: Between Asilomar, EMBO, the OECD and the European Community," *Science as Culture* 14 (2005): 325–338, esp. 327.

world. Foreign leaders first proposed a global conference on development in 1960; the State Department disliked the idea but feared bad publicity more. Eugene Skolnikoff, veteran of various White House science offices from 1958 to 1963, remembers it as an "initiative conceived by scientists with political payoffs in mind."[75] Held in Geneva in 1963, the first UN conference on science and technology led to multinational pledges to share the benefits of applied research, although disputes about intellectual property rights arose.[76] Developing countries formed the Group of 77 (G-77) the following year to represent their economic interests. Among their demands was access to technology, medicine, and basic knowledge as part of the "common heritage" of mankind; US Undersecretary of State George Ball felt the proposal was "an organized pressure campaign designed to force a massive transfer of resources from the industrialized countries to the less-developed countries."[77] The CIA worried politicization would spread from the UN general assembly to the technical agencies.[78]

Genetic engineering created conflict between the United States and developing world at the UN. UNESCO set up an international network for microbiology to encourage cooperation in 1975.[79] The same year, the Lima Declaration mandated the UN help developing nations increase production; many officials believed biotechnology could provide a critical boost.[80] However, while developing nations saw the UN as the means to access the emerging field, the United States viewed the UN as the means to globalize intellectual property rights. Determined to protect a growing domestic industry, the American delegation proposed a treaty guaranteeing global patent rights for micro-organisms (the "Budapest Treaty"), but UN support was minimal (as of this writing only 1/3 of member countries have signed).[81] Rebuffed, the Carter administration suggested an Institute

---

[75] Skolnikoff, *Science, Technology and American Foreign Policy*, esp. 165–178, quote on 173.

[76] The full name was UN Conference on the Application of Science and Technology for the Benefit of Less-Developed Areas or UNCSAT.

[77] Ball quoted in Chorev, *The World Health Organization between North and South*, 47. See also 52–54.

[78] On the CIA, see Chorev, *The World Health Organization between North and South*, 55.

[79] The network is MIRCEN. See Lynton K. Caldwell, "International Aspects of Biotechnology," *MIRCEN Journal* 4 (1988): 245–258.

[80] Raymond A. Zilinskas, "The International Centre for Genetic Engineering and Biotechnology," *Technology in Society* 9 (1987): 47–61, esp. 49.

[81] The full title is the "Treaty on the International Recognition of the Deposit of Micro-organisms for the Purposes of Patent Procedure."

for Scientific and Technical Cooperation (ISTC) to gain support in the developing world.[82]

The ISTC revived the promise of American assistance familiar since Point Four. The State Department circulated a pamphlet highlighting American funding for medical and environmental research in the developing world.[83] Officials solicited advice across the globe: Science advisor Frank Press met with leaders across Africa, while diplomats requested suggestions for projects.[84] Backed by USAID and the Ford Foundation, proponents argued the institute "would not merely funnel American technology, developed for American purposes, into foreign lands. It would help develop technology appropriate for the particular needs of foreign countries."[85] However, the American focus on commerce resurfaced as well: the State Department proposal listed the first priority as guaranteeing "private industries and organizations enjoy due protection and due returns on their investment and inventiveness."[86] Nonetheless, administration officials hoped the ISTC would win approval from the developing world at the next UN conference.

Preparations for the conference revealed potential problems. Delegates from the developing world eagerly awaited the opportunity to revisit scientific and technical assistance. The Ivory Coast ambassador, for example, remarked, "Africa looks forward with high hopes" to the meeting, considering it a step toward achieving "the advantages which science and technology can offer which will help [Africans] meet the essential needs of their populations."[87] Many American observers were

---

[82] The original term for the organization was the Foundation for International Technical Cooperation, said in a speech to the Venezuelan congress. See John Walsh, "President and Science Advisor Push for a Foundation for Development," *Science* 200 (June 16, 1978): 1252–1253. Over time the name changed: An official view of the process and proposal is found in Department of State, US National Paper prepared for the 1979 UN Conference on Science and Technology for Development, *Science and Technology for Development*, (Washington, DC: Government Printing Office, 1978). For an overview of the failure of the ISTC initiative, see Dickson, *The New Politics of Science*, 185.

[83] Department of State, *Some United States Activities using Science and Technology for Development* (Washington, DC: Government Printing Office, 1979): 1–25.

[84] On Press, see John Walsh, "Administration Backs Plan for Technology Foundation," *Science* 203 (February 2, 1979): 423. Zbigniew Brzezinski, "Science and Technology in Developing Countries," (February 16, 1978): 1–2. DNSA Presidential Directives Collection, Record # PD01568.

[85] Anon., "A Modest Aid Proposal," *Washington Post* (July 7, 1979): A12.

[86] Department of State, *Science and Technology for Development*, 32.

[87] Amoakon-Edjampan Thiemele, "Science and Technology in the Service of Development," *Technology in Society* 1 (1979): 159–164, esp. 160.

more cautious: *Science* editor William Carey predicted, "It will be a face-off between advanced and deprived societies, with science and technology as hard currencies in the new diplomacy."[88] But the American delegation arrived with more demands than money: The Senate rejected the administration's $25 million price tag for the ISTC a month before the conference opened.[89]

Held in Vienna, the United Nations Conference on Science and Technology for Development (UNCSTAD, 1979) revealed fault lines within the global community.[90] Representatives from the developing world restated their belief science and technology constituted the common heritage of mankind and demanded increased access and rights. In response, the Industrial Research Institute, a conglomerate composed of small companies conducting 85 percent of privately funded research in the United States, banded together with twenty-six multinational corporations against the idea. At the end of the conference, the United States refused to provide even $15 million for a UN center in New York City. *Science* editor Carey was candid post-conference:

If the Vienna meeting accomplished little else, it dramatized the North-South gap by pounding home the fact that three-fourths of the world's people account for only 3 percent of the world's research and development. . . . The message from the developing nations was explicit: that they will no longer accept the trickle-down method of scientific and technological transfer that suits the advanced countries. . . . In the near term, despite all that was said at Vienna, there will be no dismantling of the vast advantage in science and technology enjoyed by the advanced nations over the developing world. . . . Meanwhile, a wind is rising.[91]

The intellectual protection of biotechnology gathered the first clouds.

Legalization and Tension with Allies and the Developing World

American authorities legalized commercial biotechnology in 1980. The Supreme Court decided *Chakrabarty* vs. *Diamond* in June, ruling man-made organisms, including the oil-eating bacterium, were subject to

---

[88] William D. Carey, "Science and Technology for Development," *Science* 199 (January 20, 1979): 251.

[89] Ibid. See also Werner Fornos, "Just Another Research Subsidy," *Washington Post* (July 24, 1979): A13.

[90] For an overview of UNCNSTD, see Dickson, *New Politics of Science*, 193–202.

[91] William D. Carey, "Science and the Politics of Development," *Science* 205 (September 28, 1979): 1339.

patent laws and protection.[92] By a narrow 5 to 4 margin, the Court construed the Patent Act, originally drafted by Thomas Jefferson, to include all products of human invention, relying on a 1952 Senate report that recognized as patentable "anything under the sun that is made by man."[93] Chief Justice Warren Burger's majority opinion was strongly worded: "In short, we think the fact that micro-organisms, as distinguished from chemical compounds, are alive, is a distinction without legal significance."[94] With one ruling, the industry stood on firm legal ground. Genentech went public in October; the stock soared from $35 to $89 in the first ninety minutes of trading.[95] The USPTO granted the Cohen-Boyer patent for the recombinant DNA process two months later; at the same time, Congress passed the Bayh-Dole Act to allow universities to hold patents from federally funded research. In only six months, authorities approved the patenting of live organisms and Stanford's patent on recombinant DNA engineering.

The decisions completed the commercialization of academic biology. Within months, dozens of companies bought licenses from Stanford; by one account, the Cohen-Boyer patent generated $225 million for Stanford and resulted in more than 2,400 new products worth more than $35 billion as more than 400 companies licensed the process from Stanford during the patent's duration from 1972 to 1997.[96] Universities held around 150 patents in 1980; a decade later, the number was more than 1,500.[97] The commodification and privatization of biological knowledge concerned researchers who felt there was a conflict between the academic responsibility to disseminate information and the commercial incentive to maintain trade secrets. Others had no conflict: Harvard professor Walter Gilbert, a Nobel Prize winner in medicine (1980) and

---

[92] *Diamond* vs. *Chakrabarty*, 447 U.S. 303 (1980).

[93] Senate report is quoted in Sheila Jasanoff, *Science at the Bar: Law, Science and Technology in America* (Cambridge, MA: Harvard University Press, 1995), 144. For an excellent introduction to the legal background, see Chapter 7, "Legal Encounters with Genetic Engineering."

[94] Warren Burger and the majority opinion quoted in Nicholas Wade, "Court Says Life Can Be Patented," *Science* 208 (June 27, 1980): 1445.

[95] Special to the *New York Times*, "Harvard Considers Commercial Role in DNA Research," *New York Times* (October, 27, 1980): A1.

[96] Maryann P. Feldman, et al., "Lessons from the Commercialization of the Cohen-Boyer Patents: The Stanford University Licensing Program," in M. P. Feldman, et al., eds., *Intellectual Property Management in Health and Agricultural Innovation: A Handbook of Best Practices* (Oxford, UK: MIHR, 2007): 1797–1807.

[97] Hughes, "Making Dollars Out of DNA," 570.

founding shareholder in Biogen, stated: "Everyone is grappling with how to reconcile the objectives of an educational institutions while getting involved in a company for profit, but I welcome the initiative."[98] Not all countries agreed.

Allies refused to accept American patent policies. Though the USPTO granted the Cohen-Boyer patent on recombinant DNA techniques, the Europeans denied the patent because the process had been published first, violating EEC patent rules.[99] Nor would international collection via the Budapest Treaty guarantee US patent rights. The US Supreme Court also ruled patents on living organisms were legal, but European and Japanese laws explicitly outlawed the practice (many countries also excluded biological and medical inventions from patent protection).[100] In effect, allies accepted NIH safety regulations but rejected commercial protections, leaving diplomats determined to secure allied acceptance of American patent rulings and protect intellectual property rights overseas.

The developing world proposed an International Center for Genetic Engineering and Biotechnology (ICGEB) to benefit from the new field. The original idea, according to founder Arturo Falaschi, was to "provide developing countries with scientists trained at an international level in modern biotechnology."[101] The proposal had many advocates. UNIDO Director Abd-El Rahman Khane argued, "The applications of genetic engineering and biotechnology in vital fields such as health, energy or food are of major relevance to developing countries."[102] Nobelist David Baltimore agreed: "Despite the fact that biotechnology is a very high technology in terms of being at the forefront of basic science, it is a very appropriate technology for solving problems at all levels of development."[103] Egyptian program coordinator Wafa Kamel stressed economics: "[The developed countries] all recognize great potential in these

---

[98] Gilbert quoted in Special to the *New York Times*, "Harvard Considers Commercial Role in DNA Research," A1.

[99] See "The Global Dimensions of Intellectual Property Rights in Science and Technology," in Mitchel B. Wallerstein, et al., eds., *Global Dimensions of Intellectual Property Rights in Science and Technology* (Washington, DC: National Academy Press, 1993), 3–17, esp. 6.

[100] Ibid., 9.

[101] Arturo Falaschi, "The International Centre for Genetic Engineering and Biotechnology of UNIDO," *TIBTECH* 9 (November 1990): 314–317.

[102] Khane quoted in David Dickson, "UNIDO Hopes for a Biotechnology Center," *Science* 221 (September 30, 1983): 1351–1353, quote on 1351.

[103] Raymond A. Zilinskas, "The International Centre for Genetic Engineering and Biotechnology," *Technology in Society* 9 (1987): 47–61, quote on 49.

techniques for changing the traditional manufacturing industry and promoting economic well-being. .... The developing countries would have to buy it through patents and licenses. So we want to make investments in the brains of [our] people."[104] Championed by Boyer, Chakrabarty, and officials at NIH, the ICGEB had near universal support in the developing world; eight countries applied to host the center.[105] The United States was less enthusiastic.

The Reagan administration pressured the UN to forgo the ICGEB. Officials refused to attend planning meetings, arguing UNIDO's track record was a "mixed picture" with too much emphasis on state ownership.[106] Instead, American officials rallied their French, German, Japanese, and British counterparts against the center.[107] The French and British proposed a biotechnology network, denying any conflict with the proposed ICGEB, while the United States blocked UN incorporation of the center throughout the 1980s.[108] Ultimately, supporters established the ICGEB outside the UN system, where it maintains two centers – in Trieste, Italy, and New Delhi, India – and works on diseases such as hepatitis and malaria as well as biological research in protein chemistry and virology.[109]

Recombinant DNA engineering divided the United Nations and reoriented American diplomacy. The desire to protect private industry combined with suspicion of UN radicalism to undercut American support for the ICGEB. The UNCSTAD and ICGEB experiences reinforced American support for international patent rights and undermined support for

---

[104] Kamel quoted in Peter Humphrey, "Third World Nations Seek Gene Research Centre," *The Globe and Mail* (August 26, 1983). Accessed online.

[105] The eight countries were: Belgium, Bulgaria, India, Italy, Pakistan, Spain, Thailand, and Tunisia. For an NIH perspective, see William Gartland, "The UNIDO/Biotechnology Center" in Bendahmane and William McClintock, eds., *Science, Technology and Foreign Affairs*, 17–18. Gartland headed the Office of Recombinant Activities at NIH.

[106] Roderick Mackler, "The U.S. Government Decision on the UNIDO/Biotechnology Center," in Bendahmane and McClintock, eds., *Science, Technology and Foreign* Affairs, 19–20. Mackler was formerly the UNIDO desk officer at the State Department.

[107] Dickson, "UNIDO Hopes for a Biotechnology Center," 1351–1353.

[108] On the proposed British and French network, see David Dickson, "Biotechnology Network Planned," *Science* 221 (September 30, 1983): 1352. Raymond A. Zilinskas was the United States representative at OTA and UNIDO. For his perspective on the why the United States failed to support the ICGEB, see Raymond A. Zilinskas, "The International Centre for Genetic Engineering and Biotechnology," *Technology in Society* 9 (1987): 47–61.

[109] Arturo Falaschi, "The International Centre for Genetic Engineering and Biotechnology of UNIDO," *TIBTECH* 9 (November 1990): 314–317. See also the ICGEB website: www.icgeb.trieste.it/home.html.

scientific access based on humanity's "common heritage." Bilateral or multilateral cooperation along American guidelines became the preferred relationship, with the US government promoting international regulations and intellectual property rights.

Nor was biotechnology the only field involved. The shift toward market-based R&D ushered in a new phase in the patentization of science: Until the 1970s, research was often classified for national security reasons (nuclear physics and the bomb provide the obvious example); after, research was frequently patented and restricted for commercial reasons. Determined to capitalize on private research, American policy-makers focused on intellectual protections overseas, thus making some scientific knowledge – and lots of technical know-how – proprietary. Although biotech is the example here, the centrality of advanced technologies such as computers and satellites to research began to fuse science and technology in novel and unexpected ways, whether in genetic engineering, telecommunications or the materials sciences necessary for manufacturing semiconductors. Terms such as biotechnology and nanotechnology began to be heard, but little understood, in the halls of Congress. As private research and development blossomed, traditional Cold War alliances shifted; in the process, access to American science and technology became integral to a reorientation in American diplomacy, whether initiating détente with the Soviet Union, normalizing relations with China, or negotiating in the Middle East amid wars and an oil crisis.

5

Reorientation

From an altitude of 20,000 feet, Irving Langmuir, Nobel Prize winner in chemistry, watched as the dry ice fell through the haze below, later writing, "the whole cloud rapidly dissipated as the upper part changed to ice and the lower part rained out."[1] A leading researcher in Project Cirrus, Langmuir flew dozens of missions in the United States, Honduras, and Puerto Rico as part of a General Electric/US military experiment to weaken hurricanes by "seeding" the cloud wall with dry ice.[2] When six major hurricanes devastated the East Coast from 1954–1955, Congress appropriated millions for hurricane research at the US Weather Bureau.[3] Presumed success with Hurricane Esther in 1961 led to Operation Popeye (attempts to increase rainfall over the Ho Chi Minh trail in Vietnam) as well as Project Stormfury (a collaboration of the US Navy and Commerce Department). But weather modification programs struggled to find

---

[1] Langmuir quoted in Barrington S. Havens, *History of Project Cirrus* (Schenectady, NY: Research Publication Service, 1952), 41. The Public Relations Services Division of General Electric Research Laboratory produced the text.

[2] Cirrus was a collaboration of General Electric and the United States Army Signal Corps, the Office of Naval Research and the United States Air Force. See Havens, *History of Project Cirrus*, 15. See also "Pathological Science" in James Rodger Fleming, *Fixing the Sky: The Checkered History of Weather and Climate Control* (New York: Columbia University Press, 2010), 137–164.

[3] Research eventually evolved into the National Hurricane Laboratory in 1964 and the Hurricane Research Division at the Atlantic Oceanographic and Meteorological Laboratory in 1983. See H. E. Willoughby, et al., "Project STORMFURY: A Scientific Chronicle, 1962–1983," *Bulletin of the American Meteorological Society* 66 (May 1985): 505–514. Willoughby and co-authors were members of the Hurricane Research Division, Atlantic Oceanographic and Meteorological Laboratory, NOAA.

partners or targets; participating scientists acknowledged, "International agreements required to move Stormfury to the Western Pacific never materialized ... attempts to move Stormfury to the eastern Pacific or to Australia also met political resistance."[4] There was too much uncertainty and risk. Instead, the potential ramifications of human-caused weather events – whether hurricanes, floods, or droughts – across national borders led to calls for a world weather modification bureau and legal studies of national liability.[5]

Weather modification made a brief appearance in American diplomacy. Presidents took the field seriously in the 1960s: Johnson funded attempts to increase rainfall in India and Pakistan; Nixon required all "climate modifications (civilian or military)" get the "specific approval of the president" (NSDM 165).[6] Congress held hearings in 1975 after USAID fielded multiple requests for weather aid (the agency issued a policy to discourage such requests).[7] The Soviet Union drafted a UN resolution prohibiting "hostile use of environmental modification techniques" and American meteorologists visited Soviet weather modification facilities as part of the détente exchanges.[8] The CIA surveyed Chinese and Soviet research, reporting on experiments to clear fog from airports, moderate typhoons, and "raise the level of Lake Sevan in Armenia by increasing rainfall."[9] As late as 1983, USAID supported a $6 million dollar precipitation enhancement initiative in Morocco, which included

---

[4] H. E. Willoughby, et al., "Project STORMFURY," *Bulletin of the American Meteorological Society*, 508.

[5] On calls for an international bureau, see Georg Breuer, *Weather Modification: Prospects and Problems* (New York: Cambridge University Press, 1979). On the legalities of weather modification, see Edith Brown Weiss, "International Liability for Weather Modification," *Climatic Change* 1 (1978): 267–290.

[6] Henry A. Kissinger, National Security Council, "International Aspects of Weather Modification [NSDM 165]," (May 2, 1972): 1–2. DNSA Presidential Directives collection, Record # PD01270.

[7] Joanne Simpson, testimony to the House Subcommittee on International Organizations and Movements, reprinted as "Concerning Weather Modification," *Bulletin of the American Meteorological Society* 56 (January 1975): 47–49. See also Jack Vanderryn, "AID's Policy on Precipitation Enhancement," in Bendahmane and McClintock, eds., *Science, Technology and Foreign Affairs*, 65–68.

[8] The UN treaty is the Convention on the Prohibition of Military or Any Other Hostile Use of Environmental Modification Techniques (ENMOD, 1978). See Fleming, *Fixing the Sky*, 184–185. On the détente exchanges, see Louis J. Battan, "Weather Modification in the Soviet Union – 1976," *Bulletin of the American Meteorological Society* 58 (January 1977): 4–19.

[9] CIA, "National Intelligence Daily (Cable)," (December 16, 1978): 1–15. CIA Doc no./ ESDN CIA-RDP79T00975A030900010126-9.

ground-based radar, satellite receiving stations, training for radar technicians and pilots, and hydrological modeling (a similar program was also
underway in Panama).[10] However, after numerous tests failed to verify
results, funding and the field shrunk considerably.[11] Yet while weather
modification collapsed in the 1980s, weather prediction, like many other
environmental fields, helped knit the globe together.

Weather prediction and satellites provided opportunities for American
diplomacy. Consider the following Kennedy–Khrushchev exchange from
1962:

Perhaps we could render no greater service to mankind through our space programs than by the joint establishment of an early operational weather satellite
system.

President Kennedy

It is difficult to overestimate the advantage that people would derive from the
organization of a world-wide weather observation service using artificial earth
satellites. Precise and timely weather prediction would be still another important
step on the path to man's subjugation of nature.

Premier Khrushchev[12]

The World Weather Watch began the following year; by 1968, American and Soviet satellites provided continuous data to researchers below.
In his first year, President Nixon promised the UN General Assembly the
American space program would produce knowledge for the world; three
years later, the United States launched its first Earth Resources Technology Satellite (renamed Landsat). Landsat data, available for purchase,
immediately became a powerful tool for tracking environmental
changes, including disappearing forests and grasslands, flood erosions
and siltation, and fluctuations in water and soil temperatures. More
than 120 countries took part by 1980, with two-thirds in the
developing world.

The system fundamentally altered meteorology and global cooperation, whether on weather (short-term atmospheric conditions) or climate (long-term atmospheric conditions). A drought in the Sahara

---

[10] Vanderryn, "AID's Policy on Precipitation Enhancement," in Bendahmane and McClintock, eds., *Science, Technology and Foreign Affairs*, 67–68.
[11] For an example of a later study testing Soviet experiments, see J. Bader, et al., "Further
   Results of Grossversuch IV: The Effect of the First Rocket Launched into a Potential Hail
   Cell," *Journal of Applied Meteorology* 31 (July 1992): 700–707.
[12] Kennedy–Khrushchev exchange reprinted in Angelina Long Callahan, "Part III: NASA
   and the Soviet Union/Russia," in Krige, et al., *NASA in the World*, 125–182, quote
   on 128.

(1968–1972), for example, highlighted the need for greater coordination, spurring the Global Atmospheric Research Program (GARP), which relied on parachute-borne instruments, dozens of ships, and hundreds of buoys, along with two polar-orbiting and five geostationary satellites.[13] GARP represented another iteration of scientific internationalism and networking – a partnership between the WMO, ICSU, and dozens of participating nations (including the United States and Soviet Union). It also produced results: Weather forecasts improved for up to five days and the UN established the World Climate Research Program (1980) with US support.

Meteorology illustrated many of the tensions common to the intersection of science and politics. Until the field matured, it was difficult to distinguish worthwhile projects from those that wasted resources; weather prediction and weather modification were both emerging areas in meteorology in the 1960s and 1970s. It was also difficult to anticipate the environmental risks, legal liabilities, or military implications of rapidly evolving fields and decisions had to be made about whether meteorological data should be classified, privatized, and used in diplomacy (both prediction and modification played a part in détente with the Soviet Union). Complicating matters, the weather concerned multiple government departments and agencies, including Defense, Commerce, State, NOAA, and the EPA. At the same time, meteorology strengthened international relations, economic development, and global governance. And the field was only one of many advancing faster than Washington could keep up with.

Politicians felt insecure. Disagreements over Vietnam, supersonic transport and an anti-ballistic missile system led President Nixon to dismiss his science advisors (see Chapter 4). But hard choices on complex scientific questions remained. Representative George P. Miller (D-CA), chair of the Committee on Science and Astronautics, spoke for many of his colleagues: "One of our difficulties is how to evaluate these decisions. We have to take a great deal on faith. We are not scientists... How can we evaluate?"[14] Congress established the Office of Technology Assessment (OTA, 1972) to provide better analyses of scientific and technical issues.

---

[13] John S. Perry, "The Global Atmospheric Research Program," in Wallerstein, ed., *Scientific and Technological Cooperation among Industrialized Countries*, 149–161.

[14] Miller quoted in Peter D. Blair, *Congress's Own Think Tank: Learning from the Legacy of the Office of Technology Assessment (1972–1995)* (New York: Palgrave MacMillan, 2013), 25–26.

Early priorities included energy research, telecommunications, and applied genetics, but the list of possibilities seemed endless. The importance of international relations to research added yet another dimension: the State Department established a Bureau of Oceans and International Environmental and Scientific Affairs (OES, 1974) to stay informed and advance American interests.

Determined to get a "big picture," Congress commissioned a six-year study on "Science, Technology and American Diplomacy in an Age of Interdependence." The study's conclusions, published in 1976, are worth quoting at length:

Rarely, if ever, has U.S. foreign policy faced so many fundamental changes... Disorientation is not too strong a term for the state of U.S. foreign policy in the mid-1970s... As matters stand now, it is difficult to know what coherent and purposeful U.S. policies are being pursued in Africa or Latin America, in the export or import of technology, in the support of world health, in the export of U.S. surplus grains, in the control of multinational corporations, in the allocation of the world's resources of petroleum and minerals, in the United Nations and associated institutions, in the expansion of nuclear power with or without reference to conflict with control of nuclear proliferation, in the encouragement of the world science community, in the use of survey satellites, in global communications, and generally in defining the significance of the overused word "interdependence" as a principle of U.S. diplomacy.[15]

Even the summary could be disorienting – interdependence resisted easy assessment and policy-making. Advances in one field influenced others, research was increasingly multinational and/or private and American science priorities remained unclear. Regional geopolitics further confused the picture: European allies both cooperated with and competed against the United States, while America and its allies cooperated in response to the OPEC oil embargo and the Group of 77's calls for more access to science and technology. Aid programs, already limited by economic stagflation, guarded against radicalization at the WHO and UNESCO and worried about the costs of technology transfer. The familiar patterns of the postwar-era were coming to an end, producing a sense of disorientation among policy-makers.

In hindsight, however, a reorientation in American science and diplomacy is clear. Part of the reorientation was domestic: As noted in the

[15] US Congress, House of Representatives, Subcommittee on International Security and Scientific Affairs, Committee on International Relations, *Science, Technology and American Diplomacy in the Age of Interdependence* (Washington, DC: US Government Printing Office, 1976), quotes on 18 and 351.

previous chapter, the government encouraged private R&D, whether through grants to cooperative research between industry and academia, such as the Polymer Research Center at MIT (1973), or through legislation to lessen the time between federal research and industrial application.[16] Decentralized and networked, applied science and development took off across private industry, leading American diplomacy to support international patent rights and intellectual protections. Managing access to science and technology figured ever more prominently in American foreign relations. The other reorientation was geopolitical: The Nixon administration hoped to split the communist "bloc" and science diplomacy both signified détente with the Soviet Union and presaged normalization with the People's Republic of China. In the Middle East, multiple administrations used access to secure alliances and recycle American payments for high oil prices after the Yom Kippur War and embargo. Whether in the Middle East, Asia, or outer space, the sciences were critical to American foreign relations in the 1970s.

The "Heavenly Geopolitics" of the era reflected new political realties below. Scientific exchanges and the Apollo–Soyuz Test Project symbolized détente, as the "Handshake in Space" showcased warmer relations between the two Cold War enemies. Meanwhile, the space race morphed into a competition over commercial satellite launches and communications after the moon landing, as both superpowers offered services for their allies; NASA, for example, broadcast television to India and other developing nations to sell satellites. In the process, space communications evolved from an American system into a private, internationally regulated system, adding another layer of global governance. And when US–Soviet relations deteriorated in the late 1970s, ending most scientific exchanges, US–Chinese relations improved.

Science diplomacy played an important role in normalizing relations with the People's Republic of China. The Chinese policy of "Walking with Two Legs" sought to achieve a balance between basic and applied research, and between East and West, providing the United States an

---

[16] The legislation was the Federal Laboratory Consortium Technology Transfer Act (1974). An excellent overview is National Science Foundation, "Major Federal Policies Promoting Technology Transfer and Commercialization of R&D," in *Science and Engineering Indicators, 2014* (Washington, DC: National Science Foundation, 2014), 4–42. For an overview highlighting academia (rather than the national labs), see "Universities and Industries: Knowledge as Commodity," in Dickson, *New Politics of Science*, 56–106.

opportunity to open China.[17] Throughout the 1970s, scientific delegations maintained steady contact with the People's Republic even as the political process ebbed. For American administrations, normalization meant access to the world's largest market and possible influence through a generation of US-trained politicians; for the Chinese, normalization meant gaining access to American science and technology. By the end of Reagan's first term, the PRC was America's largest bilateral science partner, a testament to the significance of science in Chinese–American relations.

The final section considers American diplomacy in the Middle East after the OPEC oil embargo, highlighting programs with Iran, Saudi Arabia, and Israel. Although Iran graduated from Point Four in 1969, the oil embargo caused the United States to expand technical programs in the region to recycle money back into the American economy. The Saudis, for example, spent $200 million annually on a twenty-five-year program for development, which included the King Abdul University for Science and Technology, solar research and desalination, and innumerable commercial and industrial projects. At the same time, the United States and Israel pioneered three still-existing bilateral research initiatives, illustrating the key role science plays in US–Israeli relations and the longevity of American science diplomacy.

HEAVENLY POLITICS

An "artificial satellite" at the correct distance from the earth would make one revolution every 24 hrs: *i.e.*, it would remain stationary above the same spot and would be within optical range of nearly half the earth's surface. Three repeater stations, 120 degrees apart in the correct orbit, could give television and microwave coverage to the entire planet.

Arthur C. Clarke (1945)[18]

It is settled U.S. policy to encourage international cooperation in basic science ... Space is already a matter of broad international cooperation. We have some 250 agreements with 74 countries covering space

---

[17] For an introduction, see Emil Smith, Robert Coe, Alexander Tseng, and Joyce Kallgren, "Walking on Two Legs: A Panel Discussion of Science Policy in the People's Republic of China," *Bulletin of the American Academy of Arts and Sciences* 28 (November 1974): 26–41.

[18] Arthur C. Clarke, "V2 for Ionosphere Research?" *Wireless World* 51 (February 1945): 58.

cooperation. And space has already been put to the service of man in the new global communications systems and in weather prediction systems. But this is only a beginning.

National Security Council internal paper (1970)[19]

Arthur C. Clarke, best known for *2001: A Space Odyssey*, served in the Royal Air Force during World War II and helped pioneer the use of radar to land planes in poor conditions. At the end of the war, he outlined the basics of satellite communications in a letter to the editors of *Wireless World* (above), but failed to patent his idea, later writing a comic lament titled "How I Lost a Billion Dollars in My Spare Time."[20] But even Clarke could not imagine satellite communications would become a reality so fast.

The evolution of space science and technology from 1960 to 1980 illustrated the transition to globalization. Space programs were the height of bipolar Cold War competition in 1960; twenty years later, they were an aspect of global economic competition. Dwindling financial and public support required the post-Apollo program justify itself as satellite communications and launch services took center stage. NASA's achievements remained significant: Astronauts spent more than 150 days aboard *Skylab*, *Voyager* toured the outer plants, the *Vikings* (*I* and *II*) landed on Mars, and satellites relayed information to ground stations, where computer analysis refined weather prediction and environmental sciences. By 1980, well-regulated national, international, and private satellites orbited the Earth, providing television, telephone, and data transmissions linking the globe (other benefits included military reconnaissance and intelligence). The conquest of space also enabled globalization down on earth: Although globalization is often thought of as a terrestrial process, the process needed space to occur.

Satellite Communications and American Diplomacy

The United States pioneered satellite communications in the 1960s. Only a few years after Sputnik, AT&T requested permission to launch a

[19] National Security Council, internal paper, NSSM 102 related, "Science and Technology [Attribution to National Security Council Based on Format]," (December 1970): 18–24. DNSA Presidential Directives (Part II) Collection, Record # PR00639.
[20] The full title was "A Brief History of Comsats, or, How I Lost a Billion Dollars in My Spare Time." See Gerald Jonas, "Arthur C. Clarke, Author Who Saw Science Fiction Become Real, Dies at 90," *New York Times* (March 19, 2008). Archived online at: www.nytimes.com/2008/03/19/books/19clarke.html?pagewanted=all&_r=0.

satellite for a private communications system, sparking a global debate over private vs. public control of space communications. Congress passed the Communications Satellite Act in 1962 to craft a global network and established the Communications Satellite Corporation (Comsat) to represent American interests. The network went global a year later when President Kennedy invited foreign nations to contribute to Intelsat, a new International Telecommunications Satellite Consortium. By 1965, the first generation of "Early Bird" satellites circled the planet. Launched by NASA, the satellite orbited 22,300 miles over the equator in a stable geostationary position (satellites in Geostationary Earth Orbit [GEO] rotate in time with the Earth below and are thus able to receive signals from Earth stations without calculating for satellite movement). Intelsat benefits were obvious: If countries bought an Earth station, the satellite network would revolutionize their communications; Latin Americans, for example, could call Europe (and other countries in the region) without going through New York first.

American dominance worried potential foreign partners. Senator John Pastore (D-RI), for example, spoke bluntly: "*Early Bird* is orbiting and is controlled by the Comsat Corporation and can become the show window through which America will be seen throughout the world. And indeed it will be the mirror of our image."[21] But Intelsat structure was a problem: The consortium accorded voting rights on the basis of financial contribution and the United States paid for and controlled a majority of the system. Intellectual and economic benefits also stayed home; as late as 1971, only 5 percent of the contracts for satellite construction and R&D were spent outside the United States.[22] Such tight control caused European allies to coordinate national programs and hesitate before joining the post-Apollo space program (see Chapter 2). As the United States vacillated, France and West Germany collaborated on an alternative to Intelsat and the British developed Skynet (1965) for military communications.

The Soviets rallied nations against US control of Intelsat. Khrushchev denounced the consortium as a capitalist plot and introduced Interkosmos (1967) to launch Soviet-allied satellites; the following year, he

[21] Pastore quoted in Walter A. McDougall, *The Heavens and the Earth: A Political History of the Space Age* (Baltimore, MD: Johns Hopkins University Press, 1985), 359. An excellent overview of the communications debate can be found on pages 355–360.

[22] Michiel Schwarz, "European Policies on Space Science and Technology, 1960–1978," *Research Policy* 8 (1979): 204–243, esp. 216.

announced Intersputnik, a communications network open to all nations on a one-country, one-vote basis.[23] Many considered the offer a propaganda ploy highlighting Intelsat's system of voting by financial contribution; the CIA considered it less a challenge than a Soviet attempt "to link the two systems, thus gaining the use of Intelsat's global coverage while enjoying the prestige of having their own 'international' system."[24] The Soviet proposal inflamed nations unable to buy influence as well as countries with a tradition of public ownership of utilities, such as West Germany, India, and Canada.[25] Within months, sixty-five of the sixty-seven Intelsat members requested the "equivalent of a constitutional convention."[26] By the time Intelsat III beamed Neil Armstrong's "one small step for mankind" to 500 million people worldwide, the organization's charter was undergoing revision on the ground.[27] A 1973 agreement eventually resolved the impasse: Although the United States retained a plurality of votes (and costs), the next generation of satellites (Intelsat IV) muted protests by doubling the capacity for world telephone and TV transmissions.[28] As the United States renegotiated its position within Intelsat, it offered potential allies the chance to benefit as well.

Satellite communications bolstered American bilateral relations and trade. A signature initiative was the Satellite Instructional Television Experiment (SITE) with India.[29] Originally conceived at the height of the Vietnam conflict – when modernization theorists hoped to put a television in every home – the final project borrowed one of NASA's most advanced GEO satellites to broadcast television to rural villages (India,

[23] Anon., "Soviet Discloses Satellite Plans," *New York Times* (August 20, 1968): 14. Jonathan F. Galloway, "Worldwide Corporations and International Integration: The CASE of INTELSAT," *International Organization* 24 (Summer 1970): 503–519.
[24] See CIA, Directorate of Intelligence, "Central Intelligence Bulletin," (August 19, 1968): 1–8. CIA Doc No/ESDN: CIA-RDP79T00975A011900040001–5. For an earlier discussion, see also CIA, Directorate of Intelligence, "Central Intelligence Bulletin," (April 19, 1967): 1–6. CIA Doc No/ESDN: CIA-RDP79T00975A009800230001–8.
[25] Anon, "U.S. INTELSAT Role May Be Reduced," *New York Times* (April 15, 1971): 14.
[26] Anon., "Revamping Intelsat," *Science News* 95 (March 15, 1969): 256.
[27] Intelsat III had been in space for nineteen days. Virgil Labrador, *Heavens Fill With Commerce* (New York: Satnews, 2005), 58.
[28] Irving Goldstein, "INTELSAT and the Developing World," *IEEE Transactions on Communications* COM-24 (July 1976): 742–748. Goldstein was COMSAT's Director of International Affairs.
[29] See Ashok Maharaj, "Satellite Broadcasting in Rural India: The SITE Project," in Krige, et al., *NASA in the World*, 235–248.

like many developing countries, lacked a national TV system).[30] SITE required multilevel coordination: Participants included the Indian Space Research Organization, NASA, Hughes Aircraft, GE, and MIT. The experiment went live from August 1975 to July 1976, providing local agricultural, health, and family programming for four hours a day and reaching approximately 2,300 villages (about 2.8 million people). Asked about the project's impact, NASA coordinator Arnold Frutkin responded: "We took the satellite back. What was the consequence? India contracted with Ford Aerospace for a commercial satellite to continue their programs."[31] Brazil and Indonesia soon requested similar help; over the next decade, USAID enlisted more than twenty-five developing nations in bilateral programs.[32] The globe grew smaller.

Satellites and launches became bargaining chips in US diplomatic negotiations. In the Middle East, for example, American intelligence reported the Arab Satellite System (Arabsat) was "more and more politicized as time passes," but felt confident the United States would get the final contract because "the Saudis are willing to go through the motions of investigating all sources of supply for the system, but have indicated a preference for US equipment."[33] The United States launched Arabsat in 1976. In Asia, the Chinese pressured the United States to downgrade Taiwan to non-member status in Intelsat; CIA analysts warned, "the Chinese have the necessary votes to make good on their threat" and pointed out non-member status would have "very little practical effect" on Taiwan's use of the system.[34] Taiwan lost status. The free use of geostationary orbits became an issue: In 1976, eight equatorial states, including Brazil, tried to claim the GEO above their national territory.[35] In response, American officials rejected the claim as

---

[30] For an overview of the science and engineering of the system, see John E. Miller, "ATS-6 Satellite Instructional Television Experiment" *IEEE Transactions on Aerospace and Electronic Systems* AES-11 (November 1975): 1033–1037.

[31] Frutkin was the NASA Deputy Director of International Affairs. Frutkin quoted in Maharaj, "Satellite Broadcasting in Rural India: The SITE Project," 245. See also E. V. Chitnis, "The Role of Space Communication in Promoting National Development with Specific Reference to Experiments Conducted in India," *Advanced Space Research* 3 (1983): 125–132. Chitnis was the ISRO representative for the SITE project.

[32] Sheehan, *The International Politics of Science*, 69.

[33] CIA, Directorate of Operations, "Soviet Interest in MARISAT Terminals/Arab Satellite System/Compromise on INTELSAT Membership," (November 20, 1975): 1–2. CIA Doc No/ESDN: 0005903876.

[34] Ibid.

[35] The nations were Brazil, Colombia, Congo, Ecuador, Indonesia, Kenya, Uganda, and Zaire.

a violation of the UN Outer Space Treaty and pointed out Intelsat provided Brazil's first-ever national coverage and developing countries comprised one-third of the members.[36] Terrestrial politics even stretched to the moon: when the UN General Assembly passed a Moon Treaty (1979) to extend the "common heritage" concept to celestial bodies, no nation with a capable space program ratified the treaty; one example of US/Soviet agreement. Instead, developing nations require an advanced partner to access outer space, which became the site of superpower cooperation in the mid-1970s.

Science and Détente with the Soviet Union

Scientific and technical exchanges and agreements preceded President Nixon's formal détente with the communist bloc. After the isolation of Stalin's rule, Soviet participation at the UN and in the IGY offered opportunity for dialog in the mid-1950s; one of the earliest official interactions was an American medical mission in 1956.[37] The United States and Soviet Union established regular exchanges the following year (see Chapter 2), which eventually expanded to include meteorology and satellite communications. In 1959, Vice President Nixon visited Moscow and suggested a joint US–Soviet lunar mission, foreshadowing his later willingness to cooperate.[38] President Kennedy restated the offer but Congress outlawed it; Eugene Skolnikoff, a member of the president's science council, believed the State Department and NASA squashed a potential scientific détente.[39] A few years later, Chinese nuclear tests (1964) convinced both superpowers of the necessity of arms control and President Johnson reduced American production of enriched uranium as a step toward the nuclear non-proliferation treaty (1968).[40] At the same time, the space race compelled agreement: The United States and

---

[36] Thirty-three participating countries were listed as developing. See Irving Goldstein, "INTELSAT and the Developing World," *IEEE Transactions on Communications* COM-24 (July 1976): 742–748.

[37] Anna Geltzer, "In a Distorted Mirror: The Cold War and U.S.–Soviet Biomedical Cooperation and (Mis)understanding, 1956–1977," *Journal of Cold War Studies* 14 (Summer 2012): 39–63, esp. 39.

[38] Anne de Tinguy, *U.S.–Soviet Relations During the Détente* (New York: Columbia University Press, 1999), 68.

[39] Regarding Kennedy's scientific "détente," see Skolnikoff, *Science, Technology and American Foreign Policy*, 32–38. See also Wang, *In Sputnik's Shadow*, 193–194.

[40] Hal Brands, "Progress Unseen: U.S. Arms Control Policy and the Origins of Détente, 1963–1968," *Diplomatic History* (April 2006): 253–285.

Soviet Union coordinated rendezvous and docking plans for rescue missions and supported the Outer Space Treaty (1967), which limited use to peaceful purposes.[41] Finally, the two superpowers agreed to cooperate on atomic energy and medical research before the "high détente" of 1971–1974.[42]

President Nixon and Premier Khrushchev had different reasons for renewing relations, but their agreements were scientific and technical. In April and May of 1972, for example, the United States and Soviet Union agreed to limit strategic arms and anti-ballistic missile systems while expanding outer space research and scientific and technical cooperation. In addition to geopolitical benefits, détente offered the Soviets access: *Science* observed, "the Soviets are obviously interested in applied fields where the United States has a reputation for leadership."[43] CIA analysts reported computers, telecommunications, semiconductors, and industrial chemicals were among the primary attractions.[44] The Nixon administration, however, emphasized reducing Cold War tensions; American representatives wanted to focus on non-commercial research and guarantee cooperation did not violate export controls.[45] Soviet immigration policy was another area of concern.

The exchanges began amid friction over scientists' freedom to travel. In the early 1970s, the Soviet Union levied an education tax on Jewish scientists emigrating to Israel, leading many in the American scientific community to protest; more than 2,000 FAS and 150 NIH members petitioned the Soviets on their colleagues' behalf.[46] Additionally, physicist Andrei Sakharov – the father of the Soviet hydrogen bomb – began speaking out about Soviet human rights abuses and applied for a visa to

[41] Sheehan, *The International Politics of Space*, 63.

[42] For a brief look at the détente historiography, see Noam Kochavi, "Researching Détente: new opportunities, contested legacy," *Diplomatic History* 8 (November 2008): 419–425.

[43] John Walsh, "U.S.–USSR Agreement on Cooperation on Science, Technology: The Emphasis Is on Applications and Practical Results," *Science* 180 (April 6, 1973): 40–41, quote on 41.

[44] CIA, Office of Economic Research, "Soviet Economic and Technological Benefits from Détente," (February 1974): 1–37. CIA Doc No/ESDN: RDP80B01495R001000220009–1. See also Vladislav Zubok, "The Soviet Union and détente of the 1970s," *Cold War History* 8 (November 2008): 427–447.

[45] Anne de Tinguy, *U.S.–Soviet Relations during the Détente* (New York: Columbia University Press, 1999), 62–66.

[46] D. S., "2000 Scientists Petition for Soviet Colleagues," *Science* 185 (August 30, 1974): 769. See also, D. S., "U.S.–Soviet Pacts Threatened by Scientists," *Science* 180 (June 1, 1973): 934.

lecture in the United States.[47] But Soviet authorities argued the physicist was "groveling before the capitalist system" and refused to let him leave, even to receive his Nobel Peace Prize.[48] Jerome Wisner, president of MIT and Kennedy's science advisor, asked Kissinger to intervene with the Soviet ambassador and the NAS elected Sakharov an honorary member.[49] Congress passed the Jackson-Vanik amendment to deny most-favored nation status to countries limiting the right to emigrate, but maintained the détente initiatives.

Early Soviet actions and economic indicators were positive. The Communist Party used purchases of American science and technology to improve its stature at home and signed the Universal Copyright Convention in May of 1973 (previously they had copied around 70,000 pages annually from US scientific journals).[50] Dozens of American industries and universities signed cooperative agreements.[51] More than 2,200 scientists participated in the exchanges by 1975 and the United States became the Soviets second-largest Western trading partner.[52] The USSR also signed the Helsinki Final Act (1975), which supported human rights and required scientific and technical exchanges to further relations. As the delegates finalized the act in Helsinki, Soviet and American astronauts circled overhead in a symbolic gesture of friendship.

The Apollo–Soyuz Test Project symbolized peaceful coexistence in space, but it tested both sides on the ground. Linking US and Soviet

[47] D. S., "Sakharov Wants to Leave," *Science* 182 (December 14, 1973): 1116. Sakharov published *Progress, Coexistence and Intellectual Freedom* in 1968 and later argued that Soviet scientists failed to support him. See Andrei Sakharov, *Memoirs* (New York: Knopf, 1990), 516.

[48] TASS quoted in Theodore Shabad, "Soviet Disputes Sakharov Views," *New York Times* (July 13, 1973): 4.

[49] Jerome Wisner to Henry Kissinger, "[Visa for Andrei Sakharov]," (July 13, 1973): 1. DNSA Kissinger Telephone Conversation Collection, Item Number KA10418. Regarding the NAS, see Glenn E. Schweitzer, *Scientists, Engineers, and Two-track Diplomacy: A Half-Century of U.S.–Russian Interacademy Cooperation* (Washington, DC: National Academies Press, 2004), 9–10.

[50] Loren R. Graham, "Science in the Brezhnev Era," *Bulletin of the Atomic Scientists* (February 1982): 23–28. On the Soviets and copyrights, see House of Representatives, *Science, Technology and American Diplomacy*, 56.

[51] These included Bechtel, Boeing, General Dynamics, GE, Hewlett-Packard, Monsanto, Stanford, MIT, Brown, NYU, and the University of Chicago (among others). See United States Air Force Office of Special Investigations, "Soviet Overt Intelligence Collections in the United States," (January 1975): 1–31, esp. 11–12 and 16. DNSA Soviet Estimate Collection, Record # SE00481.

[52] Linda L. Lubrano, "National and International Politics in US-USSR Scientific Collaboration," *Social Studies of Science* 11 (November 1981): 451–480, esp. 457.

spacecraft in orbit required significant time and money and many American politicians, led by Senator William Proxmire (D-WI), considered the project either wasteful, a security risk or both. Soviet leaders, who normally refrained from announcing or televising launches, chafed at the access demanded by foreign journalists. The docking, which took place on July 17, 1975 and lasted for forty-four hours, proved spacecraft could rendezvous safely at speeds approaching 17,000 mph.[53] But the ASTP was always about politics: President Ford considered the famous "handshake in space" to be "indicative of the progress we have been making."[54] The experiment also humanized Cold War enemies: Soviet cosmonauts toured the Kennedy Space Center and Disney World, while American astronauts relayed the sounds of a party, complete with music, drinking, and girls, over the flight radio after disengaging.[55] Thomas Stafford, the US commander, related: "We had come a long way from flights along the Iron curtain, secret missile tests, and the moon race."[56] Yet the test project was an achievement without immediate application and the Soviets turned down a joint flight in 1976; with the United States focused on a potential shuttle, the communists had space to themselves.[57]

The Soviet Union capitalized on the lull in the American space program, which remained earthbound between the ASTP (1975) and the *Columbia* (1981). Cosmonaut Valentina Tereshkova, the first woman in space, frequented European capitals and international conferences as a goodwill ambassador. Interkosmos welcomed its first Soviet-bloc cosmonaut – a Czech who flew on the tenth anniversary of the Soviet crushing of the Prague Spring. Flights were consistently political: The Soviet program launched the first Asian in space (Tuan Pham – a North Vietnamese pilot who claimed to have downed an American B-52 during the Vietnam conflict) as well as Arnaldo Tamayo Mendez, an Afro-Cuban supporter of Fidel Castro (the Soviets pointed out the United States, with a sizable

[53] On the political nature of the ASTP, see Roger D. Launius, *NASA: A History of the U.S. Civil Space Program* (Malabar, FL: Krieger Publishing Company, 1994), 100. Launius was the Chief Historian at NASA at the time of publication. See also McDougall, *The Heavens and the Earth*, 431 and Debbora Battaglia, "Arresting Hospitality: The Case of the 'Handshake in Space'" *Journal of the Royal Anthropological Institute* (2012): S76–S89.
[54] Ford quoted in Jennifer Ross-Nazzal, "Détente on Earth and in Space: The Apollo-Soyuz Test Project," *OAH Magazine of History* 24 (July 2010): 29–34, quote on 33.
[55] Debbora Battaglia, "Arresting Hospitality," S76–S89.
[56] Stafford quoted in Ross-Nazzal, "Détente on Earth and in Space," 29.
[57] Tinguy, *U.S.–Soviet Relations during the Détente*, 70.

black population, had yet to send a black astronaut into space [African-American Guion Bluford orbited in 1983]). Brezhnev proudly stated: "the cosmonauts of the fraternal countries are working not only for science and for the national economy, they are also carrying out a political mission of immense importance."[58] As the United States prepared to launch the *Columbia*, familiar Cold War tensions resurfaced, ending most scientific exchanges.

Science and the Collapse of Détente

The Soviet approach to détente upset American businesses and worried US intelligence. At first, the desire to penetrate Soviet markets overcame intellectual property right concerns and American firms signed more than fifty agreements.[59] Within a few years, however, intelligence analysts reported many companies refused to provide the requested technology because "they regard Soviet compensation offers as unsatisfactory;" others rejected "Soviet restrictions on foreign ownership of production equipment and property."[60] Officials at all levels worried about Soviet duplicity. The NSF spoke against cooperative research in the Arctic, arguing, "The designation of such 'scientific enclaves' could provide Soviet military authorities with an excellent device for restricting the movements of US scientists."[61] The CIA closely tracked Soviet delegations "sampling" US satellite providers.[62] Air Force intelligence worried "unclassified literature" could aid Soviet programs, pointing out a Los Alamos report on electron-beam lasers could be purchased for $2.25.[63] The DOD initiated a study of technology transfer and recommended increased export controls. Congress held hearings on the dramatic increase in Soviet assistance worldwide: In the 1970s, the USSR signed

---

[58] Brezhnev quoted in Sheehan, *The International Politics of Space*, 56.

[59] de Tinguy, *U.S.–Soviet Relations during the Détente*, 67.

[60] CIA, "National Intelligence Daily Cable, CG NIDC 77–145C," (June 23, 1977). CIA Doc No/ESDN: CIA-RDP79T00975A030200010039-3.

[61] NSF Director H. Guyford Stevens, "[U.S. –Soviet Scientific and Technological Cooperation]," (April 3, 1973): 1–3. DNSA Presidential Directives (Part II), Record # PR01180.

[62] CIA, Directorate of Operations, "Soviet Interest in MARISAT Terminals/Arab Satellite System/Compromise on INTELSAT Membership," (November 20, 1975): 1–2. CIA Doc No/ESDN: 0005903876.

[63] United States Air Force Office of Special Investigations, "Soviet Overt Intelligence Collections in the United States," (January 1975): 1–31, esp. 11–12 & 16. DNSA Soviet Estimate Collection, Record # SE00481.

scientific agreements with seventy countries and established 170 technical training centers abroad, educating more than 400,000 workers.[64]

Soviet actions ended most exchanges. In 1978, the Soviet Union tried physicist Yuri Orlov and computer scientist Anatoly Scharansky for monitoring Soviet compliance with the Helsinki Accords. In response, the United States canceled four delegations and an EPA program.[65] The Soviet invasion of Afghanistan a few months later elicited a stronger response: President Carter cut back on high-tech exports, ended arms limitation talks, stopped excess grain sales, and canceled most official programs. When Sakharov publicly criticized the Soviet invasion, he was exiled to Gorky and kept under tight surveillance. Multiple scientific societies protested the punishment; polls showed many NAS and FAS members would opt out of future programs.[66] William Carey, head of the AAAS, observed, "détente through scientific exchanges is, at present, flat on its back."[67] Debate arose as to whether superpower relations should impede scientific cooperation.

The American scientific community, perhaps remembering the McCarthy era, objected to government interference with Soviet scientists' travel. At the Helsinki Scientific Forum, held only two months after the Afghan invasion, the American delegation criticized Soviet actions. Scientists and scientific societies widely supported the denunciation.[68] However, when the United States refused to give visas to Soviet physicists for a fusion conference in San Diego and pressured scientific societies to disinvite Soviet scholars, the American Physics Society and others protested US actions. A congressional resolution restricting private and industrial cooperation – rather than just official exchanges – came under fire from

---

[64] US Congress, House of Representatives, Committee on Science and Technology, Subcommitee on Science, Research and Technology, *Soviet Scientific and Technical Cooperation with Countries other than the United States*, 96th Cong., 1st sess., February 1979, 10.

[65] Lubrano, "National and International Politics in US–USSR Scientific Collaboration," 460.

[66] N. W., "Sakharov Protests Mount," *Science* 207 (March 14, 1980): 1186.

[67] Carey, quoted in US Congress, House of Representatives, Committee on Foreign Affairs, *United States Scientific and Technical Exchanges with the Soviet Union*, 96th Cong., 2nd sess., May 20, 1980, quote on 4.

[68] US Congress, House of Representatives, Committee on Foreign Affairs, *The Helsinki Forum and East–West Scientific Exchange*, 96th Cong., 2nd sess., January 31, 1980. Much of the publication consists of reprints from scientific societies and physics departments supporting US criticism of the USSR.

scientific societies, including the NAS.[69] As one scholar wrote at the time, "Scientists had accepted the use of science and technology as instruments of US and Soviet foreign policy, but only when this policy was conducted in ways that were compatible with the professional norms of American scientists."[70] The resolution never passed.

The collapse of détente temporarily ended official US/Soviet cooperation, but unofficial exchanges continued and Soviet attempts to acquire American science and technology intensified. In the early 1970s, Soviet intelligence established a directorate in New York City charged with evading export controls; KGB historian Vasili Mitrokhin identified thirty-two agents who penetrated dozens of high-tech companies and universities, including IBM, McDonnell-Douglas, Monsanto, MIT, and many others.[71] Agents also bugged communications between Brookhaven, Argonne, and other national laboratories. By the mid-1980s, the directorate proudly reported it returned 140,000 documents and 20,000 samples worth more than one billion rubles to the Soviet economy.[72] Of course, even as détente collapsed in the mountains of central Asia, the United States was in discussions with the People's Republic of China about satellites, Landsat data, recombinant DNA engineering, and a host of other benefits from their new partnership.

WALKING ON TWO LEGS

In the 1960s, the gap between the scientific and technological levels of China and those of the rest of the world was not very big. However, in the late 1960s and early 1970s, the scientific and technological levels of the rest of the world improved tremendously. All fields of science developed quickly. ... In 1975, I once said, China was fifty years behind Japan. ... If we do not take the newest scientific achievements as our starting points and create favorable conditions and try our best, I am afraid there is no hope for China.

Vice-Premier Deng Xiaoping (1977)[73]

---

[69] US Congress, *United States Scientific and Technical Exchanges with the Soviet Union*, 33. See also Anon., "Curb on Travel to Soviet Is Opposed by Scientists," *New York Times* (May 21, 1980): A16.

[70] Lubrano, "National and International Politics in US–USSR Scientific Collaboratio," 473.

[71] The activities of Directorate T are discussed in Christopher Andrew and Vasili Mitrokhin, *The Sword and the Shield: The Mitrokhin Archive and the Secret History of the KGB* (New York: Basic Books, 1999), 216–221.

[72] Ibid., 216.

[73] Deng Xiaoping quoted in Li Jie, "China's Diplomatic Politics and Sino–U.S. Relations," in Kirby, et al., *Normalization of U.S.–China Relations*, 56–89, quote on 79.

Our maturing science and technology cooperation with China, a cornerstone in our expanding relationship, is now in its eighth year and is our largest government-to-government program. Not a part of our foreign assistance program, science and technology cooperation is based on mutual benefit as are our other international exchanges. The Chinese have also added additional activities more attuned to their own interests on a reimbursable basis. We credit the doors opened by our successful science and technology program with contributing positively to the recent reforms made by the Chinese.

President Ronald Reagan (1986)[74]

US–Chinese normalization, like détente with the Soviet Union, relied upon scientific and technical exchanges. The Nixon and Carter administrations hoped exchanges could open China to American enterprise and stimulate political reform, while Chinese Communist Party members disagreed about the value of exchanges and the importation of foreign ideas and investment. Although normalization proceeded at many levels and through different emissaries, scientific exchanges provided a consistent cross-cultural bridge; even as the political path to normalization stagnated, scientific delegations and societies welcomed the People's Republic. By the late 1970s, the moderate wing of the party, led by Deng Xiaoping, opened the country to foreign commerce and research; in less than a decade, the PRC was America's largest bilateral science partner, a status maintained today. Indeed, in hindsight, the collapse of scientific relations after the communist revolution was the aberrant period in modern US–Chinese history.

The Collapse of US–Chinese Scientific Relations, 1950–1972

Science became part of Chinese and American relations in the early twentieth century. After World War I, the United States established the China Foundation with funds from the Boxer indemnity. According to the foundation's charter, the money had to be "devoted to the development of scientific knowledge and to the application of such knowledge to the conditions in China through the promotion of technical training, of scientific research, experimentation and demonstration, and training in

---

[74] Ronald Reagan quoted in *STAD (1986)*, vi.

science teaching."[75] Backed by more than $10 million, the foundation established the Academia Sinica and created ten research institutes, numerous medical colleges and the Amoy biological laboratory.[76] As the Japanese advanced on Nanking, the foundation's scholarships allowed the best students to escape to the University of California, Cornell, and other schools.[77] After the Japanese surrender, American technical programs around wartime bases in China led Zhou Enlai to request aid (see Chapter 3). Although the communist revolution ended official Chinese–American relations, researchers formed transnational communities: Chinese scientists who remained in the United States diversified the American community while Chinese scientists who returned introduced American university methods and cultures to PRC colleagues.[78]

Chinese state support for science grew steadily. The PRC established the Chinese Academy of Sciences (CAS) in 1950 and initiated scientific and technical cooperation with the Soviets shortly thereafter. Once biologists discarded Lysenkoism, the state founded institutes of genetics in Beijing and Shanghai. On the PRC's tenth anniversary, *Scientia Sinica* published a wide-ranging overview of Chinese scientific "victories," promising "China will undoubtedly emerge as one of the great nations pre-eminent in science and culture."[79] Agriculturists soon developed shorter, high-yielding varieties.[80] The Chinese space program, aided by

---

[75] Foundation charter quoted in E.V. Cowdry, "The China Foundation for the Promotion of Education and Culture," *Science* 65 (February 11, 1927): 150–151. Cowdry was an executive at the Rockefeller Institute for Medical Research

[76] A. M. Boring, "Summer Institute for Biological Research at Amoy, China," *Science* 72 (October 24, 1930): 429–430. See also Chungshee H. Liu, "The Advancement of Science in China during the Past Thirty Years," *Science* 98 (July 16, 1943): 47–51.

[77] Anon., "Scientific Awards in China," *Science* 86 (July 30, 1937): 95.

[78] Zuoyue Wang, "Transnational Science during the Cold War: The Case of Chinese/American Scientists," *Isis* 101 (2010): 367–377.

[79] Du Ruen-Sheng, "Great Progress Made in the Natural Sciences in China During the Last Decade," in *Scientia Sinica* 8 (November 1959): 1196–1217. Reprinted as Du Ruen-Sheng, "Communist Chinese Claims Regarding Scientific Progress in the Last Decade," *The Science Newsletter* 78 (December 10, 1960): 377–392, quote on 384.

[80] See Laurence Schneider, "Michurinist Biology in the People's Republic of China, 1918–1956," *Journal of the History of Biology* 24 (2012): 525–526; Laurence Schneider, "Learning from Russia: Lysenkoism and the Fate of Genetics in China, 1950–1986," in Denis Fred Simon and Merle Goldman, eds., *Science and Technology in Post-Mao China* (Cambridge, MA: Harvard University Press, 1989), 45–68. See also Jack R. Harlan, "Plant Breeding and Genetics," in Orleans, ed., *Science in Contemporary China*, 295–312., esp. 306–309.

Qian Xuesen (see Chapter 2), advanced rapidly.[81] A scientific symposium in Beijing attracted researchers from forty-four nations.[82] State support for physics resulted in a successful atomic test at Lop Nur in 1964; the following year, PRC biologists successfully synthesized bovine insulin (a first). Thus Chinese accomplishments were significant before the Cultural Revolution.

The impact of the Cultural Revolution varied among scientific fields and locations. Revolutionaries attacked the elitism of academic research and demanded "science for the masses," closing universities and forcibly transferring many researchers to the countryside.[83] Disruptions to academic work were severe. CAS institutes closed, although the state relocated defense-related research to the country's interior to guard against attack (foreign or domestic – the military defended Lop Nur from Red Guards).[84] Protected by the party, Chinese physicists beat the French to the thermonuclear bomb at the height of the revolution (1967). The Chinese Academy of Space Technology (CAST) opened in 1968 and launched the first "East is Red" satellite two years later, leading one historian to conclude, "the backbone of China's aerospace industry has roots in the Cultural Revolution."[85] Finally, mass science – connecting the laboratory to the factory and farm – may have benefited plant breeding and genetics, animal sciences, meteorology, seismology, entomology, and others.[86] Nonetheless, by the early 1970s, Mao and Zhou both

---

[81] See "Medieval Rockets to First Satellites," in Brian Harvey, *China in Space: The Great Leap Forward* (Chichester, UK: Springer-Praxis Publishing, 2013), 29–50.

[82] A participant in the 1964 conference as well as the later exchanges was Zhou Peiyuan, "Scientific and Technological Exchanges Between China and the United States," *Science* 205 (September 28, 1979): 1354–1355.

[83] Richard P. Suttmeier, "Science Policy and Organization," in Leo A. Orleans, ed., *Science in Contemporary China* (Stanford: Stanford University Press, 1980), 31–52.

[84] Defense-related fields included nuclear physics, aeronautics, electronics, optics and communications. On the relocation of defense research, see Darryl E. Brock, "The People's Landscape: Mr. Science and the Mass Line," in Chunjuan Nancy Wei and Darry E. Brock, *Mr. Science and Chairman Mao's Cultural Revolution: Science and Technology in Modern China* (New York: Lexington Books, 2014), 41–117, esp. 80–83. On the military protecting Lop Nur from Red Guards, see John Wilson Lewis and Xue Litai, *China Builds the Bomb* (Stanford: Stanford University Press, 1988), 202–204.

[85] Stacey Solomone, "Space for the People: China's Aerospace Industry and the Cultural Revolution," in Wei and Brock, *Mr. Science and Chairman Mao's Cultural Revolution*, 233–250, quote on 239.

[86] Schneider, "Science, Technology and China's Four Modernizations," 291–303. See also Suttmeier, "Science Policy and Organization," in Orleans, ed., *Science in Contemporary China*, 31–52.

worried about the impact of the revolution on Chinese education; universities reopened and Zhou reinstated all CAS members.

The CIA monitored Chinese scientific progress throughout. As the PRC approached completion of the bomb, for example, requests for intelligence on Chinese science and technology jumped from 23 to 278 per year.[87] DCI McCone instructed Albert Wheelon, a CIA deputy, to "maximize" activity on Chinese research (Wheelon requested help with translation).[88] Analysts reported the PRC had a shortage of well-trained Chinese scientists (about three thousand doctorates in total), but recounted active attempts to acquire scientific training and technology from allies, including France, Great Britain, and Japan.[89] The Chinese also launched Point Four-style assistance: Intelligence tracked Chinese aid spreading throughout the southern hemisphere, with more than 20,000 technicians working in Africa, South America, and the Middle East.[90] In 1966, the NAS and Secretary of State Rusk offered to host Chinese scholars, but the PRC rejected the offer, eventually canceling ongoing ambassadorial contacts in Warsaw.[91]

The benefits of rapprochement proved irresistible to both sides. The Nixon administration believed normalization could exploit the discord between Brezhnev and Mao and spur the Chinese to urge a compromise on their Vietnamese compatriots.[92] Normalization offered Chinese leaders a means to resolve tension over Taiwan and mitigate growing security concerns with the Soviet Union. Additionally, a moderate clique

---

[87] Albert D. Wheelon (Deputy Director for Science and Technology) to DCI, "Status of the DD/S&T Scientific and Technical Intelligence Effort on China," (October 4, 1965). CIA Doc No/ESDN: CIA-RDP80B01676R000300020005–4.

[88] Ibid.

[89] CIA, Office of Current Intelligence, "Special Report: Science and Technology in Communist China," (January 28, 1966): 1–9. Available via DDRS (Declassified Document Retrieval System).

[90] Countries included Somalia, Ethiopia, Sierra Leone, Chile and Iraq. See CIA Directorate of Intelligence, "Special Report: Chinese Aid in the Third World," (June 30, 1972):1–7. CIA Doc No/ESDN: RDP85T00875R001500040025–1.

[91] Jonathan D. Spence, *The Search for Modern China*, 2nd en. (New York: W. W. Norton & Company, 1999), 596. See also Wang, *In Sputnik's Shadow*, 193.

[92] For a recent multi-perspective work on normalization, see William C. Kirby, Robert S. Ross and Gong Li, *Normalization of U.S.–China Relations: An International History* (Cambridge, MA: Harvard University Press, 2005). See also Michael Schaller, "Détente and the Strategic Trangle, Or, 'Drinking your Mai Tai and Having Your Vodka, Too,'" in Robert S. Ross and Jiang Changbin, *Re-Examining the Cold War: U.S.–China Diplomacy, 1954–1973* (Cambridge, MA: Harvard University Press, 2001), 361–389. Finally, an excellent chronological rendering is Annapurna Nautiyal, "Kissinger and the China Opening," *China Report* 22 (1986): 463–471.

within the party, led by Premier Zhou Enlai and Deng Xiaoping, considered access to foreign science and technology essential. So, nearly twenty-five years after he first requested aid from General Marshall, Zhou indicated Chinese willingness to reopen talks after Nixon's inauguration.

Scientific Exchanges and Revolutionary Resistance

US–Chinese diplomacy proceeded step-by-step. President Nixon promised open communications with "all countries" in his inaugural address; Mao had the speech translated and reprinted in the *People's Daily*.[93] President Nixon relayed his support for rapprochement through Pakistani President Khan later that year and canceled American naval patrols in the Taiwan straits. Low-level talks resumed in Warsaw with Zhou Enlai. In his foreign policy message to Congress in February 1970, Nixon stated his desire to normalize relations with China and eased travel and trade restrictions. Mao suggested repairing relations in talks with foreign emissaries and American journalist Edgar Snow. Finally, in March 1971, Nixon removed the last restrictions on travel; a month later the PRC invited the American ping-pong team, then touring Japan, to play exhibition matches on the mainland. US corporations followed for a trade fair in Guangzhou. In October, the PRC took its seat in the United Nations without American objection; President Nixon traveled to Shanghai in February to endorse the new relationship.

Normalization altered American relations across Asia. Taiwanese nationalism (and its Intelsat member status) became subordinate to America's "One China" policy, while the Japanese realized the United States had another partner in the region. To build US–Chinese relations, the *Shanghai Communiqué* encouraged scientific and cultural exchanges:

The two sides agreed that it is desirable to broaden the understanding between the two peoples. To this end, they discussed specific areas in such fields as science, technology, culture, sports and journalism, in which people-to-people contacts and exchanges would be mutually beneficial.

Shanghai Communiqué (February 28, 1972)

---

[93] Gong Li, "Chinese Decision Making and the Thawing of U.S.–China Relations," in Ross and Changbin, *Re-Examining the Cold War*, 321–360, esp. 333. See also Margaret MacMillan, "Nixon, Kissinger and the Opening to China," in Fredrik Logevall and Andrew Preston, eds., *Nixon in the World: American Foreign Relations, 1969–1977* (New York: Oxford University Press, 2008), 107–125.

The document heralded a remarkable thawing of the Cold War. Within months, Nixon and Brezhnev signed the SALT Accords, ABM treaty, and approved the Apollo–Soyuz Test Project. Less than a year later, the Paris Peace Accords ended the American presence in Vietnam. US–Chinese trade jumped from $4.9 million (1971) to $92.5 million (1972).[94] Only Watergate and the OPEC oil embargo intruded, curtailing Nixon's plan for rapprochement. But even as the political path to normalization stagnated, scientific and technical exchanges continued the process.

Exchanges were the most tangible sign of US/PRC relations. Chinese interests were apparent from the start: Between 1972 and 1975, eighteen of twenty scholarly delegations focused on science and technology.[95] PRC representatives met with COMSAT executives and toured the Stanford Linear Accelerator, Fermilab (the National Accelerator Laboratory), Bethlehem Steel, Boeing, and other industrial research centers; delegations also met with academics in agriculture, medicine, and mathematics. US intelligence reported the Chinese desired American, rather than French or Soviet, computers.[96] American scientists, organized by the National Academy of Sciences, flocked to China, often returning impressed with mass science.[97] H. T. Tien, a Chinese-American biophysicist, was struck by the up-to-date equipment at the Institute of Molecular Biology and the upbeat attitudes of Chinese state researchers.[98] Nicholas Bloembergen, leader of the physics delegation, felt Chinese purchase of an accelerator

---

[94] Li Jie, "China's Diplomatic Politics and Sino–U.S. Relations," 121.

[95] Douglas P. Murray, "Exchanges with the People's Republic of China: Symbols and Substance," *Annals of the American Academy of Political and Social Science* 424 (March 1976): 29–42, esp. 33.

[96] CIA, Directorate of Science and Technology, Office of Scientific Intelligence, "Computer Research and Development in the People's Republic of China [OSI-STIR/75-2]," (February 1975): 1–15. CIA Doc No/ESDN: 0000712269.

[97] Between 1972 and 1979, the Committee on Scholarly Communication with the People's Republic of China (CSCPRC) sponsored 35 delegations to China and hosted 42 delegations from China. The official CSCPRC publication is Orleans, ed., *Science in Contemporary China*. For other positive contemporary analyses, Ethan Singer and Arthur W. Galston, "Education and Science in China," *Science* 175 (January 7, 1972): 15–23; Emily Smith, et al., "Walking on Two Legs: A Panel Discussion of Science Policy in the People's Republic of China," *Bulletin of the American Academy of Arts and Sciences* 28 (November 1974): 26–41 and Leo A. Orleans, "China's Experience in Population Control: The Elusive Model," *World Development* 3 (July–August 1975): 497–525. For a more recent scholarly analysis, see Sigrid Schmalzer, "On the Appropriate Use of Rose-Colored Glasses: Reflections on Science in Socialist China," *Isis* 98 (September 2007): 571–583.

[98] H. T. Tien, "Biophysical Research in the People's Republic of China," *Biophysics* 15 (June 1975): 621–631. Tien, from MSU, visited China as part of the official exchanges.

would "enable the PRC to catch up on several important high-technology developments . . . and make future exchanges with China more meaningful."[99] But the informal exchanges exacerbated domestic Chinese disagreements.

Engagement with the United States provoked intra-party factionalism. The politburo split over the acceptance of foreign influences and scientific methods: the "Gang of Four," led by Jiang Qing (Mao's last wife), wanted to reject American overtures; Zhou Enlai and Deng Xiaoping considered American science and technology useful to development.[100] Such stark divisions magnified minor incidents. Jiang Qing, for example, did not want an American satellite to broadcast Nixon's trip to China, while the acceptance of a glass snail – a gift from the Corning Glass Company – led to an official inquiry into the delegation's "slavishness" to foreign influences (the inquiry resolved to open a dialogue with Corning). At the Fourth People's Congress in 1975, Deng stressed basic research, the reinstatement of scientific professionalism, and imported technology; Zhou reintroduced his "Four Modernizations," which emphasized science and technology (the Gang of Four responded by postponing a national science conference).[101] As Deng Xiaoping replaced an ailing Zhou, reactionaries intensified their campaign, accusing him of ignoring "experience as the basis for developing science in favor of closeting scientists behind closed doors."[102] Briefly removed from power, Deng seized the opportunity that came after Mao's death and the Gang of Four's arrest.

Scientific reform and openness were central to Deng Xiaoping's critique of the Cultural Revolution. He had recognized the impact of science and technology, regardless of origin, in the early 1960s. By the 1970s, one daughter was a medical student; his youngest son, a former

---

[99] Nicholas Bloembergen quoted in Schneider, "Science, Technology and China's Four Modernizations," 300.

[100] See, for example, Li Jie, "China's Diplomatic Politics and Sino-U.S. Relations," in Kirby, et al., *Normalization of U.S.–China Relations*, 56–89, esp. 62–73. See also Gong Li, "Chinese Decision Making and the Thawing of U.S.–China Relations," in Ross and Changbin, *Re-Examining the Cold War*, 321–60 and Suttmeier, "Science Policy and Organization," in Orleans, ed., *Science in Contemporary China*, 31–52.

[101] Deng introduced "three documents" on the importance of revitalizing Chinese S&T. See "The Scientists and Deng Xiaoping," in Merle Goldman, *China's Intellectuals: Advise and Dissent* (Cambridge, MA: Harvard University Press, 1981). See also Tony Saich, *Chinese Science Policy in the 80s* (Manchester: Manchester University Press, 1989), 7–9.

[102] Jonathan D. Spence, *The Search for Modern China*, 2nd edn. (New York: W. W. Norton & Company, 1999), 611.

Red Guard, majored in physics after his father helped reopen Beijing University (and eventually did postgraduate work at the University of Rochester).[103] Rehabilitated in 1977, Deng was outspoken about accepting foreign methods and results, telling the head of the Chinese Academy of Sciences: "In the past, we did not absorb advanced knowledge from foreign countries. The developed countries attach importance to scientific achievements ... Why not absorb these results? ... For our whole country to catch up and surpass the advanced countries of the world, scientific study is the prerequisite."[104] He went further at the long-awaited National Science Conference in March 1978, opening with a pointed attack:

Science and technology are a kind of wealth created by all mankind. Any nation or country must learn from the strong points of other nations and countries, from their advanced science and technology. It is not just today, when we are scientifically and technically backward, that we need to learn from other countries; after we catch up with the advanced world levels in science and technology, we will still have to learn from the strong points of others ... The Gang of Four made the absurd statement, "The more knowledgeable, the more reactionary." They said they "preferred laborers with no culture' and they boasted as a "model of being red and expert" an ignorant counterrevolutionary clown who handed in a blank examination paper. On the other hand, they vilified as being "white and expert" good comrades who studied diligently and contributed to the motherland's cause of science and technology. This reversal of right and wrong and of ourselves and the enemy seriously muddled people's minds for a time.[105]

The intellectual counter-revolution was soon over. Science was once again a "productive force" and scientists a source of national pride (Deng suggested the PRC needed 800,000 new researchers alongside eighty-eight new universities). As a welcome aside, Chinese scientists could once again study achievements from anywhere in the world, were not beholden to political ideology and would not have their results labeled "bourgeois."[106] Eight months later, the politburo approved a

---

[103] Richard Evans, *Deng Xiaoping and the Making of Modern China* (New York: Penguin Books, 1997), esp. 223–225.

[104] Xiaoping quoted in Li Jie, "China's Diplomatic Politics and Sino-U.S. Relations," in Kirby, et al., *Normalization of U.S.–China Relations*, 56–89, quote on 79.

[105] Deng Xiaoping, "Speech at the Opening Ceremony of the National Science Conference," (March 18, 1978) reprinted in Orleans, *Science in Contemporary China*, 535–546, quotes on 539 and 540.

[106] Saich, *China's Science Policy in the 1980s*, 22. See also Spence, *The Search for Modern China*, 619.

modernization plan opening China to global commerce. The United States, and others, took notice.

Markets and Anti-communist Resistance

Competition for Chinese markets helped restart normalization. Even as interest in the political process ebbed, the Ford administration signed off on the sale of advanced computers and welcomed a delegation of Chinese scientists.[107] Chinese officials, meanwhile, visited France and West Germany in 1977 in search of advanced technology.[108] Deng signed a $10 billion dollar industrial agreement with Japan in July (putting Americans on notice about Japanese willingness to work with other regional partners).[109] The Carter administration, not yet six months old, recognized the opening and sent Secretary of State Cyrus Vance to restart normalization talks, stressing an informal and apolitical diplomatic track centered on science, technology, and commerce.[110] In January of 1978, Frank Press, Carter's science advisor, wrote the president about the potential benefits:

Chinese purchases of foreign technology are likely to begin soon that could top the billion dollar mark within the first twelve months ... [the Chinese desire a] proton accelerator ranking in size to the world's largest ... Chinese scientists being sent for training in Western European laboratories and Peking indicate that foreign training programs will be arranged in connection with purchases ... Western European Nations and Japan are actively seeking trade, training and exchange links with China that may preempt deferred US moves [underlining in original].[111]

---

[107] Henry Kissinger to United States Liaison Office (China), "CDC Computer for PRC," (October 15, 1976): 1–2. DNSA China and the U.S. Collection, Record # CH00431. On the Ford visit, see Zhou Peiyuan, "Scientific and Technological Exchanges between China and the United States," 1354–1355.

[108] Li Jie, "China's Diplomatic Politics and Sino–U.S. Relations," in Kirby, et al., *Normalization of U.S.–China Relations*, 56–89. See also Dickson, *The New Politics of Science*, 185.

[109] Spence, *The Search for Modern China*, 619.

[110] Benjamin Huberman, Associate Director of OSTP, testified that, "By 1977, it had become clear that science and technology would be a key element of China's modernization effort." The first track was normalization, the second was consultation on global matters and the third was science, technology and commerce. See Congress, House of Representatives, Committee on Science and Technology, *United States–China Science Cooperation: Hearings before the Subcommittee on Science, Research and Technology of the Committee on Science and Technology*, 96th Cong., 1st sess., May 7, 8, 10; June 22, 1979, 1–284, esp. 5.

[111] Frank Press, "Memorandum for the President: An Approach to the People's Republic of China through Science and Technology," (January 23, 1978): 1–2: Available via DDRS.

The NSC tracked PRC scientists working with a French oceanography survey and covered the National Science Conference in March.[112] Three months later, in June of 1978, Press led "the highest-level delegation of S&T officials the U.S. has ever sent abroad," as America sent its best and brightest to China.[113] Notables included the directors of the NSF, NIH, NASA, and the USGS, as well as advisors from the NSC, OSTP, State, Commerce, Agriculture, and Defense. The list of scientific enticements was equally notable: space cooperation (including Landsat data and weather satellites), energy research and assistance in genetic engineering.[114] Carter wrote to his advisor: "I do not want you to go as Santa Claus: Be sure exchanges are equitable and mutually beneficial."[115]

Scientific and technical relations expanded in the months before normalization. In November of 1978, the president charged State, Defense, and OMB with overseeing US–China relations.[116] The directive required the PRC pay for all exchanges beneficial to China and approved reimbursable launch services for a telecommunications satellite as well as genetic samples for recombinant DNA research.[117] At the same time, the USDA agreed to cooperate on "germ plasm (seed research and selection), biological control of pests, livestock and veterinary science, and agricultural education and research management methods."[118] Carter also approved China's purchase of a French nuclear reactor: Chinese refusal to sign the non-proliferation act prevented American firms from selling directly, but the French firm Faratome built plants under a licensing agreement with Westinghouse, and thus the sale required US approval.[119] In December, the United States and China announced the normalization of relations and an agreement on space technology; the two

---

[112] National Security Council, "Memorandum for Zbigniew Brezezinski," (February 7, 1978). Available via DDRS.

[113] Frank Press, "Science and Technology Delegation to China" (June 26, 1978). DNSA China and the U.S. Collection, Record # CH00444.

[114] Ibid.   [115] Carter, hand-written notes on Ibid.

[116] Presidential Directive, PD/NSC 43, "U.S.–China Scientific and Technological Relationships," (November 3, 1978). DNSA China and the U.S. Collection, Record # CH00447.

[117] Ibid.

[118] Robert Berglund, US Secretary, Department of Agriculture, "Understanding on Agricultural Exchange between the United States of America and the People's Republic of China," (November 1978): 1–3. DNSA China and the U.S. Collection, Record # CH00446.

[119] S.K. Ghosh, "Sino–French Nuclear Cooperation," *China Report* 15 (1979): 3–6.

countries signed a broader agreement on US–China scientific cooperation a month later.[120]

The dramatic turn in relations sparked congressional inquiry. The House Committee on Science and Technology held hearings on "U.S.–China Scientific Cooperation" in May and June of 1979.[121] Concern for America's international scientific standing surfaced throughout. Representative Allen Ertel (D-PA), chair of the task force, opened:

We must consider the simple fact that in the competitive international environment of the next twenty years leading up to year 2000, much of America's advantage and strength will be directly attributable to our leadership in science. Our continuing industrial innovativeness and our growing agricultural productivity, and deriving from these, our industrial productivity, our balance of trade, and ultimately our economic health and world position, will, to a very significant extent, be controlled by our leadership in science. ... Our cooperation with China should not be seen as a passing of the torch of scientific leadership. Rather, we hope it will serve to bring a new partner into the world scientific community.[122]

Of course, the PRC was part of the world scientific community with or without American blessing: The Chinese, according to OSTP associate director Benjamin Huberman, "were already reaching agreement with France, Japan and the Federal Republic of Germany."[123] Ultimately, the task force supported furthering US–Chinese scientific cooperation for a variety of reasons, including, adding "momentum to the pace of normalization, "scientific gain in certain areas," and access to "an expanding commercial market."[124] In essence, the American desire for Chinese markets overcame lingering anticommunism since the PRC already had access to most science and technology.

Scientific and Technical Relations after Normalization

Cooperation increased after normalization. In July, when the hearings concluded, the Chinese leased seismic survey ships from US oil firms,

[120] Fang Yi, Chairman, State Scientific and Technological Commission, "[Chinese Letter regarding China-U.S. Agreement on Cooperation in Science and Technology; English Language Translation Attached]" (January 31, 1979): 1–3. DNSA China and the U.S. Collection, Record # CH00451.

[121] US Congress, House of Representatives, Committee on Science and Technology, *United States–China Science Cooperation: Hearings before the Subcommittee on Science, Research and Technology of the Committee on Science and Technology*, 96th Cong., 1st sess., May 7, 8, 10; June 22, 1979, 1–284.

[122] Ibid., 2 and 3.   [123] Ibid., 5.   [124] Ibid., 5 and 6.

negotiated the purchase of a communications satellite and a Landsat ground station, and initiated cooperation in high energy physics and the construction of a synchrotron. In preparation for the first US–China Joint Commission on Scientific and Technological Cooperation, President Carter told Press to "try to obtain maximum quid pro quo."[125] Cooperation soon involved nearly every federal department: the EPA approved programs in environmental sciences (air and water quality, solid waste management, environmental health, and toxic substances control); the Department of Transportation worked on hazardous materials and marine safety; and the NSF developed "programs in basic science cooperation. Areas of special interest for programs include R&D management, science policy, astronomy, botany, natural products chemistry, and paleontology."[126] Only the Department of the Interior and the Smithsonian had yet to put forward proposals. A second wave of American scientists washed upon China's shore.

The changes impressed the visitors. The Cultural Revolution was past: William Carey, executive officer of the AAAS, reported the Chinese "were disarmingly frank in assessing the damage visited upon education and science between 1965 and 1975."[127] Philip Abelson, editor of *Science*, confirmed the experience: "At virtually every university or institute that the AAAS group visited, the briefings included a denunciation of the Gang of Four ... The Gang of Four were blamed for the severe disruption to research that occurred during the Cultural Revolution."[128] *Radio Beijing*, for example, accused the clique of having "widely disrupted scientific studies, undermined the Party's leadership over scientific and technological work, distorted the Party's principles and politics, strangled scientific experimentation and ignored basic theoretical research."[129] Abelson was also "impressed with the people we met. They were informed about their subject fields, aware of relevant material that had appeared in the

[125] Jimmy Carter, handwritten notes on Frank Press, "Memorandum for the President: Technological Relationships with China (July 27, 1979):1–3. Available via DDRS.

[126] Harold D. Neeley, USAF, Joint Chiefs of Staff, "U.S.–China Scientific and Technological Relationships," (September 5, 1979): 1–3. DNSA China and the US Collection, Record # CH00467.

[127] William D. Carey, "AAAS Board Visit to China – A Brief Report," *Science* 203 (February 9, 1979): 533–535.

[128] Philip H. Abelson, "Education, Science and Technology in China," *Science* 203 (February 9, 1979): 505–509, quote on 505.

[129] *Radio Beijing* quoted in S. K. Ghosh, "Sino-French Nuclear Cooperation," *China Report* 15 (1979): 3–6, quote on 5.

literature, and enthusiastic and energetic in their approach to science."[130] State scientists received preference in housing, were well-paid and given cash rewards for exceptional work.[131] Even though average per capita income lingered around $340 in 1979, China aimed to be a leader in global science: estimates for the Four Modernizations ranged near $600 billion; outside Beijing, near Ming dynasty tombs, a fifty-billion volt photon synchrotron was already under construction (consultants included Brookhaven, Argonne, and Lawrence-Livermore national labs).[132] Perhaps the quickest impact came in agriculture.

American and Chinese agricultural interests coincided: The PRC wanted to improve harvests and the United States had extensive experience in opening markets through agricultural assistance. In 1973, the PRC signed contracts with American and European companies to build fertilizer factories.[133] The following year, the NAS and officials from the Rockefeller Foundation, including Norman Borlaug, visited laboratories, exchanged rice samples and heard about mass science and communes.[134] The Chinese purchased 15,000 tons of Mexican seed stock and began crossing Mesoamerican and Chinese wheat.[135] A chemical industry grew alongside. The American Entomological Society visited in 1975 to learn about Chinese pest-control techniques because they minimized environmental impact; within a decade, however, the PRC was a leading importer of pesticides.[136] Normalization also led to collaboration on hybrid rice: At the urging of the World Bank and Rockefeller Foundation, the PRC joined CGIAR and established the China Rice Research Institute in Hangzhou in 1981.[137] As in India, the Green Revolution – industrial

[130] Abelson, "Education, Science and Technology in China," 506.
[131] Barbara J. Culliton, "Science in China," *Science* 206 (October 26, 1979): 426–428 and 430, esp. 427.
[132] Barbara J. Culliton, "China to Build Synchrotron Near the Ming Tombs," *Science* 206 (October 26, 1979): 428–29.
[133] Sigrid Schmalzer, *Red Revolution, Green Revolution* (Chicago: University of Chicago Press, 2016), 13.
[134] For more detail, see National Academy of Sciences, *Plant Studies in the People's Republic of China: A Trip Report of the American Plant Studies Delegation* (Washington, DC: National Academy of Sciences, 1975)
[135] Jack R. Harlan, "Plant Breeding and Genetics," in Orleans, ed., *Science in Contemporary China*, 306–309.
[136] Sigrid Schmalzer, "Insect Control in Socialist China and the Corporate United States: The Act of Comparison, the Tendency to Forget, and the Construction of Difference in 1970s U.S.–Chinese Scientific Exchange," *Isis* 104 (June 2013): 303–329.
[137] XiaoBai Shen, "Understanding the Evolution of Rice Technology in China – From Traditional Agriculture to GM Rice Today," *Journal of Development Studies* 46

agriculture, hybrid seeds, and chemical fertilizers and pesticides – reinforced US–Chinese relations.

American diplomacy aided China's entrance into the global community. Deng Xiaoping's historic 1979 visit to Washington and the Houston Space Center were the first images beamed by satellite to Chinese television.[138] Scientific and technical agreements allowed each party to determine intellectual property rights and China joined the World Intellectual Property Organization in 1980.[139] Chinese students began immigrating to the United States for education, often focusing on scientific and technical subjects – the State Department offered more than 50,000 visas in the first eight years.[140] Believing it might demonstrate a pro-American posture, the PRC launched a military campaign against North Vietnamese communists in Cambodia; remarkably, the United States counseled restraint.[141] Clearly, the world was changing.

Science was central to the shifts in American relations with China and the Soviet Union in the 1970s. Scientific cooperation created personal, institutional, and financial linkages; superpower competition in science and technology moderated military competition and established a status quo. Satellites showed launches and presidential visits, connecting citizens across national boundaries. Exchanges maintained public relations even during tense political periods. Yet science diplomacy was not limited to superpower relations; the final changes in American scientific relations occurred throughout the Middle East after the OPEC oil embargo.

PETROSCIENCE

We are heading toward the most acute shortages of energy since World War II... Let us unite in committing the resources of this Nation to a major new

---

(2010): 1026–1046. See also Schmalzer, *Red Revolution, Green Revolution* and the Institute's website: www.chinariceinfo.com/english/.

[138] Spence, *The Search for Modern China*, 624.

[139] China–United States Joint Commission on Scientific and Technological Cooperation, "Agreement between the Government of the United States of America and the Government of the People's Republic of China on Cooperation in Science and Technology," (January 31, 1979): 1–5. DNSA China and the US Collection, Record # CH00454.

[140] Leo A. Orleans, *Chinese Students in America: Policies, Issues and Numbers* (Washington, D.C.: National Academy Press, 1988), 9. In 1983, for example, 5 percent life sciences, 5 percent mathematics, 4 percent health sciences, 14 percent physical sciences, 23 percent engineering, 13 percent computer science.

[141] The PRC also strongly disliked the new Soviet–Vietnamese relationship and Vietnamese claims throughout Southeast Asia. Xiaoming Zhang, "Deng Xiaoping and China's Decision to go to War with Vietnam," *Journal of Cold War Studies* 12 (Summer 2010): 3–29.

endeavor, an endeavor that in this Bicentennial Era we can appropriately
call "Project Independence."... Let us set as our national goal, in the spirit
of Apollo, with the determination of the Manhattan Project, that by the end
of this decade we will have developed the potential to meet our own energy
needs without depending on any foreign energy sources."

President Richard Nixon, "Address to the Nation on
Energy Shortages" (1973)[142]

But at least two other important considerations weighed heavily in the
decision to launch a large-scale nuclear power program. First, Iran was
under substantial pressure – or at least it felt that it was under substantial
pressure – to "recycle" its newly increased petrodollar earnings through
purchases of massive quantities of goods and services from the major
Western industrialized countries. Second, the royal family and others close
to the regime...stood to earn substantial sums from the purchase of each of
the multi-billion dollar reactors. ...commissions on the initial reactor purchases amounted to an astounding 20% of the total contracts, or some
several hundred million dollars per reactor.

Bijan Mossavar-Rahmani, Iranian Energy Company Executive
[from abroad] (1980)[143]

The Yom Kippur War and OPEC embargo altered US scientific and
technical relations across the Middle East. Reaction to the embargo was
swift. OECD established an international agency (1974) to coordinate
energy policy, research, and technology. Additionally, because the United
States could not increase production to offset cutbacks, allies diverted
non-Arab oil to nations on the embargoed list and shipped Arab oil to
those on the preferred list.[144] At home, President Nixon announced
Project Independence, guaranteeing "Americans will not have to rely on
any source of energy beyond our own" by the end of the decade.[145] With
the Manhattan and Apollo projects as examples, the Nixon administration promised to use American science and technology to solve the energy
crisis. Domestically, independence meant passing the trans-Alaskan

[142] President Nixon, "Address to the Nation about Policies to Deal with the Energy
Shortages" (November 7, 1973). Available online at: www.presidency.ucsb.edu.
[143] Bijan Mossavar-Rahmani, "Iran's nuclear power programme revisited," *Energy Policy*
(September 1980): 189–202, quote on 192.
[144] Daniel Yergin, *The Prize: The Epic Quest for Oil, Money and Power* (New York:
Simon & Schuster, 2009), 606.
[145] President Nixon, "Address to the Nation About National Energy Policy," (November
25, 1973), available online at: www.presidency.ucsb.edu/ws/?pid=4051.

pipeline, mandates to reduce energy consumption, and a reorganization of American research. In the Middle East, the United States established research programs with Israel and initiated joint commissions to entice Saudi Arabia and Iran to purchase American assistance to recover American payments for oil.

Scientific and technical assistance promised to fix America's balance-of-payments problem after the embargo. With the price of oil quadrupled, the United States ran a significant deficit with oil-producing nations. Conversely, oil-producing nations had an abundance of riches as higher oil prices flooded their countries with dollars. Determined to recycle dollars back into the American economy, the Treasury Department created Joint Economic Commissions (JEC) to sell bonds, promote American exports, and offer scientific and technical goods and services to oil-producing states. Iran and Saudi Arabia, the American bulwarks in the region, were the primary targets. Both nations looked to modernize and so the familiar combination of arms and assistance remained an effective enticement. Only this time, host nations, rather than the United States, would pay for assistance and it would have to be sold. The leadership of the commissions reflected their financial focus: Whereas previous technical programs were run through the State Department or USAID, the Iranian and Saudi Arabian commissions were unique in being overseen by the Department of the Treasury.[146] Once again, consider Iran.

US–Iranian Relations from the Embargo to the Revolution

The embargo increased American insistence on assisting Iran. As one of the first Point Four participants, the shah recognized the value of US aid. But the State Department reported the Iranians were "not especially interested" in another agreement at their expense.[147] Financial pressure and the desire for modernization eventually won out: In 1974, a new US–Iranian joint commission funded programs in agriculture, geology, nuclear physics, and engineering. Each area involved multiple federal agencies: "agriculture," for example, included the Department of the Interior (for water management), the TVA (for fertilizer production),

---

[146] David E. Spiro, *The Hidden Hand of American Hegemony: Petrodollar Recycling and International Markets* (Ithaca, NY: Cornell University Press, 1999), 89.
[147] Robert S. Ingersoll (Deputy Secretary) to Henry A. Kissinger (Secretary of State), "U.S.–Iran Joint Commission [Organizational Chart Attached]," (November 7, 1974): 1–8, quotes on 3 and 4. DNSA Iran Collection, record # IR00912.

NASA/NOAA (for satellite data and construction of an earth station), and AID (which partnered with the USDA). The National Science Foundation worked alongside to improve technology transfer.[148]

Opportunities for American investment grew. In its first year, the commission secured $3 billion in contracts for American firms.[149] DuPont, for example, established the Polyacryl Iran Corporation, then the largest American joint venture, to develop synthetic fibers (polyester and acrylic).[150] Frederick Seitz, former president of the Rockefeller Foundation and the NAS, spent two years establishing a biomedical center in Shiraz.[151] Iran and the United States signed a $15 billion dollar, five-year bilateral trade agreement in 1975; the package included eight nuclear power plants.[152] Two years later the Commerce Department advertised opportunities in chemical manufacturing and health, such as equipment and personnel for seventy new operating rooms.[153] The *New York Times* later reported, "Few governments have spent so much so fast to shove a modest agrarian economy into the technological age."[154] It was also a martial age: By the mid-1970s, Iran consumed half of American arms sales abroad.[155]

American administrations supported Iranian nuclear development until the revolution. US/Iranian atomic cooperation began with Eisenhower's "Atoms for Peace" (see Chapters 2 and 3). Within a decade, the United States supplied a five-megawatt reactor and helped establish the Nuclear Research Center at Tehran University; Iran signed the non-proliferation treaty in 1968 and later accepted IAEA safeguards.[156] After the embargo, the commission encouraged Iran to invest millions in an American uranium enrichment project led by Bechtel.[157] Similar talks began with France and West Germany. Yet the proposed contracts, like

[148] Ibid.   [149] Spiro, *The Hidden Hand of American Hegemony*, 88.

[150] Regina Lee Blaszczyk, "Synthetics for the Shah: DuPont and the Challenges to Multinationals in 1970s Iran," *Enterprise & Society* 9 (December 2008): 670–723.

[151] Frederick Seitz, "The Role of Universities in the Transnational Interchange of Science and Technology for Development," *Technology in Society* 4 (1982): 33–40.

[152] For an excellent overview of the "boom," see "The Triumph of Repression," in Bill, *The Eagle and the Lion*, 183–215, esp. 204.

[153] Department of Commerce, Bureau of International Commerce, *Iran: A Survey of U.S. Business Opportunities* (Washington, DC: Government Printing Office, 1977), 4.

[154] Peter T. Kilborn, "Iranian Festival Is Over for American Business," *New York Times* (January 17, 1979): A10.

[155] Yergin, *The Prize*, 626.

[156] Ivanka Barzashka and Ivan Olerich, "Iran and Nuclear Ambiguity," *Cambridge Review of International Affairs* 25 (March 2012): 1–25.

[157] William Burr, "A brief history of U.S.–Iranian nuclear negotiations," *Bulletin of the Atomic Scientists* 65 (January/February 2009): 21–34.

those throughout the US–Iranian relationship, primarily benefited Iranians connected to the monarchy, thereby increasing economic inequality and validating anti-Western and anti-shah sentiments.[158]

The Islamic revolution ended American programs in Iran. No reactors shipped. Instead, a host of American corporations lost financial holdings and property.[159] The Islamic government cancelled AT&T's contract for a satellite telecommunications network. The *New York Times* announced, "The Festival is Over for American Business."[160] American diplomacy shifted from promoting Iranian access to science and technology to preventing Iranian access (see Chapter 7).[161] Yet even as the Iranian program collapsed, the Saudi program expanded.

JECOR and American Assistance to Saudi Arabia

Oil was the basis for US–Saudi relations. American geologist Karl Twitchell discovered oil on the peninsula while exploring for much-needed water; a concession to Standard Oil of California led to the creation of California Arabian Standard Oil Company (CASOC) in 1933. During World War II, FDR approved a US military mission for road repairs and airfields (much like the IIAA programs in Latin America) and considered creating a government-controlled petroleum corporation (the oil industry vetoed the idea).[162] At the end of the war, Roosevelt and Ibn Saud agreed to maintain the partnership: Saudi Arabia provided cheap oil and the United States built Dhahran airfield and established a financial system. CASOC evolved into the Arab American Oil Company (ARAMCO) and Saudi Arabia joined Iran as a bastion of American commerce and anti-communism in the Middle East for two decades.[163] Thus Saudi participation in the OPEC oil embargo shocked American leaders.

---

[158] Blaszczyk, "Synthetics for the Shah," 706 and Mossavar-Rahmani, "Iran's nuclear power programme revisited," 193.

[159] Corporations included GM, DuPont, Pfizer, Colgate-Palmolive, General Dynamics, Grumman Aircraft, Fluor, and Bechtel, among others.

[160] Kilborn, "Iranian Festival Is Over for American Business," A10.

[161] See, for example, Caspar Weinberger to Robert McFarlane, "U.S. Export Policy toward Iran," (June 5, 1984). 1–3. DNSA Iraq-Gate collection, Record # IG00205.

[162] Barry Rubin, "Anglo-American Relations in Saudi Arabia, 1941–45," *Journal of Contemporary History* 14 (1979): 253–67. On the Petroleum Reserves Corporation, see "The Policy of Solidification" in Yergin, *The Prize*, 378–381.

[163] For the official Saudi perspective see, "Oil: The Start of It All" in Royal Embassy of Saudi Arabia, Information Office, *Saudi–U.S. Relations*. This is an official publication of the Saudi Embassy, available online.

Saudi Arabia was essential to American diplomacy in the Middle East. Massive oil reserves gave the country unique bargaining power in OPEC and the United States worked to maintain its favor. After the embargo, the United States and Saudi Arabia established the Joint Economic Commission Office Riyadh (JECOR) to coordinate American government and corporate assistance. At the same time, Secretary Simon offered the Saudis an opportunity to buy US bonds outside the normal global auction (i.e. without competition) and encouraged the Saudis to pressure fellow Arab nations to continue selling oil in dollars and launch Arabsat.[164] The congressional research service summed up the program decades later: "In the wake of the embargo, both Saudi and US officials worked to re-anchor the bilateral relationship on the basis of shared opposition to Communism, renewed military cooperation, and through economic initiatives that promoted the recycling of Saudi petrodollars to the United States via Saudi investment in infrastructure, industrial expansion, and U. S. securities."[165] After meeting with a Saudi Deputy Prime Minister, for example, President Ford's press secretary reported, "at the last OPEC meeting Saudi Arabia was almost single-handily responsible for blocking a price rise . . .[and] The Kingdom is pursuing a $140 billion Five-Year-Plan aimed at improving and expanding the domestic infrastructure of the country."[166] The Five-Year Plan provided the United States an opportunity to sell the few American products more valuable than oil – arms and access to American science and technology. The prices were secondary; the Saudi's were willing to pay for the entire program, including all costs incurred to American personnel, in return for close cooperation with American industry and academia.

JECOR helped build Saudi Arabia over the next twenty-five years. The original agreement required the United States establish "a comprehensive Saudi Arabian science and technology program keyed to the national goals of the Kingdom" and highlighted solar energy and desalination.[167]

---

[164] Regarding selling oil in dollars, see Spiro, *The Hidden Hand of American Hegemony*, 91.

[165] Christopher M. Blanchard, Congressional Research Service, "Saudi Arabia: Background and U.S. Relations," (December 16, 2009), quote on 4. Footnote #6 identifies the program as JECOR. CRS 7–5700, RL33533. Available at www.crs.gov.

[166] Ron Nessen, "Meeting with Prince Abdallah Ibn Abd Al-Aziz Al Saud Second Deputy Prime Minister of Saudi Arabia," (July 9, 1976), quote on 2. Located in the Ron Nessen Papers, Box 124, folder "Saudi Arabia" at the Gerald Ford Presidential Library. Available online.

[167] Henry Kissinger and Fahd bin Abd al Aziz, "Joint Statement on Saudi Arabian-United States Cooperation," *Middle East Journal* 28 (Summer 1974): 305–307, quote on 306.

Requests poured in; David Harbison, a commission official, described the early years as a "building and buying boom."[168] Only the increased arms sales drew attention: a *Foreign Affairs* article (1977) on US–Saudi relations observed "American arms deliveries to Saudi Arabia were stepped up vastly," but failed to mention JECOR.[169] Operating largely outside the public spotlight, the commission recycled billions of dollars back into the American economy.

JECOR emphasized scientific and technical programs from the beginning.[170] There were more than twenty projects involving eleven different US departments and agencies by 1982. More than 300 fulltime federal employees lived and worked in the Islamic country (a rarity); the GAO reported, "Saudi Arabia is by far the largest [such aid program] in scope, costs and number of US personnel involved."[171] JECOR oversaw initiatives in agriculture ($88 million, USDA and Interior), water desalination (HYDROS, $34 million, Interior), medical services, access to LANDSAT data (and Arabsat), meteorology, power generation and environmental education.[172] Desalination research, for example, was a shared priority, as the United States hoped to address water quality and shortages in the Colorado River basin.[173] Treasury and the EPA collaborated on CONPROT (consumer protection), a $19 million effort to introduce food quality controls to Saudi Arabia's four major laboratories, including the "introduction of standard methods of analysis; modern and practical

---

[168] David K. Harbinson, "The US–Saudi Arabian Joint Commission on Economic Cooperation: A Critical Appraisal," *Middle East Journal* 44 (Spring 1990): 269–283, quote on 277.

[169] Dankwort A. Rustow, "U.S.–Saudi Relations and the Oil Crises of the 1980s," *Foreign Affairs* 55 (April 1977): 494–516, quote on 514.

[170] David Harbinson, an official with the JECOR program, stated, "From the beginning, commission projects have focused on technical activities." David K. Harbinson, "The US–Saudi Arabian Joint Commission on Economic Cooperation: A Critical Appraisal," *Middle East Journal* 44 (Spring 1990): 269–283, quote on 277.

[171] Comptroller General, Report to the Chairman, Subcommittee on Europe and the Middle East, Committee on Foreign Affairs, House of Representatives, *Status of U.S.–Saudi Arabian Joint Commission on Economic Cooperation*, (Gaithersburg, MD: US General Accounting Office, 1983), 1.

[172] An excellent source for the long-term impact on Saudi Arabia is Hamad I. Al-Salloom, ed., *Science and Technology in Saudi Arabia* (Beltsville, MD: Amana publications, 1995). This work is an official publication prepared by the Saudi Arabian Cultural Mission to the United States of America.

[173] Werner Luft, "Solar Energy Water Desalination in the United States and Saudi Arabia," (April 1981). Paper prepared by the Solar Energy Research Institute for conferences in Washington, DC and Atlanta, GA.

equipment for all laboratories; [and a] computerized data collection system."[174] Smaller projects included $5 million for the Saudi Arabian National Center for Science and Technology (SANCST), $5 million to establish Saudi Arabia's first National Park (Asir), and $7 million for meteorology and environmental education.[175] Of course, the commission was also responsible for an additional $350 million in direct contracts with private industry in its first five years.[176] One of the more unique programs involved constructing a "Solar Village."

Solar energy research was a focus of US–Saudi cooperation. With an average annual high temperature of 90°F, Saudi Arabia provided an ideal test-site. Experiments to use solar energy for power generation, desalination and air-conditioning began in the late 1960s and quickly expanded as part of the JECOR program.[177] The United States was most invested in SOLERAS, a cooperative effort between the DOE, Saudi Arabia, and the Solar Energy Research Institute (a non-profit division of the Midwest Research Institute in Kansas City, Missouri). Unlike other JECOR programs, the United States contributed $50 million per year to SOLERAS and funded related research in eight American universities.[178] SOLERAS constructed a "Solar Energy Village" outside Riyadh to demonstrate the use of solar energy for everyday domestic use and agriculture.[179] By the early 1980s, the GAO considered the "solar village" a "prototype facility" and "a showcase project demonstrating U.S. and Saudi cooperation and the feasibility of obtaining energy from the sun."[180] Reporters considered the village a "showpiece" because so many heads of state visited.[181]

[174] The best source on specific JECOR programs to 1981 is "Appendix IV: Description of Joint Commission Projects, Objectives, Status, Costs (note a) and Personnel (note b)" in Comptroller General, *Status of U.S.–Saudi Arabian Joint Commission on Economic Cooperation*, 29–40.

[175] Ibid.   [176] Ibid., iii.

[177] Saleh H. Alawaji, "Evaluation of Solar Energy Research and Its Applications in Saudi Arabia – 20 Years of Experience," *Renewable and Sustainable Energy Reviews* 5 (2001): 59–77.

[178] STAD 1986, 68.

[179] Midwest Research Institute, *Solar Controlled Environment Agriculture Project, Final Report: Vol. 1, Project Summary* (Kansas City, MI: Midwest Research Institute, 1986). This work was "Published for the United States–Saudi Arabian Joint Program for Cooperation the Field of Solar Energy, SOLERAS."

[180] Comptroller General, *Status of U.S.–Saudi Arabian Joint Commission on Economic Cooperation*, 12.

[181] On reporters, see John Duke Anthony, "The U.S.–Saudi Arabian Joint Commission on Economic Cooperation," in Willard A. Beling, *King Faisal and the Modernisation of*

JECOR ended its first decade a success. The GAO concluded in 1983: "The Joint Commission provided the United States with a mechanism to facilitate the flow of American goods and services to Saudi Arabia, while contributing to the economic development of that country through technical assistance."[182] More than 8,000 Saudis enrolled in US universities, while the American influence on Saudi education lasted for decades; Henry Albers, an American advisor to the Saudi University of Petroleum and Minerals, believed the school's standards and culture, "remained American ... The English language has provided an important means for scientific and technology transfer which has served Saudi Arabia well."[183] The Saudi embassy acknowledged the American role in creating the "Kingdom's modern infrastructure, schools, roads, hospitals [and] telecommunications."[184] Only arms sales and Middle East investments in American universities sparked a backlash. By 1979, nearly one third of foreign students studying in the United States came from OPEC nations (more than 80,000 in total) and nearly half were from Iran, which maintained links to more than fifty American universities.[185] The American Jewish Committee, for example, worried grants from Islamic countries could "be used to skew university curricula [or] underwrite biased anti–Israeli programs."[186] Perhaps. But American support for Israel was unrivaled.

US–Israeli Scientific Relations

The United States supported Israel since its conception. Although the country remained outside the Mutual Security Program, special legislation authorized its participation in technical aid programs.[187] Within its

---

Saudi Arabia (Boulder, CO: Westview Press, 1980), 103. On its visitation by heads of state, see Comptroller General, *Status of U.S.–Saudi Arabian Joint Commission on Economic Cooperation*, 12.

[182] Comptroller General, *Status of U.S.–Saudi Arabian Joint Commission on Economic Cooperation*, 21.

[183] Henry H. Albers, *Saudi Arabia: Technocrats in a Traditional Society* (New York: Peter Lang, 1989), 213.

[184] Royal Embassy of Saudi Arabia, Information Office, *Saudi–U.S. Relations*. This is an official publication of the Saudi Embassy, available online at www.saudiembassy.net.

[185] R. Jeffrey Smith, "Middle East Investments in American Universities Spark Campus Confrontations," *Science* 203 (February 2, 1979): 421–424.

[186] American Jewish Committee quoted in ibid., 421.

[187] Sergei Y. Shenin, *The United States and the Third World*, 100–101. The best overview of the unique level of American support for Israel is "The Great Benefactor" in John

first five years, Israel received $348 million dollars in American aid, much of it technical: Bruce McDaniel, director of US operations mission in Israel, considered technical aid the mission's cornerstone and cited successes in geology, hydrology, health, and sanitation, etc.[188] Disparities in assistance across the region drew attention. Jonathan Bingham, Point Four administrator, suggested Iranian and Arab "bitterness" had "unfortunately been accentuated by the large sums given to Israel, which have made the programs in the Arab states look puny by comparison and which have been regarded by Arabs as proof of the fact that we were not, as we claimed, impartial."[189] By the late 1950s, Israeli scientific progress meant the country could collaborate as an equal partner; one study found the US government invested around $70 million in research at Israeli institutions between 1958 and 1972.[190] Much of the research, whether in computers, aeronautics, or telecommunications, benefited the Israeli nuclear program.[191]

US diplomacy shielded the Israeli program from international oversight. Israeli leaders considered nuclear weapons critical to their state's survival; David Ben-Gurion, Israeli's first prime minister, wrote in 1956: "What Einstein, Oppenheimer and Teller, the three of them are Jews, made for the United States, could also be done by scientists in Israel for their own people."[192] He went further in a farewell address in 1963: "And the Jewish brain does not disappoint; Jewish science does not disappoint ... I am confident that science is able to provide us with the weapon that will secure the peace, and deter our enemies."[193] Aided by the French, the Israeli reactor program at Dimona drew international attention. But the United States backed the Israeli refusal to allow international inspections of the Dimona site – a special privilege infuriating to compliant states as well as those penalized for non-compliance. American

J. Mearsheimer and Stephen M. Walt, *The Israel Lobby and U.S. Foreign Policy* (New York: Farrar, Straus and Giroux, 2007), 23–48.

[188] Bruce McDaniel interviewed in Anon., "U.S. Aid to Israel Put at $348,000,000," *New York Times* (February 28, 1954): 2.

[189] Bingham, *Shirt-Sleeve Diplomacy*, 182

[190] Max Hellmann, "The U.S.–Israel Binational Science Foundation," in Wallerstein, *Scientific and Technological Cooperation among Industrialized Countries*, 111–122, esp. 111. See also Anon., "U.S. Technical Help in Israel to Change," *New York Times* (November 29, 1960): 15.

[191] Avner Cohen, *Israel and the Bomb* (New York: Columbia University Press, 1998), 342. Note that Israel Dostrovsky, Director-General of the Israeli Atomic Energy Commission, was trained for four years at Brookhaven National Laboratory.

[192] Ben-Gurion quoted in ibid., 12.   [193] Ibid., 13.

determination to maintain the Israeli narrative and downplay US visits is evident in a State Department cable to all Middle East embassies: "Department intends to initiate NO repeat NO publicity with regard visits by scientists. If leaks occur, US spokesman will say two US scientists recently in Israel in connection with US Atoms-for-Peace Agreement had opportunity to visit Dimona installation which they observed to be of nature and scope publicly disclosed by Israel Government [capitalization in original]."[194] The cover continued even after Israel refused to allow American inspections.

Military and economic cooperation cemented US–Israel relations. In 1963, for example, the United States provided $25 million for HAWK anti-aircraft missiles to guard the Dimona site, $22 million for construction through the World Bank, and numerous smaller "science grants."[195] State Department officials reported, "we are continuing, as in previous years, our multi-million-dollar support of scientific research in Israel. Contracts led by United States government agencies, paid partly in local currency and partly in US dollars, average just under $5 million annually."[196] The 1967 Six-Day War, partially instigated by conflict over water rights, encouraged US–Israel cooperation: Although the Nixon administration declined to build the requested desalination plants in Israel, Henry Kissinger promised to "press ahead with research in desalting technology and will ensure close cooperation with Israeli technicians."[197] Additional crises created additional ties.

The United States established three still-existing bilateral research programs with Israel in the 1970s. In 1972, the United States and Israel each contributed $30 million to establish a Binational Science Foundation (BSF) to fund collaborative research. A significant source of money for Israeli scientists, BSF advertises its prize-winning research and "facilitated access ... to the unrivaled infrastructure of American science."[198] The 1973 Yom Kippur War and resulting oil embargo furthered US–Israeli

---

[194] Christian A. Herter, Jr. to U.S. Embassies in Middle East, "[Department of State Press Announcement on Israeli Nuclear Reactor]," (December 22, 1960): 1. DNSA Nuclear Non-Proliferation Collection, Record # NP00725.

[195] William H. Brubeck, "Memorandum for Mr McGeorge Bundy," (1963): 1–3. Available via DDRS.

[196] Ibid., 2.

[197] Henry Kissinger, National Security Council, "Water Development and Middle East Policy," (November 6, 1969): 1–2. DNSA Presidential Directives Collection, Record # PD01192.

[198] Quote from the official BSF website: www.bsf.org.il/BSFPublic/Default.aspx.

cooperation: To stimulate economic growth, the nations created the Binational Industrial Research and Development Foundation (BIRD) in 1978 to provide "matchmaking services" between R&D firms in Israel and the United States. The final component – Binational Agricultural Research and Development (BARD) – arose the following year. BSF, BIRD, and BARD cemented American and Israeli scientific, technical, and commercial relations; by the 1980s, scientists at 124 different American institutions participated in BSF research while cumulative BIRD product sales attained $1 billion by end of the Cold War.[199] The initiatives received rare bipartisan support: During the Reagan administration, a period of tense presidential and congressional relations, the State Department summary noted, "BIRD, BSF and BARD are viewed unanimously as extremely successful. They have become the models ... for similar agreements."[200] Indeed, US–Israeli agreements, focused on private research, commerce and common regulations and protections, represented the Reagan administration's ideal of American scientific relations in a global world.

---

[199] Hellmann, "The U.S.–Israel Binational Science Foundation," in Wallerstein, *Scientific and Technological Cooperation among Industrialized Countries*, 111–122, esp. 114. For cumulative sales totals, see *STAD 1991*, 132.
[200] *STAD 1988*, 48.

6

Globalization

The concrete stopped flowing into the 54-mile circular tunnel on the Texas prairie in 1993. Known as the "Big Pour" in the construction industry, the Superconducting Super Collider (SSC) split the physics community, failed to attract international funding, and eventually died, partially completed, with the end of the Cold War. High-energy physicists designed the SSC to maintain American preeminence well into the twenty-first century and its demise led to calls for support: Nobel prizewinners Leon Lederman and Sheldon Glashow wrote, "if we forgo the opportunity that [the] SSC offers for the 1990s, the loss will not only be to our science but also to the broader issue of national pride and technological self-confidence."[1] But particle accelerators were of little congressional or public interest after the collapse of the Soviet Union. Proponents suggested the SSC would bring 8,000 jobs and aid the war on cancer, yet most remained unmoved. Senator Dave Durenberger (R-MN) spoke bluntly during hearings in 1992: "If we were engaged in a scientific competition with a global superpower like the former Soviet Union, and if this project would lead to an enhancement of our national security, then I would be willing to continue funding the project. But . . . we face no such threat."[2] Nor was high-energy physics alone; interest in space lagged as well.

The US space program suffered from the end of the Cold War. Congress proposed phasing out the shuttle over ten years. *Space Station*

---

[1] Glashow and Lederman quoted in "The Death of the Superconducting Super Collider in the Life of American Physics," in Kevles, *The Physicists*, ix–xlii, quote on xix.
[2] Durenberger quoted in Ibid., xxxvi.

Freedom, a project supporting 75,000 domestic aerospace jobs and armed with $8 billion in foreign funding, came under budget scrutiny.[3] To survive, NASA welcomed Russian hardware. After the Cold War, expensive projects in "big" astronomy and physics needed international support, as multiple administrations sought to minimize American spending. When the supercollider lost funding, for example, the State Department lobbied to increase American participation at CERN, where the Large Hadron Collider generates only a fraction of the proposed current of the SSC. Of course, different fields have different trajectories, costs, and benefactors.

US government support helped "big" biology blossom as "big" physics and astronomy stagnated. Biotechnology was cheap compared to the other fields and applications seemed to arise easily from research. Private industry was willing to invest, but the field needed government protection for the applications. As research tilted toward market-oriented production, high-tech industries demanded international protection, leading President Reagan and later administrations to leverage American economic power to globalize intellectual property rights. Such intellectual protections were part a shift toward neoliberalism.

Neoliberalism influenced American science policies at home and abroad. Historian Daniel Kevles explored the connection between the economic philosophy and domestic research, observing:

a sea change has come over American politics that has impacted the consensus for science. The shift began in the 1970s. It was heralded by, among other developments, the call for deregulation of the economy, the claims that government was the problem rather than the solution, and the renewed emphasis on privatization and entrepreneurship that was signified by the emergence of the biotechnology industry.[4]

Neoliberalism promoted private research and insisted on intellectual property rights.[5] Spurred by high-tech industries, the Reagan administration introduced the threat of economic sanctions to pressure countries

---

[3] Ibid., xxxvi.

[4] Daniel J. Kevles, "What's New about the Politics of Science?" *Social Research* 73 (Fall 2006): 761–778, quote on 770. For a critical look at the relationship between neoliberalism and science, see Philip Mirowski, *Science Mart: Privatizing American Science* (Cambridge, MA: Harvard University Press, 2011). See also Elizabeth Popp Berman, *Creating the Market University: How Academic Science Became an Economic Engine* (Princeton, NJ: Princeton University Press, 2012), 172–177.

[5] For an excellent introduction, see David Harvey, *A Brief History of Neoliberalism* (New York: Oxford University Press, 2005).

to accept American intellectual property rights. The Bush administration increased the pressure during negotiations of the GATT treaty and the TRIPS agreement completed the process under the Clinton administration. The economic philosophy also underlay the failure of G7 cooperation as well as the privatization of many American scientific and technical resources, from weather services to the Navy's NAVSTAR navigation system. Yet even as the country shifted toward private R&D, American assistance programs and trade created global networks of researchers and knowledge.

By the 1980s, most federal bureaus and departments maintained an international office overseeing multiple projects around the world. Consider the following State Department submission on the USDA from 1988:

[The USDA] continues to be very active in supporting scientific and technical exchange activities throughout the world. Many USDA employees are on loan to AID, the UN's Food and Agriculture Organization, and other agencies. ... U.S. meat inspectors are inspecting plants in many areas of the world to assure wholesome processing, foresters are attempting to control damage to watersheds and improve the environment, landscape architects are introducing new species and providing areas for recreation out of wasteland, soil conservationists are working on water and soil projects to lessen harm to the environment, and nutritionists are attempting to improve the diets of the poor.[6]

Nor was the USDA unique; US programs encouraged research and networking worldwide. The USGS and Bureau of Mines, inheritors of earlier Point Four initiatives, continued to catalog strategic minerals like cobalt.[7] The Department of Commerce, in partnership with private industry, operated the US Telecommunications Training Institute to share advances with developing countries.[8] Similar programs were underway at the EPA, Interior, and Commerce. And even as American researchers labored on the ground, NASA continued its scientific and political work in the heavens above. This chapter explores American scientific relations at the end of the Cold War in three sections.

The first, "We Are Their Allies," illustrates how the Reagan administration limited cooperation with G7 nations and within NATO. Although the United States participated in the G7 science initiative, the administration refused to engage in multiple fields and the initiative ended within a few years. Additionally, *Space Station Freedom* and SDI provoked

---

[6] *STAD* 1988, 11.   [7] *STAD* 1988, 27.
[8] Other partners included the National Telecommunications and Information Administration. See *STAD* 1988, 17.

disagreements with NATO allies, reinforcing the administration's prefer-ence for bilateral relationships and the United States expanded cooper-ation with China, Saudi Arabia, and Israel throughout the 1980s. At the same time, American policy-makers began to worry about high-tech competition from allies (and rivals) such as Japan.

"Competition Rising" considers the American response to the growth of high-tech industries overseas. The Japanese national semiconductor initia-tive produced an immediate response: Congress legalized US research consortia and President Reagan announced an American competitiveness agenda. The State Department pressed Japan to increase American access and the Department of Commerce studied the Japanese influence on American research. Meanwhile, American high-tech firms demanded intel-lectual protection overseas while petitioning to end anti-communist export controls, arguing they limited access to foreign markets.

"Soviet Fission" covers the collapse of the Soviet Union and the legacy of the Cold War on US science and diplomacy. The Soviet Union struggled to keep pace in high-tech industries at the end of the Cold War, leading to frustrations within the communist bloc. The Soviet collapse prompted multiple American attempts to integrate former weapons researchers into the larger scientific community, but the initia-tives turned toward commerce when Russia maintained large defense budgets. At the same time, the United States welcomed Russian participa-tion in a new joint space station as international competition in a global knowledge economy replaced the Cold War.

WE ARE THEIR ALLIES

Cooperation in the fields of science and technology is one of the most effective means not only for contributing to the physical welfare of people, but also of fostering the dignity and worth of every person.

Pope John Paul II (1984)[9]

Science and technology have reached such a point of importance in the roles of individual countries that I feel it is time to write down a set of generally held concepts and make sure that we have agreement on them.

OSTP Director William R. Graham (1987)[10]

[9] Paul II quoted by Ronald Reagan, in "Letter of Transmittal," *STAD* 1985, vii.
[10] Graham quoted in David Dickson, "OECD to Set Rules for International Science," *Science* 238 (November 6, 1987): 743.

Papal blessing aside, scientific cooperation proved difficult during the Reagan administration. Growing European and Japanese capabilities pushed against American preeminence and strained relations. At the same time, Reagan administration priorities – export controls, private investment, deregulation, and anti-environmentalism – undermined initiatives with allies. Neoliberalism favored private over public R&D and curbed cooperation in valuable emerging industries such as information technology and biotechnology. American management of *Space Station Freedom* and the Strategic Defense Initiative increased the friction in orbit: Europeans resented their secondary role while the Soviet Union accused the United States of militarizing space. Finally, as the United States limited cooperation throughout the 1980s, disagreements over Israel and global family-planning led the Reagan administration to withdraw American support for UNESCO.

The G7 Science Initiative

President Reagan's first budget signaled a retreat from international activities. *Science* reported: "The Reagan administration's budget has already generated considerable anger and unease in Europe and in some international scientific organizations, for it would eliminate some international projects, end U.S. participation in a few multilateral activities, and reduce U.S. support for some bilateral scientific cooperation programs."[11] The administration began by withdrawing from the International Solar Polar Mission without consulting allies.[12] Since the planned experiment required two spacecraft, the cancellation meant the Europeans "lavished a good bit of their meager space science budget on a project whose scientific value would be considerably reduced."[13] Worse, the Reagan administration wanted tougher export controls on Soviet-bloc trade just as European allies looked to expand Soviet-bloc trade. It was hardly an opportune time for increased collaboration.

The French proposed a broad program of joint research at the G7 summit in Versailles (1982).[14] European science advisors considered

---

[11] Colin Norman, "Reagan Budget Would Reshape Science Policies," *Science* 211 (March 27, 1981): 1399–1402, quote on 1400.
[12] Logsden, "25-year Perspective," 75.
[13] Norman, "Reagan Budget Would Reshape Science Policies," 1401.
[14] The G7 included the United States, Canada, France, West Germany, Italy, Japan and the United Kingdom.

collaboration critical to economic development and began meeting annually with their American counterparts in 1979.[15] French advisor Jacques Attali suggested cooperative research in more than a dozen high-tech areas.[16] Although the proposal for cooperation was not their only concern – President Mitterrand wanted to stabilize US exchange rates and discuss UN aid to developing nations – the French delegation hoped to integrate G7 research.[17] National differences arose: The Germans worried about the French penchant for institution building, while American representatives worried collaboration would encroach on private-sector firms, questioning the potential benefit.[18] Instead, the American delegation defeated proposals for a UN biotechnology center, focusing on patents, intellectual property rights, and controlling the flow of sensitive equipment to the Soviet bloc.[19] Still, the French suggestion made it to the planning stage and the assembled leaders acknowledged science and technology were fundamental components of international economic relations.[20] Over the next year, representatives met to plan the projects.

The final agreements, signed at the Williamsburg summit (1983), reflected national priorities. The collected heads of state agreed to collaborate on eighteen research projects, with the United States co-leading on six. Building on American strengths dating to the Atoms for Peace and Apollo programs, the United States co-led groups on high-energy physics, controlled thermonuclear reactors, fast breeder reactor design, solar system exploration, and remote sensing from space. The United States also co-led a group on advanced materials and standards, an area critical to the fabrication of IT products. Other countries had their own interests: Japan, for example, led projects on robotics and photosynthesis and worked with Italy on photovoltaic batteries; West Germany and France collaborated on high-speed trains.[21] Priorities could also be demonstrated negatively: The United States refused to participate in research relating to solar energy, high-speed trains, housing and urban planning in developing countries, and biotechnology.[22]

[15] *STAD* 1991, 29.   [16] Wallerstein, *Scientific and Technological Cooperation*, 11.

[17] Richard J. Cattani, "Versailles Summit – 'Toughest Yet' – Ends in Standoff," *Christian Science Monitor* (June 7, 1982). Accessed online.

[18] David Dickson, "A Political Push for Scientific Cooperation," *Science* 224 (June 22, 1984): 1317–1319.

[19] *STAD* 1986, 19.   [20] *STAD* 1987, 11.   [21] *STAD* 1986, 19–21.

[22] *STAD* 1985, 17–18. See also, Dickson, "A Political Push for Scientific Cooperation," 1317–1319.

Biotechnology and IT occupied a privileged place in American diplomacy. The United States forcefully advocated for internationalizing its patent positions and refused to cooperate in either field. American export controls proved frustrating within NATO: When the Reagan administration requested British companies apply for export licenses for information inside American researchers' heads, Brian Oakley, director of the state-sponsored British computing initiative, complained, "Damn it all, we are their allies."[23] He also warned: "By encouraging European countries to seek other partners, particularly Japan, the United States could find itself facing an anti-American alliance."[24] Although American actions did not lead to an anti-American alliance, they did stimulate allies.

With biotechnology and IT outside the G7 initiative, the Europeans increased collaboration. The European Economic Commission established a $1.3 billion program to create European IT infrastructure and promote software engineering.[25] Etienne Davignon, the EEC commissioner for industry, announced Europe was "going into a major fight with the U.S."[26] The following year, the EEC announced its first "framework program" for R&D (the eighth program – nicknamed *Horizon 2020* – began in 2014).[27] Led by French President Mitterrand, the EEC launched more than sixty product-oriented research projects (EUREKA, 1985). Even Prime Minister Thatcher, President Reagan's staunchest European ally, supported the initiatives, stating: "We face the stark prospect that the United States and Japan will monopolize world markets in high technology goods."[28] Yet the European commitment was suspect: The proposed budget was lower than R&D spending at a large corporation such as IBM or GM, exposing the limits of cooperation.[29] International scientific collaboration, whether within the EEC or the G7, remained difficult.

The Versailles initiative ended after four years. Its State Department obituary read: "The Versailles Economic Summit Science and Technology

---

[23] Oakley quoted in David Dickson, "Europeans Protest U.S. Export Controls," *Science* 224 (May 11, 1984) 579–581, quote on 579.

[24] Ibid, 580.

[25] The program was the European Strategic Program on Research in Information Technology (ESPRIT, 1983).

[26] Davignon quoted in Dickson, "Europeans Protests U.S. Export Controls," 581.

[27] The full title is the Framework Program for Research and Technical Development.

[28] David Dickson, "Europe Pushes Ahead with Plans for Joint Prospects," *Science* 233 (July 11, 1986): 152.

[29] Ibid.

Initiative formally concluded during FY 1986 its mandated tasks of fostering economic growth and employment."[30] The report observed the group was never meant to be permanent and "had accomplished its original purpose of fostering cooperation in key S&T areas and disbanded ... others will continue independently or through multilateral agreements."[31] One lasting example was the Versailles Project on Advanced Materials and Standards (VAMAS), a program the Department considered "highly successful."[32] VAMAS addressed the physical properties of the materials used in information technology, standardizing chemical analysis, spectroscopy, and fatigue testing to harmonize specifications for trade. The agency furthered American influence by allowing the United States to set the standards in a new global industry (VAMAS was formally transferred to the United States National Institute of Standards and Technology in 2005).[33]

The legacy of the G7 initiative remains unclear. The collusion among developed countries led political scientist David Dickson to conclude it facilitated "access by international capital to the basic science needed for its high technology industries, while tightening the terms and conditions under which this access would be granted to others."[34] Yet it was a European initiative with only hesitant American support and the experience illustrated the difficulty of collaboration in fields with national security or commercial implications. National priorities framed cooperation: The French wanted a more robust multilateral partnership, while the United States preferred bilateral initiatives. In hindsight, the period saw the creation of multiple European research programs while US diplomacy for science, whether in the G7 or in support of Space Station Freedom and the Strategic Defense Initiative, often divided its allies.

Space Station Freedom

Allied differences in space surfaced early in the Reagan administration. In addition to withdrawing from the joint solar mission, the United States unilaterally abrogated previous agreements on the European Spacelab. After originally supporting the project, the president curtailed flights after

---

[30] *STAD* 1987, xiii.   [31] Ibid., 12   [32] *STAD* 1987, 12.

[33] Martin Rides and Graham Sims, "VAMAS contributing to international standards in the materials sector," *ISO Focus* (October 2005): 40–41. Rides and Sims were past secretary and past chair of VAMAS respectively.

[34] Dickson, *The New Politics of Science*, 311.

FIGURE 6.1 Photo of President Reagan at London Economic Summit (1984)
This photo from the 1984 G7 London Economic Summit shows President Reagan, Prime Minister Thatcher, and Prime Minister Nakasone as well as models of *Space Station Freedom* and the space shuttle (with the Canadian arm extended). Although the president hoped for allied support, he was unable to secure commitments at the meeting.

its maiden voyage in 1982 and canceled requests for additional modules (Europeans had hoped to recoup start-up costs by selling up to six additional labs). The president also threatened to tighten export controls and announced his Strategic Defense Initiative in early 1983, disturbing allies and setting off an international debate about the militarization of space. Determined to allay allied fears, the administration suggested cooperation.

President Reagan invited Europeans to participate in a new space station – christened *Freedom* – in 1984. NASA and the State Department believed the station would demonstrate allied unity and the administration promised to exclude national security research, thereby downplaying the fears of weaponization aroused by SDI. *Freedom* was a consistent "talking point" when Reagan met privately with other leaders during the London G7 summit (1984).[35] At the administration's request, NASA Langley Research Center set up a model of the station complete with each country's proposed contribution (Figure 6.1). Yet the allies left without signing an agreement, only a vague statement each would "consider carefully the generous and thoughtful invitation."[36]

Europeans worried about the costs and control of *Freedom*. Given the experience with Spacelab, ESA members were hesitant to contribute the proposed $2 billion: Reimar Lust, the ESA director-general, stated, "we

[35] John M. Logsden, *Together in Orbit: The Origins of International Participation in the Space Station* (Washington, DC: NASA History Division, 1998), 25–26.
[36] "The London Economic Declaration," (June 9, 1984). Available at: www.g8.utoronto.ca/summit/1984london/communique.html.

could never give a blank check to NASA; we need some guarantees ...
Suddenly we were being required to make changes in Spacelab because of
alterations to the design of the shuttle, making it much more expensive
than we had planned."[37] He later added, "It is important that Europe
should accept that the United States is autonomous, and that it can
accomplish all tasks in space on its own; but it is of equal importance
that the U.S. accept the fact that Europe, too, will eventually be autonomous."[38] Further disagreement over a proposed laboratory module (the
European contribution), led one German official to grouse: "Perhaps we
will need American passports to visit parts of the space station."[39] While
Freedom did not require passports, it did require negotiation.

The final agreement contained several compromises. The United States
could not test weapons and the ESA could veto research. US patent laws
would apply, but any patents arising from microgravity experiments
would be shared.[40] Agreement also required compromises on trade policies: The Reagan administration relaxed export controls and reached six
new agreements under COCOM auspices.[41] The space station eventually
gained the support of eleven European allies; only Great Britain
abstained. But the British participated in SDI, though the initiative split
scientific communities on both sides of the Atlantic.

The Strategic Defense Initiative

The Strategic Defense Initiative politicized the scientific community.
While still a candidate in the late 1970s, Ronald Reagan befriended
retired Lieutenant General Daniel O. Graham, the founder of High Frontier – an organization dedicated to space-based defense. Sponsored by the
Heritage Foundation, Graham and physicist Edward Teller proposed a
similar system to President Reagan. Against his advisors' advice, the
president announced the Strategic Defense Initiative in his first term; the
NAS rejected the idea and thousands of scientists pledged not to take
funds. Historian Audra Wolfe concluded, "Reagan's announcement of

[37] David Dickson, "Space Station Plan Upsets Europe," *Science* 234 (December 19, 1986): 1487.
[38] David Dickson, "Europe Plans Its Own Mini Station," *Science* 232 (May 16, 1986): 816–817, 816.
[39] Dickson, "Space Station Plan Upsets Europe," 1487.
[40] Peter Coles, "ESA and NASA Get It Together," *Nature* 331 (February 18, 1988): 550.
[41] *STAD* 1988, viii.

SDI represented a complete repudiation of advice from the scientific community."[42] Instead, the president listened to outside conservative voices for scientific guidance (he also favored the teaching of creationism, denied acid rain research, and disapproved of contraception).

Reagan relied on testimony from private foundations and associated spokesmen. In the 1970s, a few well-known scientists – notably physicists Frederick Seitz and Frederick Singer – distributed millions from RJ Reynolds Tobacco Company (and others) to researchers questioning the link between smoking and cancer (known as the "Tobacco Strategy," the companies eventually settled for more than $200 billion in damages).[43] Both physicists were advocates of SDI and ardent anti-communists (Seitz, a former NAS president, broke with many of his colleagues over Vietnam). In 1984, Seitz, Singer, and others founded the George Marshall Institute to lobby for SDI, quickly partnering with Edward Teller and the Heritage Foundation. The lobby was influential: George Keyworth, Reagan's science advisor, switched positions, becoming an outspoken proponent of the program.

Politics influenced funding. Donald Hicks, the undersecretary of defense for research and engineering, stated funds would be withheld from scientists criticizing SDI: "I am not particularly interested in seeing department money going to someplace where an individual is outspoken in his rejection of department aims . . . freedom works both ways. They're free to keep their mouths shut . . . and I'm also free not to give the money. . . I have a tough time with disloyalty."[44] Eventually, more than 6,500 scientists pledged not to work on the national defense program. Politics also determined the location of the research: Given the controversies over defense research at universities during the Vietnam War, 95 percent of SDI funding went to private contractors, who began legally lobbying Congress in 1974.[45]

However, the US initiative was controversial across the Atlantic. Prime Minister Thatcher was an outspoken supporter and the UK received contracts worth $68 million within a few years.[46] But her cabinet, like many across Europe, split: Thatcher's minister of trade,

---

[42] Wolfe, *Competing with the Soviets*, 136. See also "Connecting to Politics," in Daniel S. Greenberg, *Science, Money and Politics: Political Triumph and Ethical Erosion* (Chicago: University of Chicago Press, 2001), 278–293.

[43] Oreskes and Conway, *Merchants of Doubt*.

[44] Hicks quoted in R. Jeffrey Smith, "Pentagon's R&D Chief Roils the Waters," *Science* 232 (April 25, 1986): 443–445, quote on 444.

[45] Wolfe, *Competing with the Soviets*, 132.   [46] *STAD* 1990, 187.

for example, was against participation.[47] The State Department monitored the British press, reporting: "offers of cooperation on aspects of the Strategic Defense Initiative have met mixed reactions – some are opposed to the concept, but many are enthusiastic about being involved in the leading edge of space technology."[48] Ian Chalmers, a British vacuum expert who previously failed to secure funding, agreed, stating: "SDI came along, and we were delighted."[49] Others disagreed: More than 500 university researchers, mostly computer scientists and physicists, signed a pledge to refuse SDI funds.[50] Some felt the United States was "dangling a $200 million carrot under the noses of its European allies" in an attempt to "buy all our ideas on the cheap."[51] Even as the State Department worried about "misunderstandings and misconceptions," the Soviet Union accused the United States of planting weapons in space and launched their own space station, *Mir* ("Peace," 1986).[52] As *Space Station Freedom* and SDI strained US cooperation with European allies, the Reagan administration expanded bilateral scientific cooperation with China, Saudi Arabia and Israel.[53]

Scientific Relations with China, Saudi Arabia, and Israel

US–Chinese scientific relations strengthened in the 1980s. In 1988, the State Department observed, "The U.S. and China have developed one of the largest programs of government-to-government S&T relations in the world. The umbrella S&T agreement, signed in 1979 and renewed for an additional five years in 1984, embraces sub-agreements or protocols providing for cooperation in 29 fields. The areas range from theoretical, such as high-energy physics, to practical, such as statistics, building construction and health care."[54] The Reagan administration believed scientific relations bolstered normalization: Chinese acceptance of the

[47] David Dickson, "British Cabinet Split on SDI Agreement," *Science* 230 (December 13, 1985): 1251–1252.
[48] *STAD* 1986, 8.
[49] David Dickson, "British Researchers Seek FDI Funds," *Science* 235 (February 13, 1987): 736–737.
[50] Ibid.
[51] David Dickson, "Europeans Wary of U.S. Offer on Military R&D," *Science* 232 (April 19, 1986): 314–315.
[52] *STAD* 1986, 25.
[53] *STAD* 1986, 30. See also David Dickson, *The New Politics of Science, 2nd ed.* (Chicago: University of Chicago Press, 1988), x.
[54] *STAD* 1988, 43.

non-proliferation treaty in 1985, for example, demonstrated the country's willingness to consent to international norms. Of course, research was not the only reason for cooperation; each side had clear economic and political motives as well.

The Chinese sought to trade access to Chinese markets for access to American science and technology. Historian Denis Fred Simon summarized the Chinese perspective succinctly: "the availability of technology has become the *quid pro quo* for entrance into the PRC domestic market ... technology transfer will become the 'cement' holding China's relations with West together."[55] The approach worked: the United States, Japan, and Western Europe expanded Chinese access and relaxed export controls.[56] Additional scientific resources became available; by the 1980s the PRC could access genetic resources and cross-breeding techniques from across the world.[57] From a Chinese perspective, the relationship was a success, providing access to advanced technology and research critical to national development.

American aid targeted projects of importance to the Chinese government. The Department of Energy, for example, supported the Beijing Electron-Positron Collider (BEPC), a "key project" for the Chinese. More than two dozen agencies maintained relations, including help from NOAA on comparative climate studies and the USGS on earthquake prediction.[58] Development assistance and investment went hand-in-hand: When the Bureau of Reclamation began providing technical experts for the Three Gorges Dam project in 1984, the State Department advertised "excellent opportunities for participation by the private sector in engineering, construction and equipment supply."[59] Diplomats also helped establish a center for the management of industrial science and technology in cooperation with SUNY-Buffalo.[60]

The United States reaped political and economic benefits. China opened special economic zones in 1984 and welcomed American businesses. The Chinese signed more than 1,300 technology import contacts

[55] Denis Fred Simon, "Technology Transfer and China's Emerging Role in the World Economy," in Simon and Goldman, eds., *Science and Technology in Post-Mao China*, 289–318, quote on 307.

[56] Simon, "Technology Transfer and China's Emerging Role in the World Economy," 291.

[57] Athar Hussein, "Science and Technology in the Chinese Countryside," in Simon and Goldman, eds., *Science and Technology in Post-Mao China*, 223–250, esp 236.

[58] *STAD* 1985, 24.   [59] *STAD* 1986, 94.

[60] The full title is the Dalian Center for Industrial Science and Technology Management Development.

with foreign firms during their sixth five-year plan (1981–1985), focusing on "know-how" in electronics and advanced machinery.[61] The PRC announced stricter patent laws in 1985, but foreign inventions reverted to China if the business left the mainland and the state refused to award patents on pharmaceuticals, plants, animals or chemicals.[62] Nonetheless, the Reagan administration hoped contacts would mature into political allies: The State Department hypothesized, "with China's current emphasis on selecting young leaders with technical backgrounds, it is likely that PRC scientists who have exposure to U.S. S&T will eventually be promoted to high levels of authority in China's government."[63] Like cooperation with the Soviets, US officials hoped scientific engagement with the PRC could produce political benefits, while the Chinese hoped renewed relations would allow for additional scientific and technical access. A different comity of interests bonded US–Saudi scientific diplomacy.

The Reagan years were the heyday of the JECOR program with Saudi Arabia. The initiative, like normalization with the Chinese, dated to the 1970s (see Chapter 5). In this case, Saudi Arabia purchased American scientific and technical products and services to recycle US payments for oil. By the 1980s, JECOR involved nearly every federal agency in a "low-key but wide-ranging cooperative S&T program coordinated through the US Embassy in Riyadh."[64] The annual State Department report detailed the extent of US assistance:

Under this and other bilateral programs a number of U.S. agencies provide training and assistance to their Saudi counterparts. ... In cooperative S&T activities, NSF is working with the King Abdul Aziz City for Science and Technology (KACST) to strengthen its scientific research capabilities. ... DOE is working with KACST on solar energy and basic energy research. NASA and KACST cooperate in remote sensing experiments, and NOAA is assisting in the construction of an earth station to receive satellite data. USGS has a wide range of activities in Saudi Arabia, and the Bureau of Reclamation is working on desalination. PHS is assisting in upgrading the delivery system.[65]

Much like the Iranian program, JECOR was long-term, stretching from 1975 to 2000 and influencing a wide range of Saudi society.

[61] Simon, "Technology Transfer and China's Emerging Role in the World Economy," 299.
[62] Ibid., 301. See also, Tony Saich, *China's Science Policy in the 80s* (Atlantic Highlands, NJ: Humanities Press International, 1989), 38.
[63] *STAD 1986*, 36.   [64] *STAD 1988*, 53.   [65] *STAD 1988*, 53.

The American/Saudi solar energy project illustrated the administration's politicization of science. The Reagan administration publicly opposed solar energy, going so far as to remove the solar panels from the White House, cut solar research funding, and refuse to participate in solar research with allies. Yet throughout the Middle East, solar energy provided a critical selling point for American development programs. The Reagan administration renewed SOLERAS in 1985, quietly committing $50 million to solar research in Saudi Arabia and nine universities in the United States.[66] SOLERAS brought together the variety of institutions necessary for international projects: The American project manager was Science Applications Inc. (SAIC); other team members included the Midwest Research Institute, Texas A&M Agricultural Experiment Station, the University of Texas, the McCormick Corporation, and multiple Saudi engineering firms.[67] SOLERAS eventually built the world's largest photovoltaic power station to provide electrification for two Saudi Arabian villages, leading one participant to observe, "much effort has been exerted by the United States in the use of solar energy."[68] Saudi solar energy research, like American relations with Israel, illustrated the politics of American science diplomacy in the Middle East.

US–Israeli collaboration continued to grow in the 1980s. Before Reagan's inauguration, the United States maintained a wide variety of scientific supports for Israel, including cooperation through BIRD, BARD, and BSF, offering Israeli scientists access to classified American research and shielding Israel from international nuclear inspection or sanction. The Reagan administration only furthered pre-existing policies and maintained the country's centrality in US regional relations: UNESCO's support for Palestinian nationhood and criticism of Israeli settlements was a key reason for the American withdrawal from UNESCO in 1984. SDI proved especially lucrative for Israel: Reagan funneled money to the country via contracts; during his presidency, Israel

---

[66] *STAD* 1986, 68.

[67] Midwest Research Institute, *Solar Controlled Environment Agriculture Project, Final Report: Vol. 1, Project Summary* (Kansas City, MO: Midwest Research Institute, 1986), 4. This work was "Published for the United States–Saudi Arabian Joint Program for Cooperation the Field of Solar Energy, SOLERAS."

[68] For information on the world largest generator see, *STAD* 1985, 70. The MRI participant is quoted in Midwest Research Institute, *Solar Controlled Environment Agriculture Project, Final Report*, 4.

received half the total foreign funding for SDI, through 16 contracts totaling $184 million (for comparison, the nearest countries were Great Britain with 100 contracts worth $68 million and West Germany with 33 contracts worth $65 million).[69] Legislative support for Israel was even more direct: In 1985, Congress created the US–Israel Cooperative Development Research program to provide, according to the State Department, "$5 million to promote Israeli industry in 15 developing countries."[70] Demonstrating American priorities, Congress spent as much taxpayer money to promote Israeli industry in emerging markets as on scientific and technical assistance through USAID worldwide.[71] Diplomacy clearly favored some allies; others, like Japan, could be seen as a threat.

COMPETITION RISING

If the controls are extended much further than they are at present, it could create substantial damage to fundamental science, as well as to the relationship between British and American scientists.

British Royal Society (1981)[72]

Scientists' blanket claims of freedom are somewhat disingenuous in light of arrangements made with corporate concerns. There is no problem with holding back research for trade secret reasons. This attitude is based largely on the fact that the federal government rather than corporations is the source of the restrictions. This assumes that corporate interests are at a higher level than national security concerns. I could not disagree more.

Admiral Bobby R. Inman, Deputy Director of the CIA
and former director of the NSA (1981)[73]

In February of 1980, the University of California, Santa Barbara, planned to host an International Bubble Memory Conference. Before the meeting, however, the Commerce Department sent a threatening letter to the president of the Vacuum Society requiring him to disinvite

---

[69] *STAD* 1990, 187. In total, the United States had 215 contracts for $362 million.

[70] *STAD* 1987, 186–187.

[71] USAID scientific and technical aid totaled $5.3 million for thirty-nine projects in twenty-five nations. *STAD* 1988, ix.

[72] British Royal Society quoted in David Dickson, "Europeans Protest U.S. Export Controls," *Science* 224 (May 11, 1984) 579–581, quote on 579.

[73] Inman quoted in Gina Kolata, "CIA Director Warns Scientists," *Science* 215 (January 22, 1982): 383.

conference participants from communist countries or face jail.[74] Surprised, but concerned, the society complied. Magnetic bubble memory was critical to the materials sciences involved in IT and the United States did not want advances in computer memory to fall into a rival's hands. Such policies transcended parties: The Santa Barbara conference occurred during President Carter's time in office and President Reagan extended the pre-existing regulations. Concern over foreign competition influenced American policy at home and abroad. US business leaders desired intellectual protection and patent rights overseas but questioned whether domestic export controls worked in a global economy – some argued they benefited competitors such as Japan. Meanwhile, concern over Japanese state-sponsored initiatives led to scrutiny of US–Japanese relations and a competitiveness agenda at home.

Japan, the VLSI Initiative, and American Competitiveness

The evolution of US–Japanese scientific relations concerned the Reagan administration. Formal cooperation began in 1961 and expanded rapidly; by the 1980s, the exchange program involved more than a thousand scientists, requiring NSF open an office in the American embassy in Tokyo.[75] Yet many business leaders and politicians felt Japanese success came at American expense. The State Department characterized the changing historical relationship thusly:

S&T cooperation has served an overriding US foreign policy interest for more than 30 years with Japan. The original objective was to integrate Japan into the community of advanced Western industrial democracies. That objective having been received with resounding success, efforts are now being made to refocus cooperation ... on terms appropriate to the two countries' resources and priorities today ... This has introduced a new dimension into the bilateral relationship: the importance of the U.S. remaining abreast of Japanese S&T activities."[76]

---

[74] Edward Gerjuoy, "Controls on Scientific Information Exports," *Yale Law & Policy Review* (Spring 1985): 447–478, esp. 455. See also US Congress, House of Representatives, Committee on Science, Space and Technology, *The Effect of Changing Export Controls on Cooperation in Science and Technology*, 101st Cong., 2nd sess., May 16, 1990, esp. 91.

[75] Justin L. Bloom, "The U.S.-Japan Bilateral Science and Technology Relationship: A Personal Evaluation," in Wallerstein, *Scientific and Technological Cooperation among Industrialized Countries*, 84–122, esp. 90–91.

[76] *STAD* 1987, 27.

Comparable access and reciprocity, rather than assistance, became the buzzwords in US–Japanese diplomacy.

The belief Japanese policies provided an unfair advantage spurred action. Media personalities argued Japanese prosperity resulted from stealing US know-how and products, from the transistor to robotics, computing, pharmaceuticals, and biotech.[77] The Committee on Science and Technology of the House of Representatives joined in, asserting: "Japanese maintain that, as a people, they tend to lack originality or creativity. ... However, they remain sensitive to criticism that Japan is 'living off' discoveries made elsewhere, and especially those made by US scientists and engineers."[78] Administrators mused about the Japanese influence in American academic research: Clyde Prestowitz, a Commerce Department official, pointed out Japanese corporations funded nine chairs at MIT, thus allowing them to tap "directly into the scientific source."[79] Concerned about a lack of reciprocity, the State Department launched a government-wide review of US–Japan relations in 1984 and pushed Japan to include software programming in copyright law.[80] Under pressure, Japanese universities began allowing foreigners to occupy posts and state institutes accepted thirty foreign researchers the following year.[81] But the state-sponsored semiconductor initiative caused American policy-makers the most concern.

Japan had a long history of "guided research."[82] Beginning with polymers in the 1960s, the Japanese government organized national R&D through seventy engineering research associations.[83] Concerned IBM was ahead, Japan established an association for large integrated circuits (commonly known as the VLSI program), a three-year collaboration among five Japanese companies.[84] Success with the VLSI program led to a similar consortium of Japanese firms dedicated to achieving

---

[77] For a discussion of the literature, see Drahos and Braithwaite, *Information Feudalism*, 63.

[78] Staff Report of the Committee on Science and Technology, U.S. House of Representatives, 97th Cong., 1st sess., *Survey of the Science and Technology Issues Present and Future* (June 1981) (Washington, DC: US Government Printing Office, 1981), 323.

[79] Prestowitz quoted in Adrian Johns, *Piracy: The Intellectual Property Wars from Guttenberg to Gates* (Chicago: University of Chicago Press, 2009), 456.

[80] On the State Department's review, see *STAD* 1985, 27. On the copyright law, see Drahos and Braithwaite, *Information* Feudalism, 171.

[81] *STAD* 1988, 49.

[82] Jon Sigurdson, "Industry and State Partnership: The Historical Role of The Engineering Research Associations of Japan," *Industry and Innovation* 5 (1998): 209–241.

[83] Ibid., 239–241.

[84] The companies were NEC, Toshiba, Hitachi, Mitsubishi Electric, and Fujitsu.

artificial intelligence in 1982. Although the Japanese initiative did not merge company labs or deviate from tradition, many Americans worried state-sponsored associations provided an unfair advantage to Japanese industries (at the time, similar consortiums were illegal in the United States).[85]

Americans and Europeans scrambled to counter the Japanese challenge. Congress passed the National Cooperative Research Act in 1984, legalizing the creation of American semiconductor research consortia like Sematech, a group including Sandia national laboratory and more than 130 companies.[86] The United States also required Japan change its copyright law to protect computer software.[87] Allies took steps as well. Great Britain established the Alvey Programme, a partnership with British industry and academia to work on the physics of IT, whether silicon wafer technology or conductive polymers.[88] Others included ESPRIT (Western Europe, 1983), FINPRIT (Finland, 1984), and Information-stechnik (West Germany, 1984).[89]

Concern over Japanese research occasionally descended into parody. By 1986, Japan was the world's largest producer of microchips and the US trade deficit was growing.[90] A rumor arose the Japanese were using Freedom of Information Act (FOIA) requests to obtain plans for high-tech properties, including the space shuttle, thereby saving Japan hundreds of millions of dollars.[91] Incensed, President Reagan addressed more than 2,000 American scientists and engineers involved in superconductor research (foreigners were barred from attending), proposing Congress "move to protect intellectual property and write protections into the Freedom of Information Act for scientific and technical information

---

[85] Gerald J. Hane, "The Real Lessons of Japanese Research Consortia," *Issues in Science and Technology* (Winter 1993–1994): 56–62.

[86] Herbert I. Fusfeld, "The Role of Industry in International Technical Cooperation," in Wallerstein, *Scientific and Technological Cooperation among Industrialized Countries*, 44–57. Intel led Semiconductor Research Cooperative, while the Microelectronics and Computer Corporation was a partnership between CDC, Honeywell and Sperry). See also Hane, "The Real Lessons of Japanese Research Consortia," 61.

[87] Drahos and Braithwaite, *Information Feudalism*, 171.

[88] B. W. Oakley, "The Alvey Programme: Physics in Information Technology Research," *Physical Technology* 16 (1985): 63–68. Oakley was the head of the Alvey Programme.

[89] Michael Hobday, "Evaluating Collaborative R&D Programmes in Information Technology: The case of the U.K. Alvey Programme," *Technovation* 8 (1988): 271–298.

[90] Drahos and Braithwaite, *Information Feudalism*, 125.

[91] U.S. Congress, *The Effect of Changing Export Controls on Cooperation in Science and Technology*, 92.

generated by government laboratories."[92] Yet Reagan's request proved ironic: Not only was the FOIA story untrue, but Japan's National Research Institute for Metals discovered an important new superconductor and promptly shared it with Americans.[93] Nor was that the only episode: DARPA released a bootleg copy of *The Japan that Can Say No*, a book co-authored by the founder of Sony and a prominent Japanese politician.[94] The authors suggested racism underlay American trade concerns and argued Japan should sell microchips to the Soviet Union as a show of independence, sparking a national conversation about American support for Japan and protection of American research.[95]

President Reagan outlined a variety of proposals to help the country regain a "competitive edge" in his 1987 annual address.[96] Reagan's "Competitiveness Agenda" included increased funding for research and new biotech and robotics centers.[97] Three months later, the president issued an executive order, entitled "Facilitating Access to Science and Technology," which promised to distribute foreign research to the American government, academic institutions, and private sector.[98] One program, for example, mandated NSF and ONR compile reports on foreign research for distribution by the Commerce department's technical information service; within a few years, nearly 30 percent of the material was from overseas.[99] The order also required all federal agencies and departments consider comparable access, patent rights, and export controls before engaging in international scientific cooperation.[100] Reagan's actions found congressional support: The Trade and Competitiveness Act of 1988 stated: "federally supported international science and technology agreements should be negotiated to ensure that (A) intellectual property rights are properly protected and (B) access to research and development opportunities and facilities, and the flow of scientific and

---

[92] Ronald Reagan quoted in Ibid., 93.   [93] Ibid., 93.   [94] Johns, *Piracy*, 459–460.

[95] Nicholas Wade, "America's Japan Problem: It's Race, Not Trade, Says a Japanese Book," *New York Times* (October 5, 1989): A30.

[96] Ronald Reagan, "Address Before a Joint Session of Congress on the State of the Union," (January 27, 1987). Accessed online at: www.presidency.ucsb.edu/ws/?pid=34430.

[97] Robert W. Rycroft, "The Internationalization of US Intergovernmental Relations in Science and Technology Policy," *Technology in Society* 12 (1990): 217–233, esp. 222.

[98] Executive Order 12591. See *STAD* 1996, 19–20.   [99] *STAD* 1991, 82.

[100] See Section 4, Subsections 1–3 of Executive Order 12591, "Facilitating Access to Science and Technology," (April 10, 1987). Accessed online at: www.archives.gov/federal-regi ster/codification/executive-order/12591.html.

technological information, are, to the maximum extent practicable, equitable and reciprocal."[101]

US–Japanese relations continued to receive special scrutiny. As political scientist Daniel Greenberg pointed out, "Japan demonstrated that science need not be homegrown to be commercially exploited."[102] Determined to capitalize on Japanese research, Congress passed legislation to collect, analyze, and disseminate the country's findings.[103] The State Department mounted a "Japan Initiative ... to advance the Japanese language competence of American scientists and engineers."[104] Commerce produced a series of directories on "Japanese Technical Resources in the U.S.," showcasing Japanese investments in American semiconductors, universities, and other high-tech sectors.[105] When the US international trade commission argued the United States lost $23 billion because of poor intellectual protections and cited Japan as a perpetrator, Japanese firms spent millions on more than a hundred lobbyists.[106] The uproar influenced relations: The bilateral agreement signed by President Reagan and Prime Minister Takeshita in June 1988 specified the creation of a task force aimed at maintaining equity of access between the two countries.[107] But questions regarding Japanese willingness to observe intellectual property rights and abide by export controls remained.

Protecting American R&D: Export Controls and Intellectual Property Rights

The growth of commercial R&D increased American attempts to control certain knowledge at home and abroad. At home, a revival of anti-communism during President Reagan's first term led the administration to reclassify research and introduce surveillance of electronic communications, sparking protests from scientific organizations. The Reagan administration also introduced sanctions for countries refusing to comply with American intellectual protections. Finally, many executives and

---

[101] *STAD* 1996, 23.　[102] Greenberg, *Science, Money and Politics*, 366.
[103] The legislation was the Japanese Technical Literature Act (1987), see *STAD* 1991, 82–84.
[104] *STAD* 1988, 32.　[105] *STAD* 1991, 83–84.
[106] On the U.S. trade commission, see "The Global Dimensions of Intellectual Property Rights in Science and Technology," 5. On the Japanese lobbying effort, see Rycroft, "The Internationalization of US Intergovernmental Relations in Science and Technology Policy," 229.
[107] *STAD* 1991, 18.

scientists questioned the impact of anti-communist export controls, arguing they limited American market access in the global economy.

Government oversight of domestic research and data began shortly after Reagan's inauguration. The administration pressured cryptographers to voluntarily submit research within months of taking office; simultaneously, the National Library of Medicine informed foreign distributors of MEDLINE they "should not allow any person from a Communist country to have direct, on-line computer access to the system."[108] Academic protests led CIA deputy director Admiral Bobby R. Inman to threaten more serious action if scientists refused to cooperate with national security oversight.[109] In 1982, President Reagan issued an executive order reclassifying patents whose disclosure would be "detrimental to national security" – an ill-defined standard – under the Invention Secrecy Act, which had to be appealed annually.[110] Commerce and State Department officials began warning attendees at scientific conferences; in one example, more than 150 unclassified papers were withdrawn from a public meeting.[111] Reagan administration policies also extended to broad sweeps of the telecommunications networks beginning to connect laboratories, universities and other intellectual hubs. The president signed NSDD 145 in 1984, mandating the government closely monitor all telecommunications and data, arguing, "such information, even if unclassified in isolation, often can reveal highly classified and other sensitive information when taken in aggregate."[112] Given concern over government aggregation of personal information, it is worthwhile to remember the government instituted surveillance of electronic communications – the early internet – from the beginning. The heavy-handed oversight led to push-back.

Tensions escalated between the administration and the scientific community. In 1982, the NAS reported increased scrutiny and controls harmed the scientific community. Three years later, twelve leading

---

[108] NLM quoted in Edward Gerjuoy, "Controls on Scientific Information Exports," *Yale Law & Policy Review* (Spring 1985): 447–478, quote on 457. See also Gina Kolata, "CIA Director Warns Scientists," *Science* 215 (January 22, 1982): 383.

[109] Ibid.

[110] Executive Order 12356, "National Security Information," (April 2, 1982). See also Gerjuoy, "Controls on Scientific Information Exports," 448.

[111] Colin Norman, "Administration Grapples with Export Controls," *Science* 220 (June 3, 1983): 1021–1024, esp. 1022.

[112] NSDD 145, "National Policy on Telecommunications and Automated Information Systems Security," (September 17, 1984). Available online at: http://fas.org/irp/offdocs/nsdd145.htm.

scientific organizations protested, leading the Reagan administration to exempt unclassified basic research from restriction.[113] Most scientific societies also refused administration requests to exclude foreign members: The AAAS reported more than 60 percent of societies prohibited sponsorship of closed or restricted sessions.[114] As the administration struggled to control scientific information at home, it introduced economic sanctions to enforce controls abroad.

Foreign competition in emerging high-tech industries required a new approach. In 1983, the United States linked economic sanctions and intellectual protections in the Caribbean Basin trade initiative.[115] The following year, the Congress passed the Semiconductor Chip Protection Act to confer patent rights on the topography of integrated circuits on a microchip (the "chip mask"). At the same time, the Reagan administration threatened to initiate sanctions against South Korea to enforce US intellectual property rights across a broad spectrum of high-tech industries, including IT/software, biotechnology, pharmaceuticals, and materials sciences. Indeed, these high-tech industries drove American diplomacy for science.

The growing American biotechnology industry, like the others, required intellectual protection overseas. Advances in genetic engineering led to a boom in recombinant products: Sales reached $500 million in 1987; more than fourteen new drugs became available (for ailments including heart disease, anemia, HIV-AIDS, cancers, diabetes, and others); the United States invested $3 billion (private industry contributed another $2 billion); and the USPTO issued Harvard University a patent for its "oncomouse" (a mouse genetically modified for cancer research).[116] Yet American intellectual properties, especially in genetics, were rarely respected abroad. George Rathman, the CEO of Amgen, alleged Japan allowed domestic companies to ignore patent protections, charging: "[there are] two companies in Japan enjoying the products of

---

[113] See NSDD 189. US Congress, House of Representatives, Committee on Science, Space and Technology, *The Effect of Changing Export Controls on Cooperation in Science and Technology*, 101st Cong., 2nd sess., May 16, 1990, 67.

[114] AAAS statement quoted in Ibid., 68.

[115] Drahos and Braithwaite, *Information Feudalism*.

[116] On the RDNA sales, see M. P. Feldman, et al., "Lessons from the Commercialization of the Cohen-Boyer Patents: The Stanford University Licensing Program," in A. Krattinger, et al., eds., *Intellectual Property Management in Health and Agricultural Innovation* (Oxford: MIHR, 2007), 1797–1807. On the US funding, see Wright, *Molecular Politics*, 113. On the oncomouse see, Daniel J. Kevles, "Of Mice & Money: The Story of the World's First Animal Patent," *Daedalus* 131 (Spring 2002): 78–88.

Amgen – two products approaching a billion dollars in sales, at prices two to four times that of the products in this country, guaranteeing high profits."[117] A global survey in 1988 revealed Japan was far from alone; dozens of countries refused to recognize patents on animals, pharmaceuticals, biological processes, and computer programs.[118] The widespread disdain for patent rights angered industry leaders: The president of Pfizer biopharmaceuticals accused developing countries of "intellectual socialism," arguing they used the WIPO to access patented intellectual products.[119]

Led by industry representatives, the Reagan administration supported the inclusion of intellectual property rights in the Uruguay Round of negotiations on the GATT treaty. Working with conservative think-tanks and high-tech corporations, Pfizer executives established a committee to lobby the US government for intellectual protection.[120] Sympathetic, the Reagan administration agreed to move intellectual property debates from the WIPO to the GATT, where developing countries commanded less influence.[121] According to the State Department, the United States used the GATT negotiations for a "comprehensive agreement including strong standards of protection in all intellectual property areas (patents, trademarks, copyrights, semiconductor chip mask works, trade secrets, etc.)."[122] Negotiation, backed by economic pressure, guaranteed protections globally.

Intellectual property rights rose as the Berlin Wall fell. A study of newly industrializing countries determined many adopted the American position between 1987–1990, leading the author to conclude, "There is a close relationship between this cycle of reforms in developing countries and external pressure exerted by developed countries."[123] Most newly

[117] George B. Rathman, "Biotechnology Case Study," in Wallerstein, *Global Dimensions of Intellectual Property Rights in Science and Technology*, 319–328, quote on 327.

[118] Drahos, *Information Feudalism*, 124.   [119] Ibid., 61.

[120] Conservative think-tanks included the Heritage Foundation, American Enterprise Institute, and the Hoover Institution; high-tech corporations included Du Pont, General Electric, Monsanto, and Warner Communications. See Ibid., 119.

[121] See "The Global Dimensions of Intellectual Property Rights in Science and Technology," in Wallerstein, *Global Dimensions of Intellectual Property Rights in Science and Technology*, 15. See also Isabelle Guaran, "Intellectual Property Rights: How the Developing World Is Disadvantaged by Global Government Approaches to Scientific Innovation and Intellectual Property Rights," *Australian Quarterly* 81 (September-October, 2009): 10–16.

[122] *STAD* 1991, 16–17, quote on 17.

[123] Carlos Alberto Primo Braga, "The Newly Industrializing Economies" in Wallerstein, *Global Dimensions of Intellectual Property Rights in Science and Technology*, 169.

industrializing nations adopted US policies, including Hong Kong, Malaysia, Mexico, Singapore, South Korea, Taiwan, Argentina, Chile, and Indonesia. Only Brazil briefly resisted, resulting in sanctions and the country's eventual agreement. Intellectual property rights and sanctions, more than export controls, became critical to protecting American high-tech trade.

Businessmen and scientists petitioned to end export controls. David Packard, chairman of Hewlett-Packard and a former defense secretary under Nixon, testified before Congress: "To put the matter in plain English, the current effort of the Defense Department to censor basic research in the United States is simply stupid."[124] *Science* writer David Dickson reported a "wide feeling in Europe [that] COCOM is excessively dominated by the United States," leading to a desire for an alternative organization with fewer controls.[125] The CIA pointed out export controls failed in biotechnology, observing, "The international network of suppliers that supports genetic engineering R&D is extensive. Because of this, efforts to curtail US technology transfer would not significantly impede Soviet progress ... the net effect of a singular US embargo would be the loss of an economic market...[and] Soviets have resorted to clandestine acquisition when legal means have been denied them."[126] Physicist Robert Park testified before Congress about his experience with a Chinese molecular epitaxy beam (MEB) system used in computer manufacturing: "What made this remarkable was that MEB equipment was at that time on the list of items banned for sale to China. Had the Chinese obtained the equipment illegally through some third party? Not at all. Out of necessity, they had learned to build their own. The only effect of our export restrictions was to deny some American manufacturer a sale."[127] The NAS submitted three reports in eight years on the economic harm done to the United States. In testimony before Congress, the NAS noted: "In 1980, we had a $26 billion surplus in technology trade and last year [1986], the figure will be in the red somewhere between $4 and $6

---

[124] Packard quoted in Gerjuoy, "Controls on Scientific Information Exports," 462. Packard was a deputy defense secretary.

[125] David Dickson, "Europeans Seek Technology Transfer Agency," *Science* 226 (November 30, 1984): 1057–1058, quote on 1058.

[126] Scientific and Technical Intelligence Committee, CIA, *Soviet Genetic-Engineering Capabilities* (December 1983). CIA ESDN Doc No: 0000722702, quotes on iv, 12 and 15.

[127] Park quoted in US Congress, House of Representatives, Committee on Science, Space and Technology, *The Effect of Changing Export Controls on Cooperation in Science and Technology*, 101st Cong., 2nd sess., May 16, 1990, 83.

billion ... No longer is it just the United States and Japan and West Germany that manufacture and export technology, but countries outside COCOM like Hong Kong and Indonesia, are all getting into advanced technology."[128] By 1990, research and development were going global; only the Soviet Union stood in the way.

SOVIET FISSION

Perestroika means a resolute shift to scientific methods, an ability to provide a solid scientific basis for every new initiative. It means the combination of the achievements of the scientific and technological revolution with a planned economy.

Mikhail Gorbachev (1987)[129]

In India there are more than 100 collaborative physical science projects funded either by the Indo-US Rupee Fund or the Indo-US Science and Technology Initiative ... Cooperation under the US-PRC Protocol on Basic Sciences, slowed since the trauma of 1989, continues with some bilateral projects. The successful operation of Beijing's electron-positron collider is a demonstration of advancing Chinese capability."

US State Department (1991)[130]

The collapse of the Soviet Union removed a key impediment to globalization. Although communist aggression in Afghanistan and Poland limited US/Soviet-bloc relations during Reagan's first term, scientific exchanges never entirely ceased. Indeed, overt and covert access to US research remained critical to the Soviet Union, but COCOM maintained export controls and American intelligence tracked the regime's inability to keep up with the biotechnology and computer revolutions. Even as the Soviets fell behind, the world became more interdependent – US cooperation increased in India and throughout the fall of the Berlin Wall and Tiananmen Square protests. Much like Deng Xiaoping a decade before, Mikhail Gorbachev assumed domestic institutions needed reform and

---

[128] NAS synopsis available in US Congress, House of Representatives, Subcommittee on International Economic Policy and Trade of the Committee on Foreign Affairs, *National Security Export Controls Report by the National Academy of Science*, 100th Cong., 1st sess., February 3, 1987, 2.

[129] Mikhail Gorbachev, *Perestroika: New Thinking for Our Country and the World* (New York: Harper Collins, 1987), 35.

[130] *STAD* 1991, 35.

believed foreign science and technology could help unleash productivity; he modeled *perestroika* – or restructuring – on an experimental approach. But his experiment broke an already brittle Soviet Union, leading the United States to aid the re-integration of former Soviet-bloc researchers into the global community.

Soviet Malaise, Chernobyl, and Collapse

Soviet actions limited relations during President Reagan's first term. Scientific exchanges involved eleven different fields at their height, but President Carter cancelled many after the invasion of Afghanistan (see Chapter 5). Following the declaration of martial law in Poland in 1981, Reagan declined to renew exchanges in space and energy; when the Soviets shot down a Korean airliner in 1983, the administration cancelled an agreement on transportation research.[131] By 1984, only programs of "particular interest" to the United States – health, environmental protection, and oceanography – remained.[132] Fieldwork provided a rare space for collaboration: Under the environmental agreement, Americans participated in expeditions aboard Soviet vessels to determine the composition of gases above the Bering Sea.[133] Of course, scientific cooperation also provided a subtle way to challenge Soviet rule in Eastern Europe.

Poland played a pivotal role in Reagan's approach to the Soviet bloc. A presidential defense directive mandated a more aggressive anti-communist policy in 1983, stressing economic pressure and psychological operations.[134] The directive also suggested scientific exchanges as part of an effort to promote "evolutionary change within the Soviet system."[135] When the Polish communists lifted martial law and freed political prisoners, the Reagan administration lifted sanctions on cooperation. The National Academy of Sciences met with their Polish counterparts and the State Department posted a scientific counselor to Warsaw; the annual report for 1985 concluded, "In sum, a renewed effort is underway to re-establish contacts with the scientific communities in Poland and to increase cooperative scientific activities."[136] Similar agreements followed with Czechoslovakia and Hungary, increasing pressure on the Soviet Union. American initiatives came at an opportune time: Gorbachev

---

[131] *STAD 1985*, 31–32.   [132] *STAD 1985*, 32.   [133] *STAD 1985*, 87.

[134] See National Security Defense Directive 75, "U.S. Relations with the USSR," (January 17, 1983). Available online.

[135] Ibid., 6.   [136] *STAD 1985*, 33.

wanted to reduce nuclear arms and open the Soviet system; at the 1985 Geneva summit, for example, he and Reagan promised their countries would work together against cancer.[137] Back home, Gorbachev attempted to reform an outdated and unproductive system.

The malaise among Soviet researchers had become a crisis. Success in space and advanced weaponry led to a "cult of science" among Soviet intellectuals, who often criticized the duplication of research in capitalist societies (Dow, Du Pont, Monsanto, and Union Carbide, for example, all worked on synthetic fibers in the United States, whereas the Soviet Union had one centralized synthetic fibers research institute).[138] Yet while the Soviet scientific complex was immense (more than 1.5 million official researchers), it failed to effectively link research to production or generate ground-breaking work (from 1945–1991, for example, Soviets won eight Nobel prizes; Americans won sixty).[139] By the 1980s, the failure to maintain parity in biotechnology and computer sciences highlighted weaknesses in the centralized approach; American intelligence quoted Soviet researchers complaining, "Science, like Gulliver, has wound up tied by its hair to hundreds and thousands of pegs of different instructions and decrees."[140] The mistreatment of Sakharov undermined the morale of many researchers; one of Gorbachev's first acts was to end the famed physicist's exile in Gorky. Aware the country had "lost 15 years at the very least," Gorbachev proposed a twenty-year initiative in applied research to catch up, focusing on microelectronics, computers, biotechnology, and lasers.[141] However, an accident, rather than a breakthrough, shook up the system.

Reactor four at the Chernobyl nuclear plant exploded in April 1986, sending radioactive particles across Ukraine and much of Europe. Soviet

---

[137] *STAD* 1987, 24–26.

[138] On the cult of science, see Vladislav M. Zubok, *A Failed Empire: The Soviet Union in the Cold War from Stalin to Gorbachev* (Chapel Hill: University of North Carolina Press, 2009), 180–182. For Soviet criticisms of duplication and the synthetic fibers example see, Graham, *Science in Russia and the Soviet Union* 179–181.

[139] On the number of researchers, see Loren Graham and Irina Dezhina, *Science in the New Russia: Crisis, Aid, Reform* (Bloomington: Indiana University Press, 2008), 1.

[140] CIA Office of Soviet Analysis, "Perestroyka and US–Soviet S&T Cooperation: Opportunities and Pitfalls (August 1990), 1–17, quote on 2. CIA Doc/ESDN: 00004999186.

[141] Gorbachev quoted in CIA Office of Soviet Analysis, "The Role of the USSR Academy of Sciences in Gorbachev's Modernization Campaign," (July 1987): 1–35, quote on 1. CIA Doc/ESDN: 00004999528. Note that the Soviet initiative began before Gorbachev's campaign, see Graham, *Science in Russia and the Soviet Union*, 187. See also David Dickson, "A Bleak Portrait of Soviet Science," *Science* 241 (July 15, 1988): 287.

authorities managed to contain news of the incident, but the particles proved more difficult. When Swedish researchers revealed contamination, the United Nations demanded access; IAEA officials met with a Soviet delegation in August. The worst nuclear disaster to date, Chernobyl highlighted the dangers of Cold War secrecy and the global reach of environmental catastrophe. Within a month, the UN passed conventions on notification and assistance in case of nuclear or biological accidents and emergencies. As international monitoring became more common, the meltdown remade Russian politics and American diplomacy.

The Chernobyl disaster validated Gorbachev's criticisms and weakened the Soviet Union. Reflecting on its twentieth anniversary, the former Soviet leader opined:

> The nuclear meltdown at Chernobyl 20 years ago this month, even more than my launch of *perestroika*, was perhaps the real cause of the collapse of the Soviet Union five years later. Indeed, the Chernobyl catastrophe was a historic turning point: there was the era before the disaster, and there is the very different era that has followed.[142]

The meltdown, and the government's poor response, illustrated the inefficiencies and constraints of the Soviet system Gorbachev hoped to reform. Yet bureaucracy and the costs of disaster proved difficult to overcome.

Scientific and technical sectors were among the first to undergo market reform. When the state-sponsored IT initiative failed, the Soviets faced dissent: At the last CEMA meeting in 1988, the Hungarian, Czech, and Polish delegates demanded more independence in developing their own IT initiatives.[143] Free to seek international collaboration the following year, former Soviet-bloc members sought aid and markets where they could, often in the United States. In the Soviet Union, *glasnost* ("openness") unleashed a torrent of criticism: The media attacked state controls while several thousand researchers at the Soviet Academy of Sciences protested (thousands of others simply accepted invitations to go abroad).[144] Desperate to promote innovation and access foreign capital and know-how, Soviet authorities allowed limited foreign partnerships; within two years,

---

[142] Mikhail Gorbachev, "The Nuclear Disaster That Opened Our eyes to the Truth," *The Australian* (April 19, 2006): 12.

[143] David Dickson, "East European Scientific Cooperation Seen Lagging," *Science* 241 (July 29, 1988): 524.

[144] Gina Kolata, "Soviet Scientists Flock to U.S., Acting as Tonic for Colleges," *New York Times* (May 8, 1990): A1; Sergei Leskov, "America's Soviet Scientists," *New York Times* (July 15, 1993): A25. See also Graham and Dezhina, *Science in the New Russia*, 23–24.

there were thousands of cooperatives with European, Japanese, and American partners.[145] Soviet science and technology appeared on the market: State institutes began selling products to raise hard currency, whether atmospheric data and genetically engineered microorganisms or oceanographic vessels and wind tunnels.[146]

Perestroika and Chernobyl shaped US/Soviet relations. The Soviets reconsidered patent laws before the disaster; after, the national academies of each nation collaborated on air and water pollution and soil degradation. When an earthquake devastated Armenia two years after Chernobyl, the Soviets requested American aid, even allowing the NAS to meet with Sakharov.[147] Months before the Berlin Wall fell, the United States and Soviet Union signed agreements on oceanography, atomic energy, and basic sciences (mathematics, theoretical physics, chemistry, and biology).[148] As the Soviets entered into commercial agreements abroad, export controls were one of the few remaining stumbling blocks; CIA analysts observed, "Moscow also has intensified its assault on US restrictions on technology transfer and on [COCOM]."[149]

US officials worried about Soviet acquisition of restricted science and technology. Exchanges concerned OSTP Director John McTague from the start; he testified before Congress in 1986: "The Soviets for a time were extremely successful in tapping into our R&D effort by cutting separate deals with individual agencies that were often not in the overall national interest. Among the areas the Soviets targeted and were successful in acquiring key scientific and technical knowledge, were advanced manufacturing, robotics and information technology."[150] CIA analysts agreed behind closed doors, reporting: "The KGB briefed and debriefed all Soviet scientists traveling to the West. ... The KGB in 1988 was actively recruiting scientists at the Estonian Academy's Institute of Cybernetics to act as intermediaries in the acquisition of Western communications security technology. Apparently most Soviet scientists

---

[145] Graham, *Science in Russia and the Soviet Union*, 189.

[146] CIA Office of Soviet Analysis, "Perestroyka and US–Soviet S&T Cooperation," 13.

[147] Schweitzer, *Scientists, Engineers, and Two-track Diplomacy*, 19–21.

[148] David Dickson, "U.S.–Soviets Sign Collaboration," *Science* 243 (January 13, 1989): 161.

[149] CIA Office of Soviet Analysis, "Perestroyka and US–Soviet S&T Cooperation," 3.

[150] McTague quoted in Congress, House of Representatives, Committees on Foreign Affairs and Science and Technology, *Overview of International Science and Technology Policy: The Federal Organization, Joint Hearing before the Committees on Foreign Affairs and Science and Technology*, 99th Cong., 2nd sess. (May 20, 1986): 47.

have been willing to act as intelligence collectors for materialistic, professional, and patriotic reasons."[151] American intelligence tracked Soviet inroads abroad, reporting the Soviet Academy of Sciences had entered into thirty-three joint ventures with Western firms, even as France expelled forty-seven Soviet "diplomats" for spying on their rocket program.[152] But state initiatives could not overcome the sclerotic bureaucracy or the impact of Chernobyl and the Soviet Union disappeared into history.

Re-Integrating Post-Soviet Science: The ISTC and ISS

The transition from Soviet to Russian science was an exercise in "shock therapy," providing opportunities for American firms and causing concern among administration officials. The USSR lacked patent rights or mechanisms to stimulate investment and had no tradition of innovation through market pressure; i.e., there was no "commercial culture" in Soviet research.[153] Instead, a market in Soviet scientific and technical products developed as formerly secret groups went public in the search for financing.[154] Sun Microsystems, for example, hired the top Soviet computer scientist and his entire team of fifty from the Institute for Precision Mechanics & Computing Equipment, while the Stanford Linear Accelerator Laboratory purchased ten high-precision magnets at bargain prices from the Institute for Nuclear Physics.[155] The new Russian state tried to establish a national innovation system, but the poor economy limited funding. The pitiful state of Russian science surprised many American policy-makers and scientists; a researcher at Lawrence-Livermore laboratory suggested an aid program for his Russian

---

[151] CIA Office of Soviet Analysis, "Perestroyka and US–Soviet S&T Cooperation," 10.

[152] Ibid., 4. See also David Dickson, "France Expels Soviets for Spying on Ariane," *Science* 236 (April 10, 1987): 142.

[153] See "Developing a Commercial Culture for Russian Science," in Graham and Dezhina, *Science in the New Russia*, 67–88.

[154] See Mike Berry and Lioudmila Pipiia, "Academy-Industry Relations in Russia: The Road to Market," in Etzkowitz, *Capitalizing Knowledge*, 169–186. See also Glenn E. Schweitzer, *Moscow DMZ: The Story of the International Effort to Convert Russian Weapons Science to Peaceful Purposes* (Armonk, NY: M. E. Sharpe, 1996), 17.

[155] Anon., "Soviet Scientists Sign Pact to Help Sun Microsystems," *Wall Street Journal* (March 4, 1992): A7. See also David P. Hamilton, "Piecemeal Rescue for Soviet Science," *Science* 255 (March 27, 1992): 1632–1634.

colleagues, but financial aid for world-class scientists was unknown.[156] Yet Soviet-era specialists presented a unique threat; CIA Director Robert Gates warned of a potential "brain drain" to Cuba, Syria, and others.[157] Supported by Congress, Presidents Bush and Yeltsin finalized proposals for international funding of Russian scientists at Camp David in February 1992; the following month, the administration convened a panel of 120 leading American scientists and engineers to discuss how to reorient Russian research.[158] Keeping former Soviet weapons researchers from helping rogue nations was a primary concern.

Redirecting and containing Russian scientists became a goal of US diplomacy. According to one study, the former Soviet Union became the "object of the largest international scientific aid program in history" from 1992 to 2007.[159] Congress passed the Soviet Nuclear Threat Reduction Act (1991) and the "Freedom Support Act" (1992), eventually providing more than $18 billion in total assistance, with $1.5 billion earmarked for science and technology.[160] Amid Russian media reports of five nuclear thefts, the United States partnered with the EU and Japan to open the International Science and Technology Center (ISTC, 1994) in Moscow.[161] The ISTC originally focused on converting weapons scientists to peaceful purposes: Glenn Schweitzer, its first executive director, noted nearly all his longtime Soviet colleagues came out as having been employed by the defense industry.[162] The center also provided American access to formerly closed research cities as part of the conversion process, whether Chernogolovka, an aerospace research center, or Koltsovo, where the ISTC labored to turn bioweapons research toward vaccine production.[163] Nor was the center the only game in town: The Europeans

---

[156] On the surprise of lawmakers and the Camp David meeting, see Schweitzer, *Moscow DMZ*, 18–19. On the WPA proposal, see William G. Sutcliffe, "A Soviet-Scientist 'WPA'?" *Tulsa World* (February 15, 1992).

[157] Gates quoted in David Hoffman, "Ex-Soviet Scientists To Get Aid; Center to Employ Nuclear Experts; Yeltsin, Baker Meet," *Washington Post* (February 18, 1992): A1.

[158] Frank Press, chairman, *Reorientation of the Research Capability of the Former Soviet Union: A Report to the Assistant to the President for Science and Technology* (Washington, DC: National Academy Press, 1992).

[159] Graham and Dezhina, *Science in the New Russia*, 89.

[160] The complete title of the "Freedom Support Act" was "The Freedom for Russia and Emerging Eurasian Democracies and Open Markets Support Act," see Glenn E. Schweitzer, *Experiments in Cooperation: Assessing U.S.–Russian Programs in Science and Technology* (New York: The Twentieth Century Fund Press, 1997), 26.

[161] Schweitzer, *Moscow DMZ*, 118. [162] Ibid., 13.

[163] Ibid., 155. For a detailed overview of conversion projects, see Schweitzer, *Experiments in Cooperation*, 47–74.

set up an organization to link Russian R&D to European institutions; the British Royal Society expanded programs with the Russian Academy of Sciences; and billionaire George Soros established a foundation to provide funds to Russian scientists outside the defense industry.[164]

US initiatives went beyond the ISTC and multilateral programs. According to a RAND study, the federal government spent, on average, $350 million per year in Russia (funding peaked in 1996).[165] The Department of Defense, for example, signed hundreds of agreements; the Defense Special Weapons Agency alone entered into more than 120 contracts with Russian scientific institutions.[166] The Commerce Department focused on exploiting latent capabilities in the defense industry: GE worked with Russian aerospace on helicopters and Boeing sponsored computer modeling efforts.[167] The Department of Energy received $35 million for matchmaking services for Russian researchers and American industry, listing more than 100 institutes available for cooperative arrangements with US firms and laboratories. Staff from Los Alamos even visited Sarov, a nuclear research city closed during the Soviet era (and closed again under Vladimir Putin).[168] However, as always, the single most expensive initiative circled overhead.

Freedom provided an ideal diplomatic opportunity. The difference in rhetoric before and after the Cold War was remarkable. In 1989, President Bush reminded Americans: "To this day, the only footprints on the moon are American footprints. The only flag on the Moon is an American flag, and the know-how that accomplished these feats is American know-how."[169] But Congress capped spending on *Freedom* after the expected costs soared from $8 billion in 1984 to $38 billion in 1991.[170]

---

[164] The European organization is INTAS, see http://ec.europa.eu/research/nis/en/intas.html. The Royal Society initiative is discussed in Schweitzer, *Moscow DMZ*, 177. The Soros foundation was known as the International Science Foundation until its termination in 1996.

[165] Caroline Wagner, et al., *U.S. Government Funding for Science and Technology Cooperation with Russia* (Arlington, VA: RAND Science and Technology Policy Institute, 2002), xi.

[166] Schweitzer, *Experiments in Cooperation*, 31. Schweitzer lists more than twenty different defense research units involved in Appendix C, "Department of Defense Units Involved in Joint Activities with Russian Institutions," 135.

[167] Schweitzer, *Moscow DMZ*, 185–186.

[168] Hamilton, "Piecemeal Rescue for Soviet Science," 1634.

[169] Bush quoted in Committee on Commerce, Science and Transportation, US Senate, 109th Cong., 1st sess., *Economic Development Opportunities in Nano Commercialization* (April 20, 2006) (Washington, DC: Government Printing Office, 2010), 28.

[170] Greenberg, *The New Politics of Science*, 410–417.

Just three years later, President Clinton said, "I am calling for the United States to work with our international partners to develop a reduced-cost, scaled-down version of the original *Space Station Freedom* ... there is no doubt that we are facing difficult budget decisions."[171] Meanwhile, the CIA reported the Russian Space Agency needed to market commercial launch services to survive; the NASA administrator purchased hardware and services as part of the Shuttle–Mir agreement, telling Congress, "The bargains in Russia are simply too great to resist."[172] Seizing the opportunity, President Clinton secured $400 million to entice Russian participation in *Freedom*. John Gibbons, Clinton's assistant for science and technology, outlined the project: "The big new idea was to use this as a mechanism to build international big science and technology, and to bring Russia into it rather than justify it on the basis that the 'evil empire' is out there."[173] Renamed the International Space Station (ISS), the core section is an American-financed and Russian-built module based on *Mir-2*.[174]

The International Space Station paid political and economic dividends. *Mir* and the space shuttle performed a repeat "handshake-in-space" in 1995, while the Japanese, EU, and Canadians agreed to participate a few years later (as before, the Canadians contracted to build the ISS robotic arm). Building the station turned former aerospace rivals into industrial partners: Lockheed joined with Khrunchiev-Energia to market the Proton rocket booster, while Pratt and Whitney signed an agreement with Energomash to modernize American rocket engines.[175] Once aloft, the ISS turned to corporate advertising and sponsorship to secure support. Mir-Corp ferried wealthy private citizens to the ISS for a fee; RadioShack, Lego, and *Popular Mechanics* shot TV commercials aboard the station (a paid experiment required American astronauts measure the mass of a Lego in space).[176] Many scientists questioned its value throughout: Both the American and European Physical Societies considered the ISS a drain

[171] Clinton quoted in Committee on Commerce, Science and Transportation, US Senate, 109th Cong., 1st sess., *Economic Development Opportunities in Nano Commercialization* (April 20, 2006) (Washington, DC: Government Printing Office, 2010), 29.

[172] CIA, Directorate of Intelligence, "The Russian Space Launch Vehicle Industry: Looking to Foreign Sales for Survival," (September 15, 1992). For the bargain quote, see Schweitzer, *Moscow DMZ*, 181.

[173] John Gibbons quoted in Greenberg, *The New Politics of Science*, 416.

[174] Peter Bond, *The Continuing Story of the International Space Station* (New York: Springer Publishing, 2002).

[175] Schweitzer, *Experiments in Cooperation*, 37.

[176] Bond, *The Continuing Story of the ISS*, 299.

on limited resources; physicist James Van Allen argued, "The cost of the space station is far beyond any justifiable scientific purpose or any justifiable practical purpose."[177] But space programs were never solely about science. The 1998 Commercial Space Act, for example, opened with: "The Congress declares that a priority goal of constructing the International Space Station is the economic development of Earth orbital space. The Congress further declares that free and competitive markets create the most efficient conditions for promoting economic development and should therefore govern the economic development of Earth orbital space."[178] The following year, Congress required NASA publish a price policy to eliminate "price uncertainty."[179] By 2012, the USPTO had granted more than 1,500 ISS-related patents (with another thousand applications pending).[180] Business, and research, were good.

US–Soviet engagement shifted toward commerce in the late 1990s. Russia created its first Innovation Technology Center in St. Petersburg in 1996 and the Gore-Chernomyrdin commission signed agreements to increase US investment in Russia. The following year, the ISTC increased support for the commercialization of Russian R&D; foreign partners included 3M, Lockheed-Martin, Shell, Hitachi, Samsung, Lawrence Livermore laboratory, and CERN.[181] Loren Graham and Irina Dezhina, authors of *Science in the New Russia* (2008), argue the ISTC evolved from nonproliferation efforts to global commerce after the center's administrators realized the enduring defense budgets of the United States and Russia made it difficult to convert weapons scientists.[182] As of 2014, the center's budget continues to grow, having supported more than 2,700 projects at the cost of $870 million.[183] Even during the Putin era, the ISTC and other scientific and technical cooperation maintained US–Russian relations.

[177] Van Allen quoted in Bond, *The Continuing Story of the ISS*, 279.
[178] Quote from Title 1, Section 101 of the Commercial Space Act (1998), available at: www.nasa.gov/offices/ogc/commercial/CommercialSpaceActof1998.html.
[179] L. Bush, "International Space Station Commercialization Policy," *Technology in Society* 24 (2002): 69–75, quote 71.
[180] Committee on Commerce, Science and Transportation, US Senate, 112th Cong., 2nd sess., *The International Space Station: A Platform for Research, Collaboration and Discovery* (July 25, 2012) (Washington, DC: Government Printing Office, 2013), 12.
[181] Graham and Dezhina, *Science in the New Russia*, 142–143.   [182] Ibid., 143–144.
[183] The ISTC has also worked with more than 70,000 former weapons scientists. See the ISTC website: www.istc.ru/istc/istc.nsf/va_WebPages/WhoweareEng.

The Legacy of the Cold War on American Science and Foreign Relations

Scholars debate the impact of the Soviet collapse on American science. Historian Audra Wolfe, for example, argued: "Overnight, the collapse of Communism stripped the American scientific community of much of the justification for its existence. Without a rival superpower, wars might be fought with more mundane technologies that did not require the input of the world's leading scientists."[184] This diminishing justification argument is most evident in high-energy physics, as the United States canceled the SSC and dramatically scaled back participation in the International Thermonuclear Reactor (ITER). But neither project had significant military or civilian applications and other research had a base of support beyond the fear of communism and the Cold War. Political scientist Daniel Greenberg minimized the impact of the Soviet conflict on American science funding:

> The Cold War's influence on science funding declined substantially a decade or more before the end of the Cold War; it was replaced by other motivations, principally faith in research for industrial competitiveness, good health, a clean environment, and other important, widely endorsed social and economic purposes. Government support of science has increased substantially since the demise of the Soviet Union; in recent years, the increase in some sectors, especially medical research, has accelerated.[185]

Yet while government funding for science increased, private funding increased more. Thus, over the course of the Cold War, the growth of applied research and development in the private sector shifted the balance of R&D from public to private in the United States. In fact, the trendlines were unaffected by the Soviet collapse (see Figure I.1).

The last half of the Cold War saw the consolidation of a "market logic" in American science.[186] Industry and academia funded most research in the United States from the turn of the century to World War II, but federal funding of research grew to unprecedented levels in the following three decades (see Chapters 1–3). Some policy-makers considered science a non-commercial public good in the early Cold War, but biotechnology, pharmaceuticals, and information technology (among others) challenged

---

[184] Wolfe, *Competing with the Soviets*, 135.
[185] Greenberg, *The New Politics of Science*, 7.
[186] The phrase "market logic" comes from Berman, *Creating the Market University*, 177.

that interpretation.[187] Amid the disorientation of the 1970s, the growth of commercial R&D recast science as proprietary and the engine of national economic prosperity. The following decade, the government secured private research through patent protections and encouraged university/industry partnerships, nudging some fields toward market-driven research.[188] The Clinton administration even supported the commercialization of defense research, suggesting "dual-use" items have "the potential to enhance our economic well-being."[189] The State Department listed the ways it supported the commercialization strategy: "[1] Monitor foreign science and technology developments; [2] Conduct strategic international collaboration, sometimes with the private sector, to exploit opportunities for mutual gain; [3] Seek to eliminate international barriers that impede technology development or trade."[190] Though the collapse of the Soviet Union clarified the importance of commercial research in American foreign relations, the reorientation began long before.

A focus on commerce explains American diplomacy for science. American administrations wanted to protect domestic advantages and were unwilling to collaborate on research with commercial applications, hence the limited participation in the G7 initiative, the oversight of a bubble memory conference and the use of economic leverage to enforce intellectual patent rights. Market potential was paramount in the new "knowledge economy" and the evolution of *Freedom* and passage of the Commercial Space Act provide perhaps the clearest example of this principle after the Cold War. Additionally, because Congress and multiple presidential administrations wanted to minimize spending on "basic" research, the United States did not compete to host the International Thermonuclear Experimental Reactor, instead reducing the country's status to a junior partner (France won the honor).[191] Without

---

[187] This switch from science as a public good or resource to science as an economic engine is addressed in Mirowski, *Privatizing American Science*, 57–66 and Berman, *Creating the Market University*, 3 and 177. Regarding political interpretations of science as intellectually "value-free," see Heather E. Douglas, *Science, Policy and the Value-Free Ideal* (Pittsburgh, PA: University of Pittsburgh Press, 2009).

[188] Some see the growth of "entrepreneurial science" as a "Second Academic Revolution," see Henry Etzkowitz and Andrew Webster, "Entrepreneurial Science: The Second Academic Revolution," in Etzkowitz, *Capitalizing Knowledge*, 21–46. A good overview of the federal role in creating university/industry partnerships is Berman, *Creating the Market University*, 119–145.

[189] Clinton, quoted in "Letter of Transmittal," in *STAD 1994*, vii.   [190] *STAD 1995*, 36.

[191] On the ITER, see Homer Alfred Neal, et al., eds., *Beyond Sputnik: US Science Policy in the 21st Century* (Ann Arbor: University of Michigan Press, 2008), 204–207. For an overview of Congress and science, see Greenberg, *The New Politics of Science*, 435–437.

the Cold War competition for prestige and scientific preeminence, US diplomacy for science was focused on funding relatively low-cost "basic" research and advocacy for intellectual protections for commercial research. The Human Genome Project, first proposed in the 1980s and discussed in the next chapter, was one of the last large-scale international scientific undertakings led by the United States.

Scientific cooperation was an accepted and competitive component of international relations after the Cold War. To reaffirm their long-standing relationship, for example, the United States and European Union signed several scientific and technical agreements in the 1990s.[192] At the same time, the EU Fifth Framework (1998), hoping to capitalize on innovative ideas overseas, indicated a European willingness to fund foreign research, while the United States required principal investigators to be Americans to receive funding. In 2005, a US congressional hearing on "Science, Technology and Global Economic Competitiveness" assumed "other nations have recognized the importance of innovation to economic growth, and are pouring resources into their science and technology infrastructure, rapidly building their innovation capacity and increasing their ability to compete with the United States in the global economy."[193] The same year, the National Academy of Sciences advised *Rising Above the Gathering Storm* would require more funding for American research and education.[194] As the financial implications of biological and environmental sciences grew at the beginning of the millennium, scientific relations grew ever more complex.

[192] These include the Transatlantic Declaration (1990), the New Transatlantic Agenda (1995), and the Transatlantic Economic Partnership (1998) as well as EU–US Agreement for Scientific and Technological Cooperation (signed in 1998, renewed in 2004 and expanded in 2009 to include security and space research). Gabriella Paár-Jákli, *Networked Governance and Transatlantic Relations: Building Bridges through Science Diplomacy* (New York: Routledge, 2014). The Framework Programmes for Research and Technical Development began in 1984.

[193] US Congress, House of Representatives, Committee on Science, *Science, Technology and Global Competitiveness*, 109th Cong., 1st sess., October 20, 2005 (Washington, DC: Government Printing Office, 2005), 3.

[194] National Academy of Science, *Rising above the Gathering Storm: Energizing and Employing America for a Brighter Economic Future* (Washington, DC: National Academies Press, 2005), available online.

7

The Fray

The space shuttle *Challenger* exploded on January 28, 1986, killing all seven crew members. Watched live by millions of Americans, the disaster persuaded policy-makers to use Russian and Chinese rockets rather than the shuttle for many future orbital deliveries. Presidents Bush and Clinton approved sixteen Chinese launches even as the United States sanctioned the PRC for transferring missile technology to Pakistan.[1] But the unusual arrangement struggled to get off the ground: Chinese rocket failures claimed two Hughes Space & Communications satellites in the mid-1990s. Over the next two decades, US–Chinese relations illustrated the entanglement of science, technology, and commerce in diplomacy, as the two countries cooperated, competed, and clashed. Concerns arose about exchanges, espionage, and intellectual property rights, shaping relations and research.

Events were often seen through a political lens. When a Chinese rocket exploded on February 14, 1996, destroying the Loral Space & Communications satellite aboard (Intelsat 7A), international insurers and American

---

[1] The sanctions enacted after the Tiananmen protests in 1989 required presidential approval. Sheehan, *The International Politics of Space*, 164–165; see also Marcia Smith, "China's Space Program: An Overview," (October 21, 2003), 5–6. Report for Congress, Congressional Research Service, RS21641. Available at www.crs.gov. Finally, see James Mann, *About Face: A History of America's Curious Relationship with China, from Nixon to Clinton* (New York: Vintage Books, 2000), 286–287. For contemporaneous accounts on wavers after Tiananmen, see Roberto Suro and John F. Harris, "President Overrode China Launch Concerns," *Washington Post* (May 23, 1998): A01; on Pakistani transfers, see David Jackson and Lena H. Sun, "Liu's Deals with Chung: An Intercontinental Puzzle," *Washington Post* (May 24, 1998): A01.

politicians demanded an accounting. Dubbed the "St. Valentine's Day Massacre," the accident led to dozens of deaths and a reevaluation of Chinese rockets and American policies.[2] Although poor soldering ultimately took the blame (insurers paid $219 million), critics felt the episode damaged national security and implicated President Clinton; investigating Loral, according to Representative Dana Rohrbacher (R-AZ), was "far more important than any White House sex scandals."[3]

The Loral and Hughes "scandals" revealed the politics of US–Chinese relations in space. American presidents approved satellite sales and launches to encourage PRC support for export controls; after the Cold War, President Clinton transferred approval authority from the Department of State to the Department of Commerce to avoid diplomatic sanctions. Oversight was lax and unclear at the time of the "Massacre." Complicating matters, analysts disagreed about whether the episode was harmful; some thought it strengthened US security because the Chinese provided information on their Long March rocket for the first time.[4] An expert with the Federation of American Scientists pointed out the Chinese report contained "material a spy could only dream of;" an NSA official added, "we know the frequencies, the orbits and the way to jam it if we ever went to war," and a Pentagon representative remarked, "What our military wants most is for the Chinese military to use our satellites."[5] Loral and Hughes stressed the $1.7 billion market in China and 16,000 US jobs.[6] Nonetheless, Congress voted to ban Chinese launches and created the Cox committee to investigate, which claimed the PRC had been stealing secrets from US national labs for decades.[7] Industry leaders in aerospace, communications, fiber optics, and other "dual-use" sectors worried about government oversight. American firms in Beijing quickly distanced themselves from the Cox report; a spokesman

---

[2] Unofficial reports suggested the death count might be as high as fifty-six: See "St. Valentine's Day Massacre," in Brian Harvey, *China's Space Program: From Conception to Manned Spaceflight* (New York: Springer Praxis, 2004), 122.

[3] Regarding soldering, see John Mintz, "Missile Failures Led to Loral-China Link," *Washington Post* (June 12, 1998): A20. Rohrbacher quoted in Juliet Eilperin, "GOP Says U.S. Gave China Nuclear Edge," *Washington Post* (May 6, 1998): A04.

[4] The CIA and Air Force Intelligence disagreed about the dangers of aiding Chinese launches. See Walter Pincus and John Mintz, "White House: Chinese Launches Aid U.S.," *Washington Post* (June 19, 1998): A04.

[5] Officials quoted in Walter Pincus, "U.S. Gains Intelligence in China Launches," *Washington Post* (June 13, 1998): A18.

[6] Walter Pincus and John Mintz, "White House: Chinese Launches Aid U.S.," A04.

[7] The Cox Report is available online at: www.house.gov/coxreport/cont/gncont.html.

for Lucent worried, "The potential for really screwing things up for high-tech is very high right now."[8] Chinese participation in the International Space Station was already off-limits but concern over nuclear espionage grew: Months after the Cox report, Los Alamos laboratory fired Taiwan-born physicist Wen Ho Lee.[9]

Although Lee's innocence or guilt remains unclear, his arrest upset US–Chinese relations. Lee's treatment by federal authorities – shackled in solitary confinement for months – eventually won him a large settlement from the government and press.[10] Some researchers wondered about racial profiling of Asians and the *New York Times* apologized for its coverage.[11] Yet Lee pled guilty to mishandling classified information and misled authorities repeatedly. He lied to FBI investigators twice, once in 1982 and again about meeting with Hu Side, China's primary bomb designer, more than a decade later.[12] Hearings revealed another national lab employee had confessed to disclosing a hohlraum, a key component of an atomic bomb, to the Chinese.[13] US–PRC relations became more polit-ically contentious: Senator John Warner (R-VA), chairman of the Armed Services Committee, suggested China would be "America's and the West's natural enemy in the next millennium."[14] After the settlement, Attorney General Janet Reno and FBI Director Louis Freeh maintained

---

[8] Stephen N. Brown, "Cox Report Has Fallout for Fiber Investments in China," *Lightwave* (August 1999): 14.

[9] On the ISS and China, see Emilian Kavalski, *The Ashgate Research Companion to Chinese Foreign Policy* (New York: Routledge, 2012), 403.

[10] Wen Ho Lee, *My Country Versus Me: The Firsthand Account by the Los Alamos Scientist Who was Falsely Accused of Being a Spy* (New York: Hachette, 2003). See also "Kindred Spirit: Wen Ho Lee," in David Wise, *Tiger Trap: America's Secret Spy War with China* (Boston: Houghton Mifflin, 2011), 81–98.

[11] On racial profiling, see Anon., "The Wen Ho Lee Case," *Science* 290 (December 22, 2000): 2224. Matthew Purdy, "The Making of a Suspect: The Case of Wen Ho Lee," *New York Times* (February 4, 2001), accessed online.

[12] Lee came under suspicion again in 1996 as part of Operation Kindred Spirit, an operation to determine if the Chinese had stolen plans for the W-88 warhead. For a condensed background with details about Lee's lying to the FBI, see Joint Hearing before the Select Committee on Intelligence and the Judiciary Committee of the United States Senate, 106th Cong., 2nd sess., September 26, 2000, *The Wen Ho Lee Matter* (Washington, DC: Government Printing Office, 2001), 18–19.

[13] See Committee on the Judiciary, United State Senate, 116th Cong., 2nd sess., *The Peter Lee Case* (March 29, April 5, and April 12, 2000) (Washington, DC: Government Printing Office 2001).

[14] Warner quoted in Committee on Armed Services, United States Senate, 106th Cong., 1st sess., April 12 and June 23, 1999, *Review of Alleged Chinese Espionage at Department of Energy Laboratories* (Washington, DC: Printing Office, 2000), 2.

their position against Lee; the Federation of American Scientists and officials at Los Alamos also concluded he lacked credibility.[15] While it remains unclear why Wen Ho Lee downloaded classified files, or what he did with the copies, his case altered US–Chinese nuclear relations: Danny Stillman, the former head of Los Alamos intelligence, suggested the PRC became less willing to collaborate; Hu Side declared, "You have seriously overestimated the scientific and technical capability of the United States and seriously underestimated that of China. You have insulted us ... We did not need you!"[16] However, many in Congress and American intelligence disagreed (Stillman considered Side's claims "more bunk").[17]

Oversight of US–Chinese exchanges increased. The DOE Foreign Visitors Program became a target, with opponents citing a GAO finding that Sandia and Los Alamos national labs had "largely avoided the required background checks" and monitored only 5 percent of visitors from "sensitive" countries.[18] An investigator on Lee's case argued internationalism contributed to the careless approach: "Clinton appointees opened up our nuclear weapons laboratories to scientists from our former Cold War opponents and emphasized 'international scientific collaboration,' whose major achievement was to expose our scientists to foreign intelligence collectors."[19] House members proposed legislation to restrict visits by foreign scientists; officials worried the restrictions would end reciprocal access, nuclear safety exercises, and participation in nonproliferation.[20] The director of international security at Los Alamos considered the

---

[15] Joint Hearing before the Select Committee on Intelligence and the Judiciary Committee of the United States Senate, *The Wen Ho Lee Matter*, 29–30. After extensive study, the Federation of American Scientists found Lee "lacking in credibility," see Subcommittee on Department of Justice Oversight, Committee of the Judiciary, United States Senate, *Report on the Government's Handling of the Investigation and Prosecution of Dr. Wen Ho Lee* (December 20, 2001), available online at: http://fas.org/irp/congress/2001_rpt/whl.html.

[16] Danny Stillman was the head of Los Alamos Intelligence, see "China's Decade of Nuclear Transparency," in Thomas C. Reed and Danny B. Stillman, *The Nuclear Express: A Political History of the Bomb and its Proliferation* (Minneapolis, MN: Zenith Press, 2009), 220–235. Side quoted in Dan Stober and Ian Hoffman, *A Convenient Spy: Wen Ho Lee and the Politics of Nuclear Espionage* (New York: Simon & Schuster, 2001), quote on 226.

[17] Stillman, *The Nuclear Express*, 233.

[18] General Accounting Office, *Department of Energy: Problems in DOE Foreign Visitors Program Persist* (GAO/T-RCED-99–19), released October 6, 1998, quote on 5.

[19] Notra Trulock, *Code Name Kindred Spirit: Inside the Chinese Nuclear Espionage Scandal* (San Francisco: Encounter Books, 2003), 6.

[20] Anon, "Lab access restrictions sought in wake of Chinese espionage reports," *Issues in Science and Technology* 15 (Summer 1999): 29.

proposal "a nutty idea;" the Chinese embassy hoped it would not "go so far" as to affect "normal scientific exchanges."[21] As a compromise, Congress established the US–China Economic and Security Review Commission (2000) to monitor exchanges and instituted polygraph testing for lab employees, even though the machines had proven unreliable in Lee's case and before the Supreme Court.[22] Within a few years, however, the impact on laboratory morale and staff retention convinced policymakers to limit polygraph use. But China's rise continued to test boundaries.

The People's Republic expanded domestic R&D in the early twenty-first century. The United States blocked China's entrance into the WTO until it accepted the TRIPS agreement; the PRC's 2006 plan recommended investing 2.5 percent of GDP on strategic research to end reliance on foreign high-tech goods and establish an independent intellectual property rights regime.[23] Within a few years, China supported patents on genetics and pharmaceuticals and worked within the WTO to offer drugs to the developing world.[24] A study in *Physics Today* presented the PRC perspective: "It has become increasingly obvious to them that those who own the intellectual property and who control the technical standards, enjoy privileged positions, and profit most from, international production networks."[25] Though China is often grouped with developing countries, the study noted, "by any number of indicators of scientific activity, it is not. China ranks fifth in international S&T publications, above France, Italy and Canada. It has a relatively comprehensive S&T system, if not among the world's most advanced, with indigenous R&D

[21] Director and Chinese official quoted in David Malakoff, "DOE Lab Exchanges Targeted in Wake of Espionage Claims," *Science* 283 (March 26, 1999): 1986–1987.

[22] The FBI argued Wackenhut contractors misread Lee's polygraph results and judged him innocent. Aldrich Ames bragged about beating polygraph tests and the US Supreme Court found them uncertain in 1998. See United States Senate, *Report on the Government's Handling of the Investigation and Prosecution of Dr. Wen Ho Lee.* See also, Steven Aftergood, "Polygraph Testing and the DOE National Laboratories," *Science* 290 (November 3, 2000): 939–940.

[23] On the United States, IPR and the WTO, see Charles W. Freeman III, "The Commercial and Economic Relationship," in David Shambaugh, ed., *Tangled Titans: The United States and China* (New York: Rowman & Littlefield, 2013), 181–210, esp. 196. See also Robert G. Sutter, *U.S. – Chinese Relations: Perilous Past, Pragmatic Present,* 2nd edn. (New York: Rowman & Littlefield, 2013), 220–222. China's 15-year "Medium to Long Term Plan for the Development of Science and Technology," put forward in 2006, is discussed in Cong Cao, et al., "China's 15-year science and technology plan," *Physics Today* (December 2006): 38–43.

[24] Natalie Stoianoff, "Influence of the WTO over China's Intellectual Property Regime," *Sydney Law Review* 34 (2012): 65–89, esp. 85.

[25] Cao, et al., "China's 15-year science and technology plan," 39.

in the life sciences, nanoscience, space technology, and other internationally important fields."[26] The US–China Security and Economic Review Commission concurred, adding France and Israel collaborated with the PRC in optoelectronics to access advanced Chinese capabilities.[27]

Chinese actions and espionage led to questions about US–PRC cooperation. In 2006, a Chinese Satellite Laser Ranging station illuminated an American satellite in orbit, an unprecedented act interpretable as either an example of space science (tracking space debris, geodesy, etc.) or a demonstration of Chinese anti-satellite capabilities (such lasers could "blind" or destroy their targets).[28] The following year the PRC shot down one of its own satellites in orbit (the United States carried out a similar test in 1985). Meanwhile, the Security and Review Commission warned, "Chinese espionage activities in the United States are so extensive that they comprise the single greatest risk to the security of American technologies."[29] The Commission noted the first convictions under the Economic Espionage Act were Chinese nationals, as were the subjects of many ongoing indictments (although one case involved espionage for a British company).[30] When a member from President Obama's Office of Science and Technology Policy met with PRC representatives, domestic politics intruded. Worried about a repeat of the Loral and Hughes episode, Representative Rohrbacher called the Committee on Foreign Affairs to order: "Today's purpose is to discuss the inherent dangers of transferring America's leading-edge science to China. China is an increasingly hostile and disruptive force in the world. The idea that we are cooperating with

---

[26] Ibid., 43.

[27] U.S.–China Economic and Security Review Commission, 2007 *Report to Congress*, 110th Cong., 1st sess. (November 2007), 123. Available online at: www.uscc.gov.

[28] See U.S.–China Economic and Security Review Commission, 2007 *Report to Congress*, 6. For a scientific interpretation based, in part, on a classified and leaked Chinese military space strategy text, see Gregory Kulacki, "Authoritative Source on China's Military Strategy," (March 2014). Published by the Union of Concerned Scientists, available at: www.ucsusa.org. See also, Yousaf Butt, "Effects of Chinese Laser Ranging on Imaging Satellites," *Science and Global Security* 17 (2009): 20–35.

[29] U.S.–China Economic and Security Review Commission, 2007 *Report to Congress*, 7.

[30] The Commission highlighted six cases in 2007, see Ibid., 105. For another overview of cases, see John R. Wilke, "Two Silicon Valley Cases Raise Fears of Chinese Espionage," *Wall Street Journal* (January 15, 2003): accessed online. See also Wise, *Tiger Trap*. In 2007, Gary Min plead guilty; the chemist planned to give materials to Victrex – a British rival of Du Pont, see Ericka Chickowski, "$400 million corporate espionage incident at DuPont," *SC Magazine* (February 16, 2007) and Jeff Bliss, "China's Spying Overwhelms U.S. Counterintelligence," *Bloomberg News* (April 2, 2007).

them in any capacity is alarming."[31] Of course, US cooperation in nuclear physics, among other fields, dated to the Reagan administration.[32]

Held in 2011, Representative Rohrbacher's hearing illustrated the extent of US–Chinese scientific cooperation. Ongoing programs included dozens of fields and ranged from the necessary to the exotic: The FDA, for example, operated offices in Beijing, Shanghai, and Guangzhou to inspect factories; American nuclear power companies competed in China's $100 billion dollar reactor market; American and Chinese researchers collaborated at the Clean Energy Research Center; and Chinese experience aided the American campaign against the Asian long-horned beetle, which threatened to cause $138 billion in damage to US forests.[33] Collaboration in agricultural, oceanographic, and climate research (among others) continued as well. During the hearing, OSTP director John Holdren agreed espionage and the loss of intellectual property were legitimate concerns, but added such concerns had to be balanced against the benefits of cooperation, stressing scientific knowledge accelerates and accumulates through the exchange of data and personnel. Nor would isolation increase security or prosperity, the director continued, because the United States loses contact and intelligence while forcing foreign nations to develop alternatives rather than remain "dependent upon U.S. expertise."[34] Meanwhile, American outreach helped lessen Chinese dependence: in 2014, the Security and Review Commission report observed, "the manifest asymmetry in capabilities – which characterized the relationship in 1979 – has been reduced as a result of the remarkable development of S&T in China, made possible by its own domestic policy initiatives and its strategic exploitation of international cooperative activities, especially those offered by relations with the United States."[35] American diplomacy, from the commission's perspective, enabled the country's closest scientific competitor.

---

[31] Rohrbacher quoted in Committee on Foreign Affairs, House of Representatives, 112th Cong., 1st sess., *Efforts to Transfer America's Leading-Edge Science to China* (November 2, 2011) (Washington, DC: Government Printing Office 2011), 1.

[32] Richard P. Suttmeier, U.S.–China Economic and Security Review Commission, "Trends in U.S–China Science and Technology Cooperation: Collaborative Knowledge Production for the Twenty-First Century?" (September 11, 2014), 17. Available at: www.uscc .gov/Research/trends-us-china-science-and-technology-cooperation-collaborative-know ledge-production.

[33] Committee on Foreign Affairs, *Efforts to Transfer America's Leading-Edge Science to China*.

[34] Holdren quoted in Ibid., 39.

[35] Suttmeier, "Trends in U.S–China Science and Technology Cooperation," 4.

American relations with China, although unique in duration, scope, and complexity, provide a good starting point for analysis. First, a recap: Access to American science and technology was a benefit of normalization and the PRC became America's largest bilateral science partner during the Reagan administration. Cooperation expanded steadily throughout the end of the Cold War and the Tiananmen protests, transcending American political parties. Relations engaged nearly every field, from agricultural biotechnology to nuclear physics. The Americans and Chinese had different motivations for engaging: for the United States, scientific cooperation began as an exercise in Cold War diplomacy; for the PRC, cooperation was part of a national strategy to improve scientific and technical capabilities.[36] Questions about transferring American scientific leadership and comparative advantage arose as American expertise entered China and the United States required Chinese adherence to global international property rights law. US–Japanese relations in the 1980s provide a parallel: American–Japanese scientific cooperation became more controversial as Japanese capabilities approached those of the United States (see Chapter 6).

Collaboration and competition co-existed in US–PRC scientific relations. The PRC became the United States largest research partner during the Cold War, eclipsing traditional scientific powers like Germany, Great Britain and Japan.[37] Yet China remained outside the International Space Station. Instead, the PRC orbited its own space station (*Tiangong-1* or "Heavenly") in 2011 and offered partnership opportunities to developing nations (much like the Soviets in the 1970s). Chinese espionage and cyberwarfare caused controversy: NASA briefly blocked Chinese students from a conference in 2013; members of the US Department of Agriculture passed proprietary seeds to the Chinese Academy of Agricultural Sciences; and Google contacted the National Security Agency after the company suspected the PRC launched cyberattacks on Northrop Grumman and Dow Chemical.[38] Geopolitically, both countries offered

[36] Ibid., 6.
[37] Caroline S. Wagner, Lutz Bornmann and Loet Leydesdorff, "Recent Developments in China-U.S. Cooperation in Science," *Minerva* 53 (2015): 199–214.
[38] For cyberwarfare, see Committee on Foreign Relations, House of Representatives, 112th Cong., 1st sess., *Communist Chinese Cyber-Attacks, Cyber-Espionage and Theft of American Technology* (April 15, 2011) (Washington, DC: Government Printing Office, 2011). See also Jeffrey Mervis, "NASA Meetings Bars Chinese Scientists," *Science* 342 (October 11, 2013): 177; and John Eligon and Patrick Zuo, "Designer seed thought to be latest target by Chinese," *New York Times* (February 5, 2014). See also Wise, *Tiger Trap*, 236–237.

scientific and technical assistance to the developing world, though the United States and China approached global governance, climate change and genetic engineering from different positions (at the 2009 Copenhagen climate summit, for example, the PRC represented the developing world against American interests).[39] The prominence of science in US/PRC relations was unavoidable: The two countries accounted for nearly half of global R&D in 2014.[40]

Post-Cold War international relations take place within a scientific and technical framework, increasing the American interest in scientific affairs, especially those with economic or national security implications. It also increases the geopolitical roles of scientists, scientific organizations, and scientific data. Although some argue science is apolitical, or "above the fray," two recent histories illustrate the contemporary geopolitics of science: first, climatology and genetic engineering in American diplomacy (science in diplomacy); and second, American scientific and technical outreach in the Middle East (science for diplomacy).

"National Philosophy" considers how non-scientific interests shaped American diplomacy in climatology and genetic engineering. The United States was the global climate leader during the Cold War, championing the UN climate program and the Montreal Protocol to restrict ozone-reducing chemicals. But industrial opposition to global warming, supported by the Republican party, limited American ratification of international agreements even though the country remained at the fore-front of climate research. Genetic engineering, however, had bipartisan and commercial support and the United States led a global initiative – the Human Genome Project – to provide the genetic data necessary for biomedical advances. Commercial concerns shaped diplomacy, as the United States differed with allies over genetic patenting and labeling while pursuing a market-based approach to environmentalism, whether miti-gating global warming or preserving biodiversity. As the United States advocated for its national interests within the UN, WTO, and World Bank, familiar disagreements with the developing world resurfaced, including over American positions in the Middle East.

[39] Sutter, *U.S.–Chinese Relations*, 224–228, esp. 226. On Chinese science diplomacy, see John G. Whitesides, "Better Diplomacy, Better Science," *China Economic Review* 22 (2011): 28–29.

[40] National Science Foundation, *Science and Engineering Indicators (2016)*, (Washington, DC: National Academies Press, 2016), 0-16 and 0-17.

The terrorist attacks of 2001 shook the international system and altered American diplomacy in the region. Although US commitments in the Middle East expanded after September 11th, scientific and technical programs remained "Pale Shadows" of the American military presence. Following the overthrow of Saddam Hussein, the United States initiated programs to redirect weapons scientists and rebuild Iraq's scientific and technical infrastructure, but sectarian violence limited success. American administrations also struggled to slow the Iranian nuclear program, as Presidents Bush and Obama adopted different strategies to improve relations. But regional politics complicated outreach as well as US–UN relations: After rejoining UNESCO during the Bush administration, the United States left the organization when it recognized Palestinian statehood during the Obama administration. Finally, both presidents refused to participate in multilateral scientific initiatives in the region as American assistance, following Cold War precedents, remained a fraction of national security spending.

NATIONAL PHILOSOPHY

All evidence evaluated to date indicates that unexpected and unintended compositional changes arise with all forms of genetic modification, including genetic engineering. Whether such compositional changes result in unintended health effects is dependent upon the nature of the substances altered and the biological consequences of the compounds. To date, no adverse health effects attributed to genetic engineering have been documented in the human population.

National Academy of Sciences (2004)[41]

CLIMATE CHANGE IS ONE OF THE DEFINING ISSUES OF OUR TIME. It is now more certain than ever, based on many lines of evidence, that humans are changing Earth's climate. The atmosphere and oceans have warmed, accompanied by sea-level rise, a strong decline in Arctic sea ice, and other climate-related changes... The evidence is clear. However, due to the nature of science, not every single detail is ever totally settled or completely certain. [*capitalization in original*]

National Academy of Sciences (2014)[42]

---

[41] National Academy of Sciences, *Safety of Genetically Engineered Foods: Approaches to Assessing Unintended Health Effects* (Washington, DC: National Academies Press, 2004): 8.

[42] National Academy of Sciences, *Climate Change: Evidence and Causes* (Washington, DC: National Academies Press, 2014): 1.

When is something known? Though the answer may not be knowable, this classic epistemological question remains key to American diplomacy involving the natural world. The first NAS quote, for example, came after Zambia and Zimbabwe refused American aid because it might contain unlabeled genetically modified food. The capitalization in the second quote hints at the frustration felt by NAS members at the misrepresentation and politicization of climate research. Of course, the NAS does not make policy, and so the public understanding of science plays a role in policy-making, debate, and diplomacy. And while our abilities have progressed far beyond the natural philosophy of Newton's day, our understanding remains immature. Genetic engineering and climatology, for example, are complex, touching upon our most basic biological self and our relations to the world (with all the economic impact entailed), leaving them susceptible to politicization. An admission of uncertainty at the margins can be taken as a reason to criticize the whole and the risk created by uncertainty provides the space for political interpretation. Although the NAS intent was to be as clear as possible, individuals and nations can frame the language as calling for more or less action as they see fit. In the process, climatology and genetic engineering illustrate how natural philosophy can become national philosophy.

The two fields have similar histories. The United States took the lead in both in the 1970s and American industries helped shape American diplomacy in both afterward. As nations adopted different policies, beliefs about the appropriateness of genetic patents, the application of the "precautionary principle" or what constitutes our "common heritage" entered geopolitics. Because genetic engineering and climatology are global in scope and influence, familiar tensions between the United States and developing worlds reappeared, whether in intellectual property rights, biodiversity conservation, or global warming mitigation. American leadership was also questioned: Though American researchers, funded by federal monies, remained at the forefront of climate science, the United States abdicated its traditional role in climate diplomacy while advocating a commercial approach to genetics and conservation. This set America at odds with its European allies, who accepted anthropogenic global warming and preferred a precautionary approach to genetic modification, mandated labeling and limited patenting. Two histories illuminate the geopolitics of the natural world: first, American diplomacy on ozone regulations and climate research; second, American genetic diplomacy, especially regarding patenting and the UN Convention on Biological Diversity. Indeed, a final similarity is the uncertain US/UN relationship, as the United States has signed but failed to ratify the major environmental treaties in both fields.

US Climate Leadership: The Ozone Hole and Montreal Protocol

The United States and United Nations Environmental Program (UNEP) began campaigns to remove ozone-depleting chemicals like chlorofluorocarbons (CFCs) from the atmosphere in the 1970s. Building upon a consumer awareness campaign, the United States took the remarkable step of banning domestic aerosol use (a major source of CFCs), but the rest of the world refused to act; instead, European politicians and industries argued ozone-depletion was a minor concern. With the global market intact, major American producers such as DuPont felt little pressure to innovate and reduced research on CFC substitutes.[43] However, as CFC use increased globally, the UNEP launched a second round of international negotiations in 1981. The United States was the primary sponsor: NASA's satellites confirmed the loss of atmospheric ozone and American researchers forecast an increase in ultraviolet (UV) radiation and skin cancer.[44] In response, the UN and the United States mounted another campaign to reduce ozone-depleting chemicals worldwide.

A variety of agencies and interests shaped the American position on ozone.[45] NASA collected the scientific data, which was interpreted in collaboration with the EPA and NOAA and relayed to the White House Office of Science and Technology Policy (OSTP). For the next few months, more than a dozen different agencies, such as the Office of Management and Budget (OMB), as well as multiple departments (Agriculture, Commerce, Education, Energy, Health and Human Services, Interior, and Treasury) determined American policy. The Defense Department and intelligence services considered the national security implications and the Office of US Trade Representatives weighed the impact on American commerce. Finally, State Department officials represented the American position at the United Nations and international environmental

---

[43] Karen T. Litfin, *Ozone Discourses: Science and Politics in Global Environmental Cooperation* (New York: Columbia University Press, 1994), 71.

[44] James Gustave Speth and Peter M. Haas, *Global Environmental Governance* (Washington, DC: Island Press, 2006), 93.

[45] See "Forging the U.S. Position," in Richard Elliot Benedick, *Ozone Diplomacy: New Directions in Safeguarding the Planet, enlarged edition* (Cambridge, MA: Harvard University Press, 1998), 51–67.

conferences. Thus, though inspired by atmospheric data, American policy evolved through negotiation among officials with different concerns.

Led by the State Department and EPA, the United States took the lead in promoting worldwide reductions in ozone-depleting chemicals. Sixty embassies received talking-points to explain why action was necessary; others received personal visits: Richard Benedick, the State Department negotiator, led multiple scientific missions to European allies, Japan, and the Soviet Union (all previous holdouts).[46] The US/UN campaign achieved its first success in 1985, when most nations signed the Vienna Convention acknowledging an ozone problem (although without naming CFCs) and promised to establish national action plans. The next step was a clear timetable and emission reduction targets for an upcoming conference in Montreal. However, as regulations moved closer to reality, members of the administration began to question the American position.

The Reagan administration's dislike of environmentalism was clear from the start. Historian Benjamin Kline argued the President oscillated between "indifference and hostility"; Samuel Hayes highlighted the administration preference for executive authority and industry appointments.[47] Secretary of the Interior James Watt, for example, compared environmentalists to Nazis and communists and eleven of the first fifteen EPA appointees hailed from industry.[48] Officials also resisted advice from the scientific establishment, undermined regulations by questioning research, and meddled in the scientific review process.[49] After public criticism, the administration modified its approach during Reagan's second term, attempting to "halt emerging action on such matters as acid rain, toxic air emissions, indoor air pollution, and hazardous waste, rather than abandon established programs."[50] Nor was the approach

---

[46] Ibid., 58.

[47] Benjamin Kline, *First along the River: A Brief History of the U.S. Environmental Movement*, 4th edn. (New York: Rowman & Littlefield, 2011), 114. See also "The Reagan Antienvironmental Revolution," in Samuel P. Hayes, *Beauty, Health and Permanence: Environmental Politics in the United States, 1955–1985* (Cambridge: Cambridge University Press, 1989), 491–526.

[48] For Watt's comments, see Anon., "Watt Says Foes Want Centralization of Power," *New York Times* (January 21, 1983). Regarding the appointments, see the "Reagan Reaction, 1980–88," in Kirkpatrick Sale, *The Green Revolution: The American Environmental Movement 1962–1992* (New York: Hill & Wang, 1993), 51.

[49] See "Sowing the Seeds of Doubt: Acid Rain," and "Constructing a Counternarrative: The Fight Over the Ozone Hole," in Oreskes and Conway, *Merchants of Doubt*, 66–106 & 106–135.

[50] Hayes, *Beauty, Health and Permanence*, 511.

limited to domestic affairs; the use of uncertainty to justify inaction shaped American diplomacy: The United States disputed Canadian acid rain findings, was the only country to vote against an updated UN list of hazardous chemicals, and the only country not to sign the UN World Charter for Nature.[51] Cracks also appeared in the American position on ozone.

After a decade of global leadership, American support for reductions wavered. Administration officials first disputed the data in 1984, leading EPA staffers to leak an internal report to the NRDC, which eventually sued the government under the Clean Air Act.[52] A few years later, the Commerce Department, OMB, and OSTP reopened questions about the science and emphasized the impact of regulations on the US economy. Congressional opposition congealed along now-familiar lines: Senator Symms (R-ID) characterized ozone research as a two-sided debate where "the other side is that the ozone around the world really has not changed" and worried US policy was set by "some emotional viewpoint," arguing the government could not protect citizens who insisted on exposing themselves to "excessive sunshine."[53] Even the revelation of the ozone hole failed to persuade doubters: Secretary of the Interior Donald Hodel suggested hats and sunglasses to combat the increased UV radiation, leading the NRDC to point out, "crops, fish and wildlife aren't going to wear hats and sunglasses," while a spokesman for Ray-Ban joked, "we'll be happy to set [Hodel] up as a distributor."[54] Though the public lampooned the secretary's response, by the time the American delegation

---

[51] On acid rain and hazardous chemicals, see *Hayes, Beauty, Health and Permanence*, 509–511. On the world charter see Sale, *The Green Revolution*, 51 and Harold W. Wood, Jr., "The United Nations World Charter for Nature: The Developing Nation's Initiative to Establish Protections for the Environment" *Ecology Law Quarterly* 12 (September 1985): 977–996.

[52] Neil E. Harrison and Gary C. Bryner, eds., *Science and Politics in the International Environment* (Lanham, MD: Rowman & Littlefield, 2004), 111. David Doniger was one of the NRDC lawyers, see David Doniger and Michelle Quibell, "How the NRDC Helped Save the Ozone Layer," available at www.nrdc.org. The NRDC briefly delayed the suit after the EPA argued it might undermine international negotiations, see Litfin, *Ozone Discourses*, 72.

[53] Symms quoted in Benedick, *Ozone Diplomacy*, 62.

[54] Hopgood, *American Foreign Environmental Policy and the Power of the State*, 227. Ray-ban spokesman quoted in Robert Gillette, "Suggests Wearing Hats, Sunscreen Instead of Saving Ozone Layer: Hodel Proposal Irks Environmentalists," *Los Angeles Times* (May 30, 1987).

arrived in Montreal, only the scientific agencies, State, and the EPA supported the original American position.[55]

The Montreal Protocol (1987) is often considered the gold standard in global environmental governance, but atmospheric data was not sufficient to compel an American signature. Instead, non-scientific actors played necessary roles. DuPont, for example, announced it could create CFC substitutes within five years, shifting the position of American industry: The protocol promised to phase out CFCs while creating a market for substitutes, leading DuPont and the Alliance for Responsible CFC Policy (an industry association of more than 500 producers and users) to support curbing emissions; Brian Fay, head of the Alliance, argued without global regulations, the "U.S. [would] go its own way and commit industrial suicide."[56] At the same time, European allies and Japan began moving forward with plans for reductions and substitutes, threatening US leadership. In the public sphere, the ozone hole acted as a "forcing event" and environmental NGOs maintained media coverage, causing citizens to demand action.[57] Negotiator Benedick also highlighted the importance of the personal, whether the relationship between Reagan and Secretary of State Schultz (a prominent supporter of the protocol) or Reagan's experience with skin cancer (the President had growths removed in 1985 and 1987).[58] For a variety of reasons, President Reagan became the first head of state to sign the Montreal Protocol, which reduced use of ozone depleting chemicals by more than 90 percent and led to predictions of a recovery to 1980 levels.[59] However, while the protocol provided an example of environmental science translated into policy, climate research opened an ongoing divide in American international relations the following year.

Domestic Politics vs. Global Science: The Case of Global Warming

Research on the relationship between carbon dioxide (CO_2) and the climate was a century old before it became a political topic in the late 1980s. It was also a product of the scientific cooperation discussed in previous chapters: IGY researchers measured CO_2 at more than sixty

[55] Benedick, *Ozone Diplomacy*, 65.   [56] Fay quoted in Litfin, *Ozone Discourses*, 93.

[57] Speth, *Global Environmental Governance*, 95.

[58] Benedick, *Ozone Diplomacy*, 65–67.

[59] Mike Hulme, *Why We Disagree about Climate Change* (New York: Cambridge University Press, 2009), 291.

locations around the world, including initiating Charles Keeling's critical measurements at Mauna Loa Observatory in Hawaii and the South Pole; the IBP continued the work a decade later.[60] Scientists, rather than environmentalists, pushed the issue throughout the 1970s and the Department of Energy asked the JASONS (of Vietnam-era infamy) to consider the impact of CO_2 in 1977. The United States passed the National Climate Act in 1978 (the same year it outlawed aerosols), establishing a national climate program and the United Nations followed suit. As with CFCs and ozone depletion, the United States was the undisputed climate leader and there were few skeptics.

Climate research became more institutional and political at the end of the Cold War. In 1988, James Hansen, a NASA climatologist, testified before Congress human activities accounted for an increase in CO_2 and subsequent global warming. Within months, the United Nations, with US support, created the Intergovernmental Panel on Climate Change (IPCC) to consolidate research around the world. The IPCC represented a milestone in collaborative global research, representing the work of thousands of scientists from more than 150 countries. Although such a broad sample of scientists and peoples could be seen as value-neutral, opponents of global regulations criticized the IPCC for being political.[61] Industries opposed to CO_2 regulations, including British Petroleum, Shell, DuPont, Ford, General Motors, and others, formed the Global Climate Coalition to dispute climate research. Other conservative organizations – including the George Marshall Institute and Heritage Foundation – followed. The new lobby had an impact: In 1990, the Bush administration organized a White House conference to delay action on climate change and hesitated to support updated ozone restrictions for fear of setting a precedent on CO_2 emissions.[62]

Scholars have suggested numerous theories as to why climate science became politicized in the United States. Joyeeta Gupta pointed out climate change entered the political arena as a theory rather than through the actual experience of consequences, creating space for politicization.[63]

---

[60] See Howe, *Behind the Curve*, 22; Hulme, *Why We Disagree about Climate Change*, 55; and Spencer R. Weart, *The Discovery of Global Warming* (Cambridge, MA: Harvard University Press, 2008), 35–37.

[61] Douglas, *Science, Policy and the Value-Free Ideal*, 131.

[62] Hopgood, *American Foreign Environmental Policy and the Power of the State*, 156. See also Joyeeta Gupta, *The History of Global Climate Governance* (Cambridge: Cambridge University Press, 2014), 154.

[63] Gupta, *The History of Global Climate Governance*, 41.

Mike Hulme argued the Montreal Protocol offered a false sense of optimism because the causes and consequences of ozone depletion were limited and arose from only a fraction of economic activity, conditions which do not apply to global warming.[64] Joshua Howe proposed the protocol "put powerful corporate and governmental bodies on notice," pitting naïve scientists believing in the "forcing function" of knowledge against the lobbying power of the Global Climate Coalition.[65] Taking a wider perspective, Naomi Oreskes and Eric M. Conway consider climate change only the newest iteration in a long history of industry questioning research and emphasize the impact of misinformation campaigns on a poorly educated populace.[66] Others suggest the "green scare" replaced the "red scare" after the fall of the Berlin Wall and Soviet Union, with environmentalists compared to watermelons ("green on the outside, red on the inside") as well as communists and Nazis (later terrorists and the Taliban using the argument that "greens" are waging environmental "jihad").[67] For all of these reasons, combined with the general fear combatting climate change would undermine the economy, the United States began to backtrack on CO_2 regulations. The tensions would be clear at the Rio conference, held on the twenty-year anniversary of the Stockholm Conference (see Chapter 4).

The UN "Earth Summit" in Rio (1992) addressed the major environmental issues of the day. Representatives from more than 150 countries attended, including dozens of world leaders. More than 7,000 NGOs also attended, testifying to the explosive growth in environmental activism since Stockholm. One outcome of the conference was Agenda 21 – a sustainable development manifesto enshrining goals desired by the developing world since Stockholm, including increased access to science and technology and more control over multinationals and property rights. However, the agenda was non-binding; the American focus was on the two potentially binding pieces of legislation: the UN Framework Convention on Climate Change (UNFCCC) and the Convention on Biological Diversity (CBD).

---

[64] Hulme, *Why We Disagree about Climate Change*, 292.

[65] Howe, *Behind the Curve*, 153–154 and 186–187.

[66] Oreskes and Conway, *Merchants of Doubt*. See also Naomi Oreskes and Erik M. Conway, "Challenging Knowledge: How Climate Science Became a Victim of the Cold War," in Robert N. Proctor and Londa Schiebinger, eds., *Agnotology: The Making and Unmaking of Ignorance* (Stanford, CA: Stanford University Press, 2008).

[67] Mark Romeo Hoffarth and Gordon Hodson, "Green on the Outside, Red on the Inside: Perceived Environmentalist Threat as a Factor in Explaining Political Polarization of Climate Change," *Journal of Environmental Psychology* 45 (2016): 40–49.

The Bush administration opposed CO_2 regulations in the UNFCCC. Before the conference, officials publicly questioned whether the president needed to go to Rio, causing more than 700 members of the National Academy of Sciences to urge his attendance. However, the hesitation was merely an act to gain leverage: To secure American participation and financing, the administration received guarantees specific emissions targets were off the table.[68] As such, the UNFCCC became an agreement in principle (much like the Vienna Convention) with no enforcement mechanisms; the development of specific emission reduction targets would be addressed at a later conference in Kyoto (like the Montreal Protocol). The United States also rejected requests for more funding; EPA administrator William Reilly argued the UNFCCC "was not a Marshall Plan for the developing world."[69] Instead, the Bush administration promoted market-based solutions through the Global Environmental Facility (GEF).

The GEF represents a neoliberal response to global warming. In 1989, the French delegation to the World Bank proposed a pilot program to respond to environmental criticism of the Bank's programs.[70] The idea was to use Bank financing to introduce environmentally friendly technologies to the developing world; many environmental NGOs hoped the program would "green" the Bank, others worried it would merely "greenwash" a sick system.[71] The United States forcefully advocated for the GEF at the Rio conference while the G77 protested the market-based proposal and World Bank's management, angling for a UN-managed "green fund" instead.[72] To win support, the United States pledged $250 million but critics pointed out the amount was small for the world's largest polluter (Japan, hoping to position itself as an environmental leader, pledged $2.5 billion).[73] Restructured after Rio, the

[68] Hopgood, *American Foreign Environmental Policy and the Power of the State*, 167. See also Sale, *The Green Revolution*, 86–88.

[69] Reilly quoted in Howe, *Behind the Curve*, 181.

[70] Lin Gan, "The Making of the Global Environmental Facility: An actor's Perspective," *Global Environmental Change* (September 1993): 256–275, esp. 257.

[71] Zoe Young, "NGOs and the Global Environmental Facility: Friendly Foes?" *Environmental Politics* 8 (1999): 243–267.

[72] The G77 also protested that the conference language of costs, stakeholders and pricing of nature was part of the problem. See *ibid.*

[73] Howe, *Behind the Curve*, 178–182. Sale, *The Green Revolution*, 86–88. Japan's MITI initiated a "Green Aid Plan" to provide technical support to developing countries in Asia in 1992, see Maaike Okano-Heijmans, "Japan's 'Green' Economic Diplomacy: Environmental and Energy Technology and Foreign Relations," *The Pacific Review* 25 (2012): 339–364.

Global Environmental Facility began with four primary areas: ozone protection, climate change, biological diversity, and international waters.[74] Projects began almost immediately: The People's Republic of China, for example, submitted proposals before the Rio conference and received more than $450 million, much of it devoted to climate change.[75] The GEF remains a primary legacy of the Rio conference and American diplomacy: One researcher suggested the facility was "the only game in town" within a decade while others point out the limited NGO participation.[76]

Domestic politics have constrained American global warming diplomacy in the quarter-century since Rio. After a brief honeymoon during the first Clinton administration, the United States walked away from its traditional role as a climate leader. By a vote of 98–0, the United States Senate passed the Byrd-Hagel resolution (1997) to reject any climate treaty with the potential to undermine the American economy; as such, the United States refused to ratify the Kyoto Protocol and the Bush administration withdrew in 2001; EPA administrator Christine Todd Whitman told reporters, "No, we have no interest in implementing that treaty."[77] Historian Joshua Howe argued "maintaining the flexibility and profitability of U.S. business interests" defined American international climate change policy after Rio, as the United States promoted a cap-and-trade system, essentially a compliance regime based on the "right to pollute."[78] Nor has the situation changed: After President Obama signed the Paris Climate Accord in 2015 – an international accord composed of non-mandatory pledges lacking enforcement mechanisms – Senate Majority Leader Mitch McConnell (R-KY) countered, "The climate proposal announced today represents nothing more than a long-term planning document. The president is making promises he can't keep ... this is an unattainable deal based on a domestic energy plan that is likely illegal,

---

[74] For an excellent, albeit brief, overview of GEF programs, see Laurence D. Mee, Holly T. Dublin, and Anton A. Eberhard, "Evaluating the Global Environmental Facility: A Goodwill Gesture or a Serious Attempt to Deliver Global Benefits?" *Global Environmental Change* 18 (2008): 800–810.

[75] Gorild Heggelund, Steinar Andresen and Sun Ying, "Performance of the Global Environmental Facility (GEF) in China: Achievements and Challenges as Seen by the Chinese," *International Environmental Agreements* 5 (2005): 323–348, esp. 339.

[76] Young, "NGOs and the Global Environmental Facility," 246.

[77] Whitman quoted in Julian Borger, "Bush Kills Kyoto Treaty," *The Guardian* (March 29, 2001), accessed online.

[78] Howe, *Behind the Curve*, 185.

that half the states have sued to halt, and that Congress has already voted to reject."[79] Environmentalists hoped Paris leads to "an upward spiral of [national] ambition over time," but the history of climate diplomacy is uninspiring.[80] Instead, the politics and emphasis on markets remain consistent: at Paris, EPA administrator Gina McCarthy stated, "we are unequivocally sending market signals that are spurring US action, and unleashing businesses to think creatively and seize the opportunity."[81] Though the rejection of global warming, especially given the progress in climate science, was unprecedented in the history of American diplomacy, the emphasis on markets was equally apparent in genetics.

The Human Genome Project and Convention on Biological Diversity

Markets were critical to the rise of American biotechnology. Markets fueled the original rush to use recombinant genetic engineering to synthesize proteins like insulin, interferon and growth hormone, while the United States refused to cooperate through the UN or G7, using access to genetic materials as part of diplomacy in the 1970s (see Chapters 4 and 5). As the first generation of biotech companies received legislative and judicial support throughout the 1980s (see Chapter 6), another generation focused on decoding the human genome with the hope of marrying genetics, biomedicine, and pharmaceuticals. The second group relied heavily on information technology, a story dating to the late 1970s (when the United States banned aerosols and initiated a national climate program).

The potential for genetic databases became clear as computers revolutionized data-processing. In 1965, Margaret O. Dayhoff, a pioneer in bioinformatics, published the first compendium of nucleic acid and protein sequences (the *Atlas of Protein Sequence and Structure*).[82] Fourteen years later, Dayhoff and a few dozen molecular biologists and computer scientists met at Rockefeller University to discuss creating a computerized

[79] McConnell quoted in Charles S. Clark, "EPA and Energy Department Leaders Hail Paris Climate Accord," *Goverex.com* (December 14, 2015) accessed online.
[80] David J. C. MacKay, "Price Carbon: I will if you will," *Nature* 526 (October 15, 2015): 315–316.
[81] McCarthy quoted in Clark, "Epa and Energy Department Leaders."
[82] Margaret O. Dayhoff, *Atlas of Protein Sequence and Structure* (Silver Springs, MD: National Biomedical Research Foundation, 1965).

genetic database.[83] Sponsored by NIH, NSF, and the European Molecular Biology Laboratory (EMBL), the meeting was encouraging, but the participants adopted different strategies afterwards: EMBL announced the creation of public database in 1980; Dayhoff organized private companies such as Genex, Merck, Eli Lily, and DuPont for a subscription service in 1981; and NIH searched for funding.[84] By 1983, NIH, Los Alamos laboratory, and Bolt, Beranek & Newman (a DARPA contractor) agreed to establish a free nucleic acid database distributed using the Defense Department's Arpanet (the early internet).[85] Named GenBank, the computerized database became the first federally funded public repository for basic genetic data, serving as a model for the Human Genome Project (and PubMed Central online later).[86]

Advocates within the government, business community, and State Department argued mapping the genome was in the national interest. Within the government, researchers at NIH and the Department of Energy advocated for a large-scale program in human genetics (the DOE's life sciences division at Los Alamos researched mutation). Many corporate backers saw the Human Genome Project (HGP) as a bridge between basic biology and biomedicine; when polled, members of the Industrial Research Institute desired the genome project more than any other.[87] As the national laboratories established mapping centers, industry spokesmen testified to Congress the HGP was necessary to remain globally competitive.[88] Japan appeared to be preparing for a major

[83] Bruno J. Strasser, "Collecting, Comparing and Computing Sequences: The Making of Margaret O. Dayhoff's 'Atlas of Protein Sequence and Structure,' 1954–1965," *Journal of the History of Biology* 43 (Winter 2010): 623–660.

[84] Bruno J. Strasser, "The Experimenter's Museum: GenBank, Natural History, and the Moral Economies of Biomedicine," *Isis* 102 (March 2011): 60–96, esp. 76–80.

[85] Bruno J. Strasser, "GenBank: Natural History in the 21st Century?" *Science* 322 (October 24, 2008): 537–38.

[86] Strasser, "The Experimenter's Museum," 94–96. See also Eugene Thacker, *The Global Genome: Biotechnology, Politics and Culture* (Cambridge, MA: Massachusetts Institute of Technology Press, 2005), 108–115; for a chronology of bioinformatics, see 333–338.

[87] Christopher Anderson, "Genome Project Goes Commercial," *Science* 259 (January 15, 1993): 300. As many as 185 companies were interested in the genome mapping, see Subhajit Basu, "Human Genome and Patent," *International Review of Law, Computers and Technology* 16 (2002): 339–357. On the IRI, see Rodney Loeppky, *Encoding Capital: The Political Economy of the Human Genome Project* (New York: Routledge, 2005), 105.

[88] In 1987, mapping centers opened at Los Alamos, Livermore and Berkeley. See Daniel J. Kevles, "Big Science and Big Politics in the United States: Reflections on the Death of the SSC and the Life of the Human Genome Project," *Historical Studies in the Physical and Biological Sciences* 27 (1997): 269–297, esp. 275–276.

research initiative (comparable to the VLSI initiative) and the Soviet Union initiated a genome program as the Berlin Wall fell. Ultimately, the "Human Genome Project" consolidated the work of multiple countries, providing access for all and depositing data in both GenBank and the EMBL database.[89] More than 400 private companies, based on access to the scientific commons and public funding, grew alongside.[90] As the value of genetic information rose, the UN addressed biotechnology at the Rio Conference in the Convention on Biological Diversity (CBD).[91]

Originally focused on forests and endangered species, the CBD evolved into an instrument of national sovereignty over biological resources, endorsing the commercial vision of nature favored by the United States and the biotech industry.[92] As such, the CBD recognizes three kinds of rights: national sovereign rights, rights held by private or public institutions, and the property rights of local communities.[93] The convention also reframed market-oriented exploitation, previously seen as a threat to biodiversity, as the means to save it: Biodiversity, properly valued, would be preserved – i.e., proper assessment of nature would reconcile profits and environmental protection.[94] Examples include ecotourism and Payments for Ecosystem Services, a program to use markets to generate quantifiable benefits from soil stability, forest health, water quality, and the like.[95] Finally, the CBD replaced the "common heritage" argument

---

[89] The international effort was coordinated by the Human Genome Organization or HUGO (1988), see www.hugo-international.org/. The American effort was coordinated by the Human Genome Research (NCHGR, est. 1989); renamed the National Human Genome Research Institute (NHGRI) in 1997. See www.genome.gov/27534788/about-the-institute/. The primary countries responsible for the mapping were the United States, France, Germany, Great Britain, Japan, and China.

[90] One study found 470 genomics companies in more than 25 countries related to the HGP, see Ilse R. Wiechers, Noah C. Perin, and Robert Cook-Deegan, "The emergence of commercial genomics: analysis of the rise of the biotechnology subsector during the Human Genome Project, 1990–2004," *Genome Medicine* 5 (2013) accessed online.

[91] See "Article 15. Access to Genetic Resources" and "Article 19. Handling of Biotechnology and Distribution of its Benefits." Convention text available at: www.cbd.int/convention/text/default.shtml.

[92] See "Article 3. Principle" for national sovereignty. See also Hopgood, *American Foreign Environmental Policy and the Power of the State*, 170.

[93] Valérie Boisvert and Franck-Dominique Vivien, "Towards a Political Economy Approach to the Convention of Biological Diversity," *Cambridge Journal of Economics* 36 (2012): 1163–1179, esp. 1167.

[94] Ibid.

[95] For an analytical overview of the program, Kemi Fuentes-George, "Neoliberalism, Environmental Justice, and the Convention on Biological Diversity: How Problematizing the

preferred by the G77 (common heritage as equal access for all) with an interpretation where resources are available to all on a commercial basis.[96] Scholars disagree about the impact: Some argue this transforms public goods into private commodities; others argue the CBD forces companies to cooperate with national governments for the first time.[97]

US support for "bioprospecting" or "biopiracy" exemplified the controversial nature of the CBD. The convention emphasized equitable sharing of benefits and attributed property rights to the nation (or peoples), thus requiring companies negotiate for use of genetic resources. Merck, for example, paid Costa Rica, a country with 4 percent of the world's biodiversity, one million dollars plus future royalties for the initial surveying rights to all indigenous animal and plant material.[98] As Merck representatives fanned out across the countryside, the United States established International Cooperative Biodiversity Groups (ICBG) to "stimulate the field of bioprospecting" and "provide models for the sustainable use of biodiversity."[99] Run by three agencies (NIH, NSF, and USAID), the ICBG established groups in nearly a dozen countries; Costa Rica eventually received more than $60 million in the 1990s.[100] Exploration continued for more than a decade: In 2009, the editors of *Pharmaceutical Biology* pointed out natural products accounted for nearly half of all contemporary drugs and yet less than 1 percent of global species had

Commodification of Nature Affects Regime Effectiveness," *Global Environmental Politics* 13 (November 2013): 144–163. For an overview of projects in more than a dozen countries, see Pieter J. H. van Beukering, et al., eds., *Nature's Wealth: The Economics of Ecosystem Services and Poverty* (Cambridge: Cambridge University Press, 2013).

[96] Thomas Pogge and Doris Schroeder, "Justice and the convention on biological diversity," *Ethics & International Affairs* 23 (Fall 2009). Accessed online. See also Pilar N. Ossorio, "The Human Genome as Common Heritage: Common Sense or Legal Nonsense?" *Journal of Law, Medicine and Ethics* (Fall 2007) 425–439.

[97] Laurelyn Whitt, *Science, Colonialism and Indigenous Peoples: The Cultural Politics of Law and Knowledge* (New York: Cambridge University Press, 2009), 15; Peter Andrée, *Genetically Modified Diplomacy: The Global Politics of Agricultural Biotechnology and the Environment* (Vancouver: University of British Columbia Press, 2007), 109–123; Pogge and Schroeder, "Justice and the Convention on Biological Diversity."

[98] Merck & Co., Inc., statement in Committee on Foreign Relations, United States Senate, 103rd Cong., 2nd sess., *The Convention on Biological Diversity* (April 12, 1994) (Washington, DC: Government Printing Office, 1994), 27–32. See also Hardy, "Patent Protection and Raw Materials," 299–326.

[99] Joshua P. Rosenthal, "The International Cooperative Biodiversity Group Programs," 3. This is a case-study prepared for the CBD through the Fogarty International Center at NIH, available at: www.fic.nih.gov.

[100] Rex Dalton, "Cashing in on the Rich Coast," *Nature* 442 (June 1, 2006): 567–569.

been investigated.[101] Although Merck and the pharmaceutical industry argued in favor of the CBD, and the United States initiated many CBD-compliant projects after Rio, the United States has been unable to ratify the Convention for twenty years.[102]

American administrations preferred the commercial protections of the WTO to those of the UN and CBD. The country led on endangered species protection throughout the 1970s and originally argued for the biodiversity treaty. However, the Clinton administration initially objected to articles requiring technology transfer and sided with the Association of Biotech Companies (ABC), refusing to sign (the ABC president branded the Rio conference a "diplomatic mugging").[103] After Rio, EPA administrator Reilly observed, "The U.S. early on supported the need for a biodiversity convention so it was a perverse twist that we alone rejected it."[104] Senator Nickles (R-OK) identified weak intellectual property rights as the reason, stating ratification would "throw away" years of negotiating stronger protections through the GATT treaties.[105] To gain support, the administration added "Seven Understandings" to protect American intellectual property rights, winning the support of the biotechnology industry and signing the CBD in 1993, but the opposition refused to reconsider. Some feared the CBD would override American law, others had broader fears: Senator Malcolm Wallop (R-WV) worried, "The biodiversity treaty, as written, would give the Clinton administration even greater authority to accomplish suspect environmental goals."[106] The United States has not taken up ratification since. Instead,

---

[101] Editors, "The International Cooperative Biodiversity Groups," *Pharmaceutical Biology* 47 (2009): 754.

[102] See William J. Snape, III, "Joining the Convention on Biological Diversity: A Legal and Scientific Overview of Why the United States Must Wake Up," *Sustainable Development Law & Policy* (Spring 2010): 6–47. Snape was the US representative for the Center for Biological Diversity; see also Robert F. Blomquist, "Ratification Resisted: Understanding America's Response to the Convention on Biological Diversity, 1989–2002," *Golden Gate University Law Review* 32 (2002): 493–586.

[103] The Administration cited articles 16(3) and 19(2) (on technology transfer) as the reason. See Cheryl D. Hardy, "Patent Protection and Raw Materials: The Convention on Biological Diversity and Its Implications for U.S. Policy on the Development and Commercialization of Biotechnology," *Journal of International Law* 15 (1995): 299–326, esp. 317. On the "mugging," see Hopgood, *American Foreign Environmental Policy and the Power of the State*, 183.

[104] Reilly quoted in Blomquist, "Ratification Resisted," 533.

[105] Nickles quoted in Ibid., 527.

[106] Wallop quoted in R. Daniel Kelemen and Tim Knieval, "The United States, The European Union, and International Environmental Law: The Domestic Dimensions of Green Diplomacy," *I.Con* 13 (2015): 946–965, quote on 962.

the country successfully advocated for the passage of the TRIPS agreement (1994), in which existing US laws formed much of the GATT (later WTO) standards.

International Tensions over Genetic Patents and Labels

American and British attempts to patent genetic material generated controversy. In 1992, NIH and the Medical Research Council (MRC) in Great Britain filed patent applications on genetic sequences whose purpose remained unknown (the Expressed Sequence Tags or ESTs). The action split the scientific community: Bernadine Healy, the head of NIH, supported the patent applications, while James Watson, the genetic pioneer, and the American Society of Human Genetics did not.[107] Revelations the Commerce Department had applied for a patent on a Panamanian woman's cell line without her knowledge and consent led to protests at the UN.[108] NGOs also claimed a US-financed ICBG offered Peruvians one quarter of one percent for products based on their knowledge and resources.[109] Eventually, NIH, Commerce, and the MRC withdrew the patent applications, but the episodes concerned researchers working on the human genome.[110]

The national scramble for patents and profits led to a reassertion of the public ideal. Organized by the Wellcome Trust, scientific leaders from twenty-two nations met in Bermuda (1996) to set the guidelines for sharing HGP data; the resulting "Bermuda Principles" required release of all sequences more than 1,000 nucleotides in length within twenty-four hours.[111] The Principles won support from computer scientists involved in a parallel open-source software movement as well as President Clinton,

[107] Hubert Curien, "The Human Genome Project and Patents," *Science* 254 (December 20 1991): 1710 and 1712.

[108] Kara H. Ching, "Indigenous Self-Determination in an Age of Genetic Patenting: Recognizing an Emerging Human Rights Norm," *Fordham Law Review* 66 (November 1997). Accessed online.

[109] Cindy Hamilton, "The Human Genome Diversity Project and the New Biological Imperialism," *Santa Clara Law Review* 41 (2001), 9.

[110] Robert Cook-Deegan and Christopher Heaney, "Patents in Genomics and Human Genetics," *Annual Review of Genomics and Human Genetics* 11 (2010): 383–425, esp. 401.

[111] National Research Council, *U.S. and International Perspectives on Global Science Policy and Science Diplomacy*, 21.

who urged "all nations, scientists and corporations to adopt this policy and honor its spirit."[112] The following year, UNESCO passed the Universal Declaration on the Human Genome and Human Rights (1997) to discourage bioprospecting and many journals required submission to GenBank before publication.[113] As researchers reaffirmed the commitment to open science, the United States Patent and Trade Office (USPTO) reversed its position, reopening a genetic market.[114]

American patent offices and companies embraced genetic commerce. The USPTO began by granting patents on ESTs in 1996, leading to a rush of applications. Two years later, Celera genomics, a private company, established a rival initiative to the HGP. However, Celera (meaning "speed") used a method HGP researchers considered too mistake-prone and the company refused to make sequences public, offering a subscription service instead.[115] Ultimately, the Celera challenge caused the public project to commit more resources while the commercial service failed after *Science* published more fine-scale maps (all data went to GenBank). Only a few years later, *Science* reported more than one-fifth of human genes, regardless of whether their purpose was known, were patented.[116]

Genetic patenting exposed divisions in the EU. European nations agreed to TRIPS, but internal disagreements remained. In 1995, the European Parliament voted against the patenting of life forms; Hiltrud Breyer, a German Green leading the opposition, argued patents make

[112] Regarding open source software, see Hallam Stevens, "The Politics of Sequence: Data Sharing and the Open Source Software Movement," *Information and Culture* 50 (2015): 465–503. Clinton quoted in Jorge L. Contreras, "Bermuda's Legacy: Policy, Patents and the Design of the Genetic Commons," *Minnesota Journal of Law, Science and Technology* 12 (2011): 61–125, quote on 61.

[113] Mohamad A.F. Noor, Katherine J. Zimmerman and Katherine C. Teeter, "Data Sharing: How Much Doesn't Get Submitted to GenBank?" *PLoS Biology* 4 (July 2006): e228.

[114] For the Ft. Lauderdale Principles (2003), see Yann Joly, Clarissa Allen, and Bartha M. Knoppers, "Open Access as Benefit Sharing? The Example of Publicly Funded Large-Scale Genomic Databases," *Journal of Law, Medicine and Ethics* (Spring 2012): 143–146. For the Amsterdam Principles (2008), see Henry Rodriguez, et al., "Recommendations from the 2008 International Summit on Proteomics Data Release and Sharing Policy: The Amsterdam Principles," *Journal of Proteome Research* 8 (2009): 3689–3692.

[115] Maynard V. Olson, "The Human Genome Project: A Player's Perspective," *Journal of Molecular Biology* 319 (2002): 931–942, quote on 934. See also Mirowski, *Science Mart*, 290–294. Heidi L. Williams, "Intellectual Property Rights and Innovation: Evidence from the Human Genome Project," *Journal of Political Economy* 121 (2013): 1–27.

[116] Sam Kean, "The Human Genome (Patent) Project," *Science* 331 (February 4, 2011): 530–531.

"human life... fodder for economic interests," but others worried about American competition.[117] When the European Parliament reversed itself, passing the "Legal Protection of Biotechnological Inventions" (1998), enforcement proved difficult: the Netherlands filed an annulment and Germany and seven other countries refused to go along; neither the European Patent Office nor biotech directive provided much clarity.[118] As the patent debate stalled, the Parliament mandated labeling of genetically modified foods, a policy popular across the EU.

Philosophical differences underlay US and European disagreements over genetic labeling. European nations adopted a "precautionary principle," which limited genetic research and products even in the absence of clear harm. However, the United States FDA argued there was a "substantial equivalence" between genetically modified and non-modified products, reducing the need for labeling unless products contained a known allergen.[119] When the USDA approved recombinant bovine growth hormone in American cattle, the EU, dealing with an outbreak of mad-cow disease, applied the precautionary principle and refused American beef. International institutions split over labeling and market-access: The WTO Agreement on Sanitary and Phytosanitary Measures (1995) required a clear scientific risk to limit imports, but the Cartagena Protocol on Biosafety (2000), an addendum to the CBD, required labeling and allowed countries to refuse GM imports.[120]

The Cartagena Protocol separated NATO allies and international institutions. The EU and UN were enthusiastic supporters; the United States and WTO were not. As political scientists R. Daniel Kelemen and Tim Knieval observe:

---

[117] Rory Watson, "Brussels Rejects Biotechnology Directive," *British Medical Journal* 310 (Mar 11, 1995): 619–620, quote on 619. See also R. Stephen Crespi, "The European Biotechnology Patent Directive is Dead," *Trends in Biotech* 13 (May 1995): 162–164.

[118] Katerina Sideri, "Practical Reasoning, Impartiality and the European Patent Office: The Legal Regulation of Biotechnology," *European Law Journal* 18 (November 2012): 821–843. Emilie Cloatre, "From International Ethics to European Union Policy: A Case Study on Biopiracy in the EUs Biotechnology Directive," *Law & Policy* 28 (July 2006): 345–367. Kara H. Ching, "Indigenous Self-Determination in an Age of Genetic Patenting: Recognizing an Emerging Human Rights Norm," *Fordham Law Review* 66 (November 1997) accessed online.

[119] Felicia Wu and William P. Butz, *The Future of Genetically Modified Crops: Lessons from the Green Revolution* (Santa Monica, CA: Rand Corporation, 2004).

[120] The best book-length treatment of the Cartagena Protocol is Andrée, *Genetically Modified Diplomacy.*

The Protocol promised to protect the EU's GMO regulatory regime from attack under world trade rules by allowing signatories to invoke the "precautionary principle" in absence of scientific certainty to restrict imports of GMOs. The United States viewed EU GMO regulations as non-tariff barriers to trade that violated world trade rules, and from the late 1990s, the U.S. threatened to take legal action against the EU before the World Trade Organization.[121]

Although American negotiators secured an exemption for pharmaceuticals during the Cartagena negotiations, they were unable to secure a similar exemption for GM foods and so refused to commit.[122] Once enacted, the Protocol had an impact: Zambia and Zimbabwe rejected unlabeled American food aid because Europeans threatened to deny them market access, while the United States filed suit against its allies at the WTO, winning judgments on growth hormone and labeling.[123] However, though the WTO affirmed the United States on GM crops, its ruling has not changed the position of most European governments or opened markets.[124] Instead, the US/UN genetic dispute continued over the relative enforcement powers of the WTO and UN, with the developing world caught between (Norman Borlaug and Jimmy Carter described the clash as "a rich-world argument that is hurting the poor").[125]

Genetic labeling remains a contested area of American diplomacy. Sheila Jasanoff suggested the United States and its European allies see genetic engineering through different lenses, with the United States focused on biotechnology as products and the Europeans focused on the risks involved in the process and the need for state control.[126] Additionally,

---

[121] Kelemen and Knieval, "The United States, The European Union, and international environmental law," 946–965.

[122] Blomquist, "Ratification Resisted," 499. The Cartagena Protocol is the only enforceable part of the CBD, see also Kemi Fuentes-George, "Neoliberalism, Environmental Justice, and the Convention on Biological Diversity: How Problematizing the Commodification of Nature Affects Regime Effectiveness," *Global Environmental Politics* 13 (November 2013): 144–163.

[123] Robert Paarlberg, *Starved for Science: How Biotechnology Is Being Kept Out of Africa* (Cambridge, MA: Harvard University Press, 2008).

[124] Debra M. Strauss, "Feast or Famine: The Impact of the WTO Decision Favoring the U.S. Biotechnology Industry in the EU Ban of Genetically Modified Foods," *American Business Law Journal* 45 (Winter 2008): 775–826, esp. 805.

[125] Norman E. Borlaug and Jimmy Carter, "Foreword," in Paarlberg, *Starved for Science*, viii.

[126] Jasanoff argued the British are more focused on process and the Germans as state control. Sheila Jasanoff, *Designs on Nature: Science and Democracy in Europe and the United States* (Princeton, NJ: Princeton University Press, 2005), 79–97. For coverage of the OSTP focus on products not process see also Wu and Butz, *The Future of Genetically Modified Crops*, 54.

the structure of the EU gives environmental groups more influence and "resistance to biotechnology became almost a surrogate for resisting America's imperial power" after the Cold War.[127] In the developing world, there is no strong political motivation for GM crops (unlike the Green Revolution of the 1960s and 1970s).[128] Nor do developing nations have an incentive to enforce intellectual property rights since profits accrue outside the country.[129]

Genetic patenting also remains contested. From one perspective, the "American" position is dominant: Biotech, pharmaceutical, and other industries desired twenty-year patent protection on genetic resources and products and this is codified in WTO agreements and European directives. However, legislative and judicial success has not brought clarity or coherence to global governance.[130] Instead, geneticists complain of "patent thickets" preventing research; one stated, "it's extraordinarily hard to argue that when we sequence an individual's genome we aren't violating patents right and left."[131] Additionally, many of the original patents from the 1990s were based on a one-gene, one-disease concept that went quickly out-of-date; nonetheless, the DNA Patent Database at Georgetown University received its 50,000th patent in 2010.[132] The USPTO began limiting patents, but a study in 2011 concluded genetic patents were still issued faster in the United States than in Europe or Japan and for broader claims (Europeans, for example, have a tougher definition of utility).[133]

Fragmentation and rivalry in intellectual property spurred innovative public/private partnerships. Merck, Lawrence Livermore Laboratory, and Washington University developed the Merck Gene Index (a publicly available database) and deposited 800,000 sequences into GenBank

[127] Jasanoff, *Designs on Nature*, 8.

[128] Wu and Butz, *The Future of Genetically Modified Crops*, xxii.

[129] James D. Gaisford, Jill E. Hobbs, and William A. Kerr, "Will the TRIPS Agreement Foster Appropriate Biotechnologies for Developing Countries?" *Journal of Agricultural Economics* 58 (2007): 199–217.

[130] For an exhaustive study of the various over-lapping and competing treaties, see Catherine Rhodes, *International Governance of Biotechnology: Nature, Problems and Potential* (New York: Bloomsbury Academic, 2010).

[131] Sam Kean, "The Human Genome (Patent) Project," *Science* 331 (February 4, 2011): 530–531, quote on 530.

[132] For 2010 numbers, see Cook-Deegan and Heaney, "Patents in Genomics and Human Genetics," 384.

[133] Ibid.

(thereby undermining future patent claims by rival companies).[134] The Wellcome Trust and ten pharmaceuticals patented sequences but withheld enforcement to deny other claims.[135] Public programs proliferated as well, whether the 1000 Genomes Project (research from 2008 to 2015 to find variations in population genetics) or ENCODE (a US-funded follow-up to the HGP to map functional genes).[136] "Biobanks" and "genebanks" range from open-science initiatives to closed commercial endeavors: Some require a data-sharing plan, but others do not, and many restrict public access.[137] This commercial value distinguished genetics from climatology for decades, as market forces supported genetic research while often opposing climate research. US diplomacy followed the market.

Genetic Engineering, Climate Change, and American Foreign Relations

Commercial concerns influenced American diplomacy in genetic engineering and climatology. American genetic diplomacy, for example, was conditioned by "substantial equivalence" because the industries threatened by global regulation persuaded Congress to accept and promote the concept.[138] Opposition to climate change was driven largely by industrial lobbies. In the United States, political persuasion influenced the

---

[134] Deegan and Heaney, "Patents in Genomics and Human Genetics," 402. Figure of sequences from Contreras, "Bermuda's Legacy," 95.

[135] Subhajit Basu, "Human Genome and Patent," *International Review of Law, Computers and Technology* 16 (2002): 339–357, 343. See also Bita Amani and Rosemary J. Coombe, "The Human Genome Diversity Project: The Politics of Patents at the Intersection of Race, Religion and Research Ethics," *Law & Policy* 27 (January 2005): 152–188.

[136] For the 1000 Genomes Project, see www.1000genomes.org/. For ENCODE, see https://genome.ucsc.edu/ENCODE/.

[137] See Yann Joly, Clarissa Allen, and Bartha M. Knoppers, "Open Access as Benefit Sharing? The Example of Publicly Funded Large-Scale Genomic Databases," *Journal of Law, Medicine and Ethics* (Spring 2012): 143–146. For an in-depth introduction, see Giovanni Pascuizzi, Umberto Izzo and Matteo Macilotti, eds., *Comparative Issues in the Governance of Research Biobanks: Property, Privacy, Intellectual Property and the Role of Technology* (New York: Springer, 2013). See also Bartha M. Knoppers, "A Human Rights Approach to an International Code of Conduct for Genomic and Clinical Data Sharing," *Human Genetics* 133 (2014): 895–903; Timothy Caulfield, Shawn HE Harmon, and Yann Joly, "Open Science versus Commercialization: A Modern Research Conflict?" *Genome Medicine* 4 (2012): 17.

[138] Francis Garon and Eric Montpetit, "Different Paths to the Same Result: Explaining Permissive Policies in the USA," in Eric Montpetit, et al., eds., *The Politics of Biotechnology in North America and Europe: Policy Networks, Institutions and Internationalization* (New York: Lexington Books, 2007), 61–82.

reaction to NAS pronouncements: Many on the political right disliked NAS support for global warming, but many on the political left disliked NAS support for genetic modification. At the same time, US diplomacy was not simply beholden to markets and industry forces: the pharmaceutical and biotech industries supported ratification of the CBD and most sectors of the American economy accepted global warming. Instead, ratification was not a priority.

Whether on the Kyoto Protocol or the CBD, the de facto American policy has been to sign but not ratify, a common diplomatic stance: Between 1776 and 1976, the United States signed but failed to ratify more than 400 treaties.[139] Some negotiations lasted generations: the country took fifty years to ratify the protocol on bacteriological warfare, forty-one years for the convention against genocide and has yet to ratify the convention on discrimination against women (signed in 1980).[140] Though impasses in the twenty-first century may be the consequence of divided government, the result was familiar: The United States received concessions during negotiations but escaped regulation, liability, or penalty. Negotiators, for example, shaped both the CBD and Kyoto Protocol in the absence of ratification, including exempting pharmaceuticals in the Cartagena Protocol and requiring the GEF (and World Bank) serve as the primary funding source for UN programs on climate change and biodiversity. Nor has the failure to ratify retarded American research or meant non-compliance: The country remained a leader in both fields, was generally compliant with the CBD, and financed climate research around the world. Indeed, failure to ratify did not lessen American influence or banish American science from the global stage. Instead, signing but not ratifying provided the United States leverage on global regulations and preserved the freedom of action traditionally sought in American diplomacy.

---

[139] See Curtis A. Bradley, "Unratified Treaties, Domestic Politics, and the U.S. Constitution," *Harvard International Law Journal* 48 (Summer 2007): 307–336, number on 309.

[140] The complete names of the conventions mentioned are: "The Geneva Protocol for the Prohibition of the Use in War of Asphyxiating, Poisonous or Other Gases, and of Bacteriological Methods of Warfare," (signed in 1925, ratified in 1975 (post-Vietnam)); "The Convention on the Prevention and Punishment of the Crime of Genocide," (signed in 1948, ratified in 1989 (post-Civil Rights era); "The Convention on the Elimination of All Forms of Discrimination against Women," (signed 1980, unratified). See Bradley, "Unratified Treaties," 309–310.

Genetics and climatology extended global governance into previously sovereign areas, but scientific knowledge could not compel national agreement or action. On one hand, data – whether on gene sequences or atmospheric ozone and CO_2 levels – provided a foundation for international agreement and cooperation. International institutions provided structure and funding, treaties established targets, and collaborative research demonstrated a shared commitment. On the other hand, genetic data did not guarantee political accord: The surge in genetic research and biotechnology resulted in ongoing disagreements over definitions of property, national sovereignty, and trade. International institutions adopted different positions: The United Nations and WTO cooperated on climate research and biodiversity but were at cross-purposes over genetic patenting and labeling. Finally, politics overruled environmental data at first: Increased ozone and CO_2 levels alone were insufficient to compel international action. Nor did ratification of a treaty guarantee national compliance or signify a successful global response.

The debates over genetic engineering and climate change also highlighted global inequality. Environmental disputes between developed and developing nations date to the Stockholm conference, while the disparity between genetic donors, often located in developing nations, and genetic patent holders, often located in developed nations, became obvious. The leading countries of the Human Genome Project – the United States, Germany, Great Britain, France, and Japan – represented the global scientific elite. Intellectual property rights accrued to those countries: any patent of significance is a "triadic" patent, i.e. a patent accepted by the US Patent and Trade Office, the European Patent Office, and the Japanese Patent Office. Yet research in genetics, climate change, and biodiversity – often funded by the United States – created networks linking the developed and developing worlds. In an era of historic scientific concentration and inequality, these networks provided a channel and opportunity for American diplomacy, whether as part of the war on terror or development programs to lessen global inequality.

PALE SHADOWS

It could be argued that our inability to continue our investment in human capital on a scale that we did in the 1960s and 1970s is a factor that has contributed in some measure to instability in many places today and hostility to the United States. ... The United States was the key influence in developing the Indian agricultural university system, the key contributor to the African agricultural universities, and to Asian and Latin American agricultural universities as well. But such US programs are now a pale shadow of what they once were. Science has disappeared. Human capital development has disappeared. And the investments for long-term institution building have nearly disappeared.

Former CIA Director and future Secretary of Defense
Robert Gates (2006)[141]

America's scientific leadership has always been widely admired around the world, and we must continue to expand cooperation and partnership in science and technology. We have launched a number of Science Envoys around the globe and are promoting stronger relationships between American scientists, universities, and researchers and their counterparts abroad. We will reestablish a commitment to science and technology in our foreign assistance efforts and develop a strategy for international science and national security.

US National Security Strategy (2010)[142]

Robert Gates has been an important voice for science diplomacy in the twenty-first century. As President of Texas A&M University he lamented "science [had] disappeared" in American diplomacy; a few years later, as Secretary of Defense, he promoted a National Security Strategy to "reestablish a commitment to science and technology in our foreign assistance efforts." His recognition of the American role in foreign agricultural colleges and candor about US funding being reduced to a "pale shadow" demonstrated a welcome awareness. In the 1980s, for example, USAID educated close to 20,000 students a year, many of whom became officials in their home countries; by 2008, the number of students was less than 1000.[143] Yet Cold War programs were not ideal: while aid demonstrated American benevolence and garnered positive public relations, the scientific and technical component was always a "pale shadow" of its military counterpart and the funds often reinforced authoritarian regimes, exacerbating pre-existing inequalities (see Chapter 3). Gates also ignores the role the United States played in protecting commercial research through

---

[141] Gates quoted in Nina V. Federoff, "Science Diplomacy in the 21st Century," *Cell* 136 (January 9, 2009): 9–11, quote on 10.

[142] US National Security Strategy, 2010, available online at: http://osce.usmission.gov/national_security_strategy_2010.html.

[143] Committee on Science and Technology, House of Representatives, 110th Cong., 2nd sess., *International Science and Technology Cooperation* (April 2, 2008) (Washington, DC: Government Printing Office, 2008), 56.

intellectual protections and limiting UN assistance. Still, the attention to science diplomacy is rare.

The attacks of September 11th shaped American diplomacy in the developing world. Both the Bush and Obama administrations engaged in scientific outreach, sponsoring programs in Iraq and other Middle Eastern countries, though the administrations adopted a different approach to nuclear diplomacy with Iran. The Palestinian/Israeli conflict continued to disrupt US/UN relations: The United States left UNESCO over recognition of Palestine, an act celebrated in Israel but damaging to America's image as a neutral patron of science. During an era of historic inequality, emerging nations such as China and Brazil encouraged scientific cooperation in the southern hemisphere while the United States remained absent from UNESCO and the World Academy of Science (TWAS), organizations whose missions include the cross-cultural dialogue and science education essential to ending terrorism.

The War on Terror and Iraqi Assistance

As in the Cold War, science played multiple roles in the War on Terror. Within months of the attacks, congressional hearings recounted new initiatives in detecting and preventing bioweapons, improving laboratory security and information-sharing, preventing cyberwarfare, and understanding the psychology of terrorism.[144] Overseas, advanced technology – from night vision goggles, drones, and stealth transport to real-time GPS coordination – protected and aided American troops, while the CIA ran a vaccination program to obtain DNA samples of Osama Bin Laden's relatives for genetic identification.[145] As before, scientific and technical aid complemented the military strategy – the United States would prevent extremism by raising living standards. The environment seemed promising: A poll found fewer than 20 percent of Moroccans and Jordanians (an

---

[144] Joint Hearing, Committee on Science (House) and Committee on Commerce, Science and Transportation (Senate), 107th Cong., 2nd sess., June 25, 2002, *Science and Technology to Combat Terrorism* (Washington, DC: Government Printing Office, 2002). The United States also worried about Iraqi smallpox warfare and initiated a vaccination program for 500,000 troops, see also "The Department of Homeland Security," in Jeanne Guillemin, *Biological Weapons: From the Invention of State-Sponsored Programs to Contemporary Bioterrorism* (New York: Columbia University Press, 2005).

[145] Saeed Shah, "CIA Organized Fake Vaccination Drive to Get Osama bin Laden's Family DNA," The *Guardian* (July 11, 2011).

ally) had a positive view of the United States, but more than 80 percent had a positive view of American science and technology.[146] At the same time, politicizing medicine carried risk: Revelation of the CIA's vaccination scheme landed a Pakistani doctor in jail and jeopardized the region's anti-polio campaign when extremists began assassinating vaccinators as foreign agents.[147]

The Bush administration introduced scientific and technical aid programs to Iraq. The country was a leading Arab scientific nation until the Iran–Iraq war and exodus of thousands of scientists after the first Gulf War.[148] Saddam's rule, sanctions, and the invasion of 2003 meant the country had to start over.[149] To assess Iraq's needs, Sandia national laboratory partnered with the Arab Science and Technology Foundation (ASTF), meeting with hundreds of scientists and eventually concluding healthcare, water resources, and energy were among the most critical.[150] Iraqi delegations also traveled to Washington to meet with officials at NSF, NIH, and other departments for help with decommissioning nuclear facilities and environmental research. The State Department established an Iraqi International Center for Science and Industry (IICSI, 2003) to redirect weapons scientists and the Departments of Energy and Defense launched similar initiatives.[151] The State Department modeled the new

[146] Dafna Linzer, "Poll Shows Growing Arab Rancor at U.S.," *Washington Post* (July 23, 2004): A26. See also David Kramer, "Science diplomacy enlisted to span US divide with developing world," *Physics Today* (December 2010): 28–30.

[147] For an in-depth, multi-part look at this story, see Alexander Mullaney, "He Led the CIA to bin Laden: And Unwittingly Fueled a Vaccine Backlash," (February 27, 2015), available online at: http://news.nationalgeographic.com/2015/02/150227-polio-pakistan-vaccination-taliban-osama-bin-laden/.

[148] See Abdalla Alnajjar, Ammar M. Munir, Arian Pregenzer and Adriane Littlefield, *International Initiative to Engage Iraq's Science and Technology Community: Report on the Priorities of the Iraqi Science and Technology Community* (May 2004), 14–16. This a joint publication of Sandia National Laboratories and the Arab Science and Technology Foundation, SAND 2004-2223.

[149] Jeffrey Mervis, "Science Minister Starts from Scratch," *Science* 302 (November 21, 2003). UN sanctions limited the import of most scientific equipment into Iraq, see Sam Jaffe, "Rebuilding Iraqi Science," *The Scientist* (July 14, 2003): 22–25.

[150] Abdalla Alnajjar, et al., *International Initiative to Engage Iraq's Science and Technology Community*.

[151] The Coalition Authority created the Iraqi Nonproliferation Programs Foundation in June 2003, endowed with $37.5 million. See Richard Stone, "New Initiatives Reach Out to Iraq's Scientific Elite," *Science* 303 (March 12, 2004): 1594; see also, Richard Stone, "Coalition Throws 11th-Hour Lifeline to Iraqi Weaponeers," *Science* 304 (June 25, 2004): 1884. The original State Department press release is available at the Federation of American Scientists, see http://fas.org/nuke/guide/iraq/dos121803.html. See Christina Asquith, "A $20 Million Carrot to Keep WMD Scientists in Iraq," *Christian Science*

Iraqi institution after the Russian ISTC (see Chapter 6) but building a scientific infrastructure during an insurgency proved difficult.

Violence and a lack of funding slowed Iraqi scientific reconstruction. The United States and ASTF hoped to raise $50 million for cancer treatment, water monitoring, and other needs, but neighboring states offered little help.[152] Instead, insurgents began targeting Iraqi researchers as American agents, including staff at the only remaining cancer center.[153] The systematic murder of hundreds of Iraqi scientists, especially physicists at the new IICSI, led to pleas from the scientific community to increase visas for Iraqi researchers as well as hesitation among Iraqis to collaborate.[154] Amid the bloodshed, the State Department and DOD set up an Iraq Virtual Science Library in 2005 (the DOD set up a similar library in Pakistan the previous year).[155] The library eventually provided students and researchers access to journals and instruction from faculty from more than thirty universities; by 2010, library users downloaded nearly 30,000 articles a month.[156] Federal agencies also partnered with American universities and industries in market-based initiatives: Texas A&M, for example, managed a "$6 million USDA program to improve agricultural extension in Iraq, and a $10 million USAID subcontract on the Inma Agribusiness program to build agribusiness in Iraq."[157]

Monitor (December 22, 2003). For the DOE's National Nuclear Security Administration program, see https://nnsa.energy.gov/mediaroom/pressreleases/nnsa-program-engage-iraqi-scientists.

[152] Richard Stone, "Priorities for Rebuilding Civilian Iraqi Science," *Science* 304 (May 14, 2004): 943–944.

[153] Richard Stone, "In the Line of Fire," *Science* 309 (September 30, 2005): 2156–2159.

[154] Yudhijit Bhattacharjee, "Groups Urge Easing of Restrictions on Visa Policies Affecting Scientists," *Science* 304 (May 14, 2004): 943; see also, Bengt Gustafsson, "The science community must unite over Iraq," *Nature* 444 (November 23, 2006): 422.

[155] The DOD Defense Threat Reduction Agency (DTRA) funded the library. See Jim Dawson, "US is Creating a Virtual Library to Help Restore Iraqi Science," *Physics Today* (November 2005): 24–26.

[156] Numbers of downloads from Anon, "Five Years Later, An Ambitious Virtual Library is Turned Over to Iraqi Researchers and Scholars," *States News Service* (June 10, 2010), accessed online. For an overview of the program, see Cathleen A. Campbell, "US Science Diplomacy with Arab Countries," in Davis and Patman, eds., *Science Diplomacy: New Day or False Dawn?* (Hackensack, NJ: World Scientific Publishing Co., 2015), 27–44. See also, Richard Stone, "Throwing a Lifeline to a One-Time Arab Science Power," *Science* 347 (January 16, 2015): 223–224.

[157] Texas A&M Prepared Statement, Committee on Science and Technology, Subcommittee on Research and Science Education, House of Representatives, *The Role of Non-Governmental Organizations and Universities in International Science and Technology Cooperation*, 110th Cong., 2nd sess. (July 15, 2008) (Washington, DC: Government Printing Office, 2008), 45.

As in India and China, commercial agriculture played a prominent role in American efforts.

Scientific outreach reflected American diplomatic goals throughout the region. To highlight women's empowerment, for example, the State Department, AAAS, and Kuwait sponsored a conference of female scientists and engineers, with nearly 200 attendees representing more than a dozen Arab countries.[158] The NAS organized a hydrological study on the Jordan valley with Palestinian and Israeli academies to demonstrate cooperation on a shared concern.[159] The AAAS even reached out to Syrian President Bashar al-Assad. After a ninety-minute meeting, AAAS official Norman P. Neureiter left thinking "a closer relationship in science may be possible," but US/Syrian relations soon soured.[160]

The Obama administration expanded diplomatic efforts in the region. A few months after taking office, the president spoke at Cairo University about American re-engagement. His remarks included a familiar refrain:

> On science and technology, we will launch a new fund to support technological development in Muslim-majority countries, and to help transfer ideas to the marketplace so they can create more jobs. We'll open centers of scientific excellence in Africa, the Middle East and Southeast Asia, and appoint new science envoys to collaborate on programs that develop new sources of energy, create green jobs, digitize records, clean water, grow new crops [and]... eradicate polio.[161]

The passage touched upon themes echoed by every president since Truman: the importance of science and technology to economic development, the creation of new research centers, the use of market forces to stimulate innovation, and the promise of better crops, water, energy, and health.

The Obama administration maintained Bush-era initiatives and NGOs responded positively to the President's announcement: The FAS worked in Yemen and the AAAS sent representatives to Syria, Cuba, North Korea, and Myanmar.[162] In 2011, the State Department inaugurated

---

[158] AAAS Prepared Statement, Ibid., 13.

[159] Michael Clegg, Foreign Secretary of the NAS, Ibid., 15–16.

[160] Norman P. Neureiter, Director, Center for Science, Technology and Security Policy, AAAS prepared statement, Ibid., 17.

[161] President Barack Obama, "Remarks by the President at Cairo University, 6–04–09," available at: www.whitehouse.gov/the-press-office/remarks-president-cairo-university-6-04-09.

[162] David Kramer, "Science diplomacy enlisted to span US divide with developing world," *Physics Today* (December 2010): 28–30; see also, Anon., "2014 AAAS Award for Science Diplomacy," *Bioterrorism Week* (January 19, 2015): 3; Campbell, "US Science Diplomacy with Arab Countries," 27–44.

the Global Innovation through Science and Technology (GIST) program, later expanded by additional corporate sponsors. The program focused on commercial research: By 2016, the GIST Network had generated more than $80 million in revenue through more than 4,500 start-ups in 86 emerging economies.[163] However, Obama's boldest initiative focused on Iran, a country whose nuclear program resulted from American assistance and tested American diplomacy.

US–Iranian Scientific Relations in the War on Terror

It is worth revisiting the history of American–Iranian scientific relations. Iran signed one of the first Point Four agreements and the United States maintained a wide-ranging aid program for nearly three decades (see Chapter 3). The Atoms for Peace program introduced nuclear research to Iran, though neither country advertises this inconvenient fact: an official Iranian website proudly cites the five-megawatt reactor at the University of Tehran as the "foundation of Iran's nuclear science and technology," but fails to name the United States as its donor.[164] Americans also helped establish the Atomic Energy Organization of Iran (AEOI) in 1974, while the Islamic revolution (1979) closed universities and scrapped nuclear research as "junk" forced upon the country by the "Great Satan."[165] The United States classified Iran as a "state sponsor of terror" in 1984; the country restarted its nuclear program shortly thereafter and Sharif University took applications for its first PhD program, in physics, in 1988.[166]

Scientific outreach provided the rare US/Iranian interaction. In 1997, newly elected Iranian President Mohammad Khatami spoke to The World Academy of Sciences about cooperation between nations "without amicable relations."[167] Khatami also suggested his country join CERN and a proposed synchrotron in the Middle East. The following

---

[163] See the GIST Network press release available at: www.gistnetwork.org/content/press.

[164] See the Atomic Energy Organization of Iran's official website: www.aeoi.org.ir/. Accessed April 25, 2016.

[165] Siegfried S. Hecker and Abbas Milani, "Ending the Assassination and Oppression of Iranian Nuclear Scientists," *Bulletin of the Atomic Scientists* 71 (2015): 46–52.

[166] On terror, see Kenneth Katzman and Paul K. Kerr, Congressional Research Service, *Iran Nuclear Agreement* (May 31, 2016), 24. Congressional Research Service, 7–5700, R43333, available at www.crs.gov. On the PhD program, see Richard Stone, "An Islamic Science Revolution?" *Science* 309 (September 16, 2005): 1802.

[167] Robert Koenig, "Iran's Scientists Cautiously Reach Out to the World," *Science* 24 (November 24, 2000): 1484–1487.

year, the NAS sent the first American delegation to Iran in over a decade. Neither country promoted the visit (the Iranians kept it secret). Over the next two years, the AAAS and FAS made initial forays while the National Academies arranged a series of workshops with faculty at Sharif and the Iranian Academy of Science.[168] But even as NGOs began to re-engage, the American and Iranian governments remained at odds over nuclear enrichment, a critical component of weapons-related research.

US/Iranian scientific relations grew in the absence of political relations. The attacks on September 11th complicated American science diplomacy with Iran: President Bush included the country in an "axis of evil" even as the National Academies and AAAS maintained talks. When American troops occupied Baghdad, the Swiss relayed an Iranian offer to cooperate on terrorism and the nuclear program.[169] The administration rejected the offer, but the Iranians signed an agreement with European nations to suspend enrichment at Natanz and avoid confrontation.[170] In 2005, the Iranian government announced a twenty-year plan for national scientific advancement and welcomed collaboration; the same year, the United States became Iran's largest foreign research partner, with ongoing projects in astronomy, seismology, and medicine.[171] However, Iran restarted enrichment and ended the European agreement, amassing more than 8,000 centrifuges and achieving low-level enrichment by 2008.[172] Relations with the United States and American scientific organizations grew strained.

Iranian politics shaped the reception to American scientific outreach. The Iranian scientific community encouraged exchanges, but engagement met opposition among conservatives. William Wulf, an official with the

---

[168] Richard Stone, "Iran's Trouble with Molybdenum May Give Diplomacy a Second Chance," *Science* 311(January 13, 2006): 158.

[169] Trita Parsi, *A Single Roll of the Dice: Obama's Diplomacy with Iran* (New Haven, CT: Yale University Press, 2012), 5–7. See also Barbara Slavin, *Bitter Friends, Bosom Enemies: Iran, the U.S. and the Twisted Path to Confrontation* (New York: St. Martin's Press, 2007), 212–215.

[170] The countries were France, Germany and the UK, see Katzman and Kerr, *Iran Nuclear Agreement*, 1.

[171] On the Iranian plan, see National Academies of Sciences, Engineering and Medicine, Glenn E. Schweitzer, ed., *U.S.–Iran Engagement in Science, Engineering, and Health (2010–2016): A Resilient Program but an Uncertain Future* (Washington, DC: National Academies Press, 2017), 24–27. See also Stone, "An Islamic Science Revolution?" 1802.

[172] Stone, "Iran's Trouble with Molybdenum May Give Diplomacy a Second Chance," 158. See also David E. Sanger, et al., "Around the World, Distress Over Iran," *New York Times* (November 28, 2010): accessed online.

Civilian Research and Development Foundation, testified to Congress: "I [was] in Iran in October, and it is just hard to explain how much the faculty at places like Sharif University, which is sort of their MIT, like Americans, understand our values, admire our values, and are some of the best ambassadors that we have in the entire world. So this is gold."[173] However, Michael Clegg of the NAS cautioned against "politicizing" scientific outreach:

For instance, the National Academy of Sciences (NAS) has been involved in a series of mutually beneficial scientific workshops with Iran over the past eight years, achieving a remarkable level of engagement with Iran's science community. AAAS has also been involved in this activity. But one must be careful what funds are used for such programs and what rhetoric accompanies them. When State declared that it had funds for NGOs to focus on fostering democracy in Iran, it resulted in the arrest and detention in Iran of a number of Iranians and Iranian-Americans suspected of using State Department money to conspire against the Iranian government.[174]

Nor did things improve after the election of Barack Obama. Iranian officials detained NAS administrator Glenn Schweitzer for eight hours, jeopardizing future exchanges (Schweitzer vowed not to go back).[175] But President Obama, who campaigned on speaking with enemies, restarted negotiations.

The president relied on two hidden strengths: knowledge of the secret Iranian nuclear facility at Fordow and a computer virus capable of disrupting enrichment.[176] In a speech broadcast online, the president spoke directly to the Iranian people on Nowruz (the Iranian New Year),

---

[173] William Wulf quoted in Committee on Science and Technology, *The Role of Non-Governmental Organizations and Universities in International Science and Technology Cooperation*, 48.

[174] Michael Clegg quoted in Ibid., 18.

[175] Richard Stone, "Tehran Incident Threatens U.S.–Iran Project," *Science* 323 (January 2, 2009): 23.

[176] The CIA provided the intelligence on the Fordow plant. The Bush administration authorized $400 million for covert operation against the Iranian nuclear program in 2007. One aspect was cyberwarfare led by the NSA and US Cyber Command, which developed multiple versions of Stuxnet (in cooperation with Israelis). See Kim Zetter, *Countdown to Zero Day: Stuxnet and the Launch of the World's First Digital Weapon* (New York: Broadway Books, 2014), 309–315. See also Ralph Langner, "Stuxnet: Dissecting a Cyberwarfare Weapon," *IEEE Security & Privacy* (May/June 2011): 49–51; and P. W. Singer, "Stuxnet and Its Hidden Lessons on the Ethics of Cyberweapons," *Case Western Reserve Journal of Law* 47 (2015): 79–86

stating his administration was "committed to diplomacy."[177] The surprise appeal was unsuccessful: A few months later, Iranians elected Mahmoud Ahmadinejad, who opposed engagement and promoted a nuclear nationalism.[178] Alex Dehgan, State Department science advisor, publicly argued relations could be repaired, pointing out "Iranian scientists publish more papers with Americans than with any other country in the world."[179] Covertly, the President authorized deployment of the virus later known as Stuxnet, infecting Iranian computers the day after Ahmadinejad's confirmation.[180] Behind the scenes, administration officials, with the help of Brazil and Turkey, hoped to break the impasses, tacitly accepting low-level uranium enrichment even though many allies and administration officials did not.[181] The exposé of the Fordow nuclear plant undermined Iranian claims and reversed the position of anti-sanction holdouts like Russia and China (some American allies recommended bombing the plant).[182] When the Iranians balked at conditions in 2010, the stuxnet virus caused centrifuges at Natanz to malfunction, adding a new twist to science in diplomacy: a cyberwarfare attack on a nuclear program to force a diplomatic resolution. But the Iranians refused to suspend enrichment and the UN Security Council imposed sanctions (Res. 1929).

US sanctions made Iranian research more difficult. The cost of medical drugs and scientific equipment increased, and most publishing houses blacklisted Iranian researchers.[183] International collaborations declined. Unknown assailants killed five physicists between 2010 and 2013, leading the American Physical Society (APS) to protest that assassinations were

---

[177] President Obama, "Videotaped Remarks by The President in Celebration of Nowruz," (March 20, 2009): available at: www.whitehouse.gov/the-press-office/videotaped-remarks-president-celebration-nowruz.

[178] Ahmadinejad's crowds often chanted for nuclear power. See Joseph Cirincione, "Controlling Iran's nuclear program," *Issues in Science and Technology* (Spring 2006), accessed online.

[179] Dehgan quoted in David Kramer, "At Work in the Trenches of Science Diplomacy," *Physics Today* (December 2010): 30–31, quote on 31.

[180] Zetter, *Countdown to Zero Day*, date of release on 303, n30; date of infection 339.

[181] Parsi, *A Single Roll of the Dice*, 58–60.

[182] Allies included Israel, Saudi Arabia, Bahrain, Oman. See David E. Sanger, James Glanz and Jo Becker, "Around the World, Distress Over Iran," *Washington Post* (November 28, 2010).

[183] Warren E. Pickett, et al., "Science Diplomacy in Iran," *Nature Physics* 20 (July 2014): 465–467. Mehdi Aloosh, "Iran deal and Global Health Diplomacy," *The Lancet* 3 (December 2015): e744.

not the road to peace (a number of Syrian physicists were killed as well).[184] However, the president of AEOI, physicist Ali Akbar Salehi, proposed the deaths inspired "students who were studying in other fields [to change] to nuclear science" (he also felt the physicists "gained martyr-dom").[185] Iranian Minister of Science Mohammad Farhadi felt the "sanctions ... forced scientists to work more creatively," although the Iranian government jailed an aspiring young physicist for refusing to contribute to the nuclear program and conservative newspapers tarred foreigners as part of an anti-government conspiracy.[186] Nonetheless, Iranian scientific publications grew (by one measure, Iran placed seven-teenth in the world) and the NAS maintained engagement.[187]

The sanctions era (2010–2015) complicated American science diplo-macy with Iran. Politicians and researchers in both countries desired engagement and the NAS looked for proposals capable of producing results.[188] Joint areas of interest included research into earthquake pre-paredness, solar energy, climate change, wetlands conservation, and wild-life conservation. But collaboration proved complex: foreign participants struggled to attain visas and travel, US participants feared for their safety and restrictions limited research in fields with national-security dimen-sions, including avionics, electronics, chemicals, energy, and many other

---

[184] Siegfried S. Hecker and Abbas Milani, "Ending the Assassination and Oppression of Iranian Nuclear Scientists," *Bulletin of the Atomic Scientists* 71 (2015): 46–52. Four of the physicists were killed by magnetic bombs attached to moving cars by motorcyclists, a trademark one work attributes to the Israelis: See Dan Raviv and Yossi Melman, *Spies against Armageddon: Inside Israel's Secret Wars* (Sea Cliff, NY: Levant Books, 2014), 10–11.

[185] Salehi quoted in Richard Stone, "Iran's Atomic Czar Describes the Art of the Deal," *Science* 349 (August 14, 2015): 674–675, quote on 675.

[186] Mohammad Farhadi, "Iran, Science and Collaboration," *Science* 349 1029 (September 4, 2015): 1029. Elise Auerbach, "Iran Needs to Present a United Front on Science" *Nature* 508 (April 24, 2014): 433. On Omid Kokabe, the aspiring physicist, see Anon., "Kokabee Has His Kidney Removed," *Iran Times International* (April 29, 2016), accessed online; Herbert L. Berk, "Free Omid Kokabee, Another Iranian Prisoner of Conscience," *Washington Post* (Oct 30, 2015), accessed online; Steven T. Corneliussen, "Not Much Media for Physicist Omid Kokabee, Imprisoned in Iran," *Physics Today* (Nov 19, 2015), accessed online.

[187] Shahin Akhondzadeh, "Iranian Science Shows World's Fastest Growth Rates: Ranks 17th in Science Production in 2012," *Avicenna Journal of Medical Biotechnology* 5 (July/September 2013): 139. See also Szczepan Lemańczyk, "Science and National Pride: The Iranian Press Coverage of Nanotechnology, 2004–2009," *Science Communications* 36 (2014): 194–218.

[188] National Academies of Sciences, Engineering and Medicine, *U.S.–Iran Engagement in Science, Engineering, and Health (2010–2016)*, 2–4.

dual-use areas. Additionally, the US State Department had to engage in domestic diplomacy, reassuring American researchers engagement was in the American national interest. Eventually, more than 1,500 scientists from 120 different universities and institutions took part between 2000 and 2016, while another 12,000 Iranian students attended US universities (with restrictions on nuclear-related courses).[189] Iranian Minister Farhadi believed the sanctions and collaboration changed Iran's internal dialogue, stating, "The environment further spurred science-driven political discourse in the country. A prominent example is the role of the scientific community in the recent negotiations on Iran's nuclear program."[190] Indeed, aided by physicists, the Obama administration achieved international agreement.

The United States and Iran signed an agreement on the Iranian nuclear program in 2015. Two years before, Iranian elections brought Hasan Rouhani, a supporter of NAS collaboration and future negotiator of the nuclear deal, to the presidency.[191] As talks moved forward, President Obama again took to the airwaves on Nowruz, promising "more partnerships in areas like science and technology and innovation."[192] In the United States, Congress held hearings opposing Obama's resumption of negotiations, arguing the sanctions worked and the administration ignored Iranian anti-American actions around the world.[193] When Iranian officials asked Ali Akbar Salehi to help negotiate, he suggested his counterpart, Secretary of Energy Ernest Moniz, a friend from the 1970s when both were at MIT.[194] The physicists played critical roles in the negotiations and public relations, appearing in multiple interviews to explain the deal. The eventual agreement dismantled much of the Iranian nuclear infrastructure, promising to convert Fordow into an international physics center while permitting Iran to continue low-level enrichment and limited research.[195] Additionally, in return for allowing

---

[189] Ibid., 2, 44 and 75.   [190] Farhadi, "Iran, Science and Collaboration," 1029.

[191] Abbas Milani and Michael A. McFaul, "A Chance for Iranian Reform," *Hoover Digest* (Winter 2016): 97–105.

[192] President Obama, "Remarks by President Obama on Nowruz," (March, 19, 2015). Available online.

[193] Committee on Foreign Affairs, House of Representatives, 113th Cong., 2nd sess., November 18, 2014, *Iranian Nuclear Talks: Negotiating a Bad Deal?* (Washington, DC: Government Printing Office, 2014).

[194] Richard Stone, "Iran's atomic czar describes the art of the deal," *Science* 349 (August 14, 2015): 674–675.

[195] The complete text of the JCPOA (Joint Comprehensive Plan Of Action), with commentary, is available at: www.state.gov/e/eb/tfs/spi/iran/jcpoa/. The Iranians, for example,

IAEA inspections, the country would receive relief from international economic sanctions. Experts immediately disagreed about its long-term prospects and US–Iranian relations remained politically contentious; following the election of Donald Trump in 2016, the NAS halted cooperation, as scientific engagement and the nuclear deal faced an uncertain future.[196] Although it is too soon to determine the legacy of US/Iranian scientific cooperation, the history suggests cooperation was not a cure-all for political disagreement, atomic assistance may have unforeseen consequences, and scientific relations were shaped by the domestic politics of participating nations.

UNESCO and Science Diplomacy in the Middle East

American diplomacy in the Middle East followed Cold War precedents. Scientists and scientific organizations maintained contact during periods of political discord and were among the first Americans "in country," whether Iraq or Iran, while foreign scientists encountered difficulties obtaining visas and attending conferences in the United States. Presidential elections were pivotal to US/Iranian engagement and both leaders faced opposition at home. Nuclear programs took center stage and negotiations required physicists act as diplomats. In Iraq, physicists and former weapons scientists became targets of American redirection campaigns and assassins (a novelty). At the UN, American support for Israel continued to cause tension within UNESCO.

The tensions between the United States and UNESCO have a long history. American anti-communists disliked the organization's neutralism in the early Cold War (see Chapters 1 and 2) and the United States withdrew in 1984 because of its support for family planning, birth control, and criticism of Israel. Scientific societies, including the American Physical Society, American Chemical Society, and AAAS (among others) urged the first President Bush to rejoin, but the administration refused.[197] Assistant Secretary of State John Bolton felt the country maintained more

---

must refrain from laser enrichment and research on uranium or plutonium metals (which have useful non-nuclear military properties like armor-piercing). See also Katzman and Kerr, *Iran Nuclear Deal*.

[196] For an excellent introduction to the debate, see the ten experts consulted by the *Bulletin of Atomic Scientists* at John Mecklin, "The Experts Assess the Iran Deal of 2015," *The Bulletin* (July 14, 2015). Available online.

[197] William Sweet, "Science Societies Press for Review of US Position on UNESCO Membership," *Physics Today* 43 (Feb 1990): 111–112.

influence from outside: "Bluntly stated, UNESCO needs the United States as a member far more than the United States needs UNESCO," adding the United States had more leverage "as a sought after non-member ... than we would wield simply being one vote among 161 others."[198] Throughout the 1990s, Republican opposition in Congress kept the United States out.[199] However, President George W. Bush, concerned about collapsing American support in the region, surprised his UN audience by announcing the United States would rejoin UNESCO in 2003.[200] American membership lasted less than decade.

The first misunderstanding arose quickly. In 2006, the American ambassador sent the UNESCO Director General a directive requiring the organization consult with US officials before partnering with American citizens or societies. Irving Lerch, an official with the American Physical Society, argued it signaled researchers "need to be vetted by the U.S. government."[201] After officials at the National Academies and UNESCO expressed concern, the United States loosened the policy.[202] But a far greater problem was UNESCO's support for Palestinian statehood.

Israeli opposition to Palestinian statehood overrode US support for UNESCO. In 2011, UNESCO members voted 107 to 14, with 52 abstentions, to accept Palestine as a nation. Palestinian President Mahmoud Abbas celebrated a "victory for rights [and] for justice," while Israeli Ambassador Nimrod Barkan accused UNESCO of adopting "a science fiction version of reality by admitting a non-existent state to the science organization."[203] American officials worried Palestine might try for membership in other UN scientific bodies like the WMO, WHO, or IAEA: State Department spokeswoman Victoria Nuland stated, "We

---

[198] Bolton quoted in Irwin Goodwin, "Bush Administration Dashes Hopes for U.S. Rejoining UNESCO: For Now," *Physics Today* 43 (June 1990): 61–62, quote on 62.

[199] Great Britain rejoined UNESCO in 1997 and thirty-seven Nobel Laureates urged the Clinton administration to rejoin as well. Ehsan Masood, "Budget Block on US Bid to Rejoin UNESCO...," *Nature* 396 (December 17, 1998): 606.

[200] Rodney W. Nichols, "UNESCO, US goals, and international institutions in science and technology: what works?" *Technology in Society* 25 (2003): 275–298.

[201] Lerch quoted in Yudhijit Bhattacharjee, "U.S. Rules Could Muffle Scientific Voices," *Science* 309 (July 22, 2005): 544.

[202] Yudhijit Bhattacharjee, "U.S. Loosens Policy on Ties to UNESCO," *Science* 313 (August 18, 2006): 900.

[203] Abbas quoted in Anon., "US cuts UNESCO funding after Palestinian admitted," *Bangladesh Government News* (November 2, 2011), accessed online. Barkan quoted in Anon., "US Quits UNESCO Over Palestine," *The Jewish Advocate* (November 4, 2011): 20.

don't see any benefit, and we see considerable potential damage, if the move is replicated in other UN organizations."[204] As required by law, the Obama administration immediately cut off funding, even failing to pay the expected American contribution at the end of the year.[205] Since the United States accounted for 22 percent of UNESCO's budget (around $65 million), this severely curtailed activities. The Arab League president expressed his "extreme astonishment" at American actions and argued it bode ill for American diplomacy in the region.[206] UNESCO Secretary General Irina Bokova, a Bulgarian politician, reasoned, "I think UNESCO was caught in the middle of this political turmoil of Middle Eastern conflict, and I think this is unfair." She pointed out that UNESCO provided education to deprived students throughout the developing world and thus the United States was "losing [its] credibility."[207] After being in arrears for two years, the United States and Israel lost voting privileges in 2013. The situation continued as of this writing, with the American commitment to development and science education questioned by the country's absence from UNESCO and other scientific organizations.

The United States, for example, paid little attention to TWAS and SESAME. In 1983, Pakistani Nobel Laureate Abdus Salam and forty-two other researchers established the Third World Academy of Sciences or TWAS (today known as The World Academy of Sciences).[208] Abdus wanted TWAS to pressure southern hemisphere governments to invest in science and technology.[209] Based in Italy, the organization became affiliated with UNESCO in 1991, receiving a $15 million dollar endowment. In 2012, China overtook Brazil as the largest contributor to the organization and Chinese President Hu Jintao opened the twenty-third general

[204] Anon., "US halting UNESCO Payments Following Palestine Vote," *Kuwait News Agency* (October 31, 2011), accessed online.
[205] John R. Crook, "UNESCO Admits Palestine; United States Cuts UNESCO Funding," *The American Journal of International Law* 106 (January 2012): 138–175.
[206] Nabil al-Arabi quoted in Anon, "US UNESCO Freeze Bad for Middle East Peace," *Daily Times (Lahore, Pakistan)* (November 2, 2011), accessed online.
[207] Anon., "Without US Funds, UNESCO Struggling to Stay Afloat," *States News Service* (October 11, 2012).
[208] Robert Koenig, "The New Groove in Science Aid: South-South Initiatives," *Science* 322 (November 21, 2008): 1176–1177.
[209] P. Balaram, "Science and the Third World," *Current Science* 81 (October 25, 1991): 865–866.

meeting in Tianjin with a donation of $1.5 million.[210] China also expanded its doctoral fellowship program, providing residence in Beijing along with language and cultural training.[211] Among advanced nations, only Italy provided an annual payment to Academy; the United States was uninvolved. Nor has the United States provided funding for the Synchrotron-light for Experimental Science and Applications in the Middle East (SESAME), an accelerator in Jordan.[212] Modeled after CERN and launched by UNESCO in the early 2000s, the synchrotron has struggled to secure funding; although the United States donated spare equipment and trained a few scientists, the country failed to match the $11 million contribution from the EU (instead, American laboratories and scientific societies participated independently).

In terms of global scientific illumination, the United States could be a brighter "city on the hill." On one hand, the United States funds dozens of scientific initiatives in the developing world annually, increasing science education and building research networks. On the other hand, the United States has been unwilling to commit significant resources to science in the developing world, especially after the Cold War. National security interests outweigh foreign development, as one would expect, but the disparity in funding demonstrates the wide difference in priority: Programs to redirect Iraqi weapon scientists received tens of millions of dollars; the Iraqi Virtual Science Library cost $340,000 to set up and most of the money was spent in the United States.[213] A single stealth bomber cost far more than all American aid programs in the region combined. Limited by funding, scientific and technical assistance in the War on Terror remained a "pale shadow" in American diplomacy.

---

[210] Donation figure in Anon., "Development Boost," *Nature* 489 (September 7, 2012): 479; see also Robert Koenig, "The New Groove in Science Aid: South-South Initiatives," *Science* 322 (November 21, 2008): 1176–1177.

[211] Anon., "Science Fellowship Programme for PHD Candidates from Developing Countries," *States News Service* (April 4, 2013), accessed online.

[212] See the official website for current information: www.sesame.org.jo/sesame/about-us/information-material/brochures.html. See also Chris Llewellyn Smith, "Synchrotron Light and the Middle East," *Science & Diplomacy* (November 16, 2012), accessed online; Nageen Ainuddin, "Science Diplomacy around the World," *Technology Times* 6 (November 15, 2015), accessed online; Chris Llewellyn Smith, "SESAME Moves towards Commissioning," *APS Newsletter* (September 2015), accessed online.

[213] Richard Stone, "In the Line of Fire," *Science* 309 (September 30, 2005): 2156–2159.

8

The Laboratory of Diplomacy

The empires of the future are the empires of the mind.
Prime Minister Winston Churchill, "The Gift of
a Common Tongue," (1943)[1]

Winston Churchill spoke at Harvard after helping craft the D-Day strategy. Although the prime minister suggested the "gift of a common tongue" would help the United States and Great Britain establish an "empire of the mind" after the war, influence over science provided the basis for a different empire of the mind. Scientific knowledge empowered nations and competition in scientific and technical achievement shaped postwar relations. Research with national security and commercial applications became integral to power and nations strove to acquire, profit from and hoard certain knowledge. Knowledge was not equal, nor were nations: instead, the United States was in the best position.

Scientific preeminence and influence enabled an American "empire of the mind" in the decades after World War II. American preeminence and global reach dated to the war: at home, government contracts funded and networked academic and industry researchers; overseas, the United States offered agricultural, industrial and medical aid to allies, incorporating scientific and technical assistance into diplomacy. By the end of the conflict, policy-makers understood science was critical to national security, prosperity, and prestige; every administration thereafter sought to

---

[1] Winston Churchill, "The Gift of a Common Tongue," address given at Harvard University (September 6, 1943). Available online.

maximize American scientific potential and understood the advantages of scientific leadership overseas. Throughout the Cold War, the United States relied on science and technology for hard power and diplomatic leverage, as the country provided access to allies while denying access to adversaries. American assistance and sponsorship of global science created soft power, demonstrating the country's benevolence and deflecting criticism, as scientists, engineers, and doctors operated as goodwill ambassadors abroad. The United States also played the dominant role in the UN technical system, while American accomplishments, whether the Green Revolution and malaria eradication or Intelsat and the moon landing, advertised American capabilities to the world. A national laboratory of diplomacy, the United States experimented with different approaches to science in foreign relations after World War II.

Science for diplomacy was extensive and effective in the developing world, especially in the early Cold War. The United States pioneered the use of scientific and technical assistance to secure geopolitical cooperation, fortify alliances, and open economies. Aid packages included research for local needs, science education and medical care along with loans, military equipment, and industrial and technical support. However, nations often worried assistance was the vanguard of exploitation: Beginning with Point Four, foreigners wondered if aid was "a way to save American private enterprise" and a few withdrew when aid was tied to anticommunism. Of course, without the fear of communism, US administrations had less reason to assist; by the end of the Cold War, American programs had shrunk to a "pale shadow" of their former selves. Additionally, science for diplomacy had political limitations in the developing world: Cooperation often bolstered authoritarian regimes and did little to promote democratization, whether in Iran, Brazil, and Argentina or Saudi Arabia and the People's Republic of China.

American diplomacy for science initially focused on international cooperation. The United States was the primary supporter of global scientific initiatives after World War II, including the International Geophysical Year, World Weather Watch, and the UN Environmental and Climate Programs. The country also funded medical research, exchanges, and foreign agricultural colleges, but development programs never prioritized research and cooperation was always on American terms. The Truman administration, for example, curtailed French proposals for international laboratories following World War II and later presidents refused multiple requests from NATO allies and others to collaborate during the Cold War.

Foreign competition in research-driven industries and domestic constraints on spending reoriented American diplomacy for science away from international cooperation and toward commercial safeguards. National and economic interests governed American diplomacy, whether the response to global warming or the demand for intellectual property rights. Biotechnology was a key test-case, as the United States discouraged cooperation, pressuring the UN to forgo the ICGEB and refusing to collaborate with G7 partners. Large-scale projects required clear justification to receive funding: Supporters argued SDI was necessary for national security and sold the Human Genome Project and International Space Station on diplomatic and commercial grounds. Projects and fields without national security or economic benefit struggled. Consider high-energy physics: The United States abandoned the Superconducting Supercollider and did not compete to host the next generation reactor. Instead, the de facto policy has been American physicists will go abroad for cutting-edge research, a predicament inconceivable during the Cold War. Yet cost is only one indicator of national focus: The cost of operating in space meant space programs command the most funding, but American administrations also spent considerable energy and political capital advocating for intellectual property rights. From 1987 to 1994, for example, more effort went into promoting the TRIPS agreement than went into promoting the Supercollider or *Space Station Freedom*, one reason neither exists.

Alliances, commerce and global services wove science further into American diplomacy. Military cooperation during World War II included sharing research with allies (or not) and the NATO science council facilitated coordination after Sputnik. Collaboration on national security research represented the pinnacle of US relations and allies lobbied the United States for increased status and access. At the same time, arms control required agreement with adversaries on scientific and technical specifications, measurement, and testing. As such, one of the most visible roles for scientists in American diplomacy was in arms negotiations. Scientific and technical standardization was also essential to international trade: Commerce demanded the National Bureau of Standards at the beginning of the twentieth century; eighty-years later, the United States led the G7 project on computer science and information technology. Finally, global services such as weather prediction, satellite communications, and disease prevention relied on American participation, confirming the benefits of global scientific cooperation.

American diplomacy aided scientists and scientific societies, providing funding and the opportunity to establish long-lasting relations with global colleagues. But defense research divided the scientific community: while many researchers enjoyed the work, others felt driven to political activism or to defend the independence of science. In retrospect, the breadth of activism during the Cold War was remarkable, including the "Scientists movement" for civilian control of atomic energy, organizing against the Baruch plan, resisting McCarthyism, supporting the Johnson presidential campaign, criticizing US use of herbicides in Vietnam, protesting travel restrictions on Soviet scientists, and refusing to participate in SDI. At the end of the Cold War, environmental issues, whether ozone or climate change, took center stage and members of the scientific community began to pressure countries into environmental compliance, a new approach to global governance.

American national interests often conflicted with the traditional openness and internationalism of science. Defense required government jurisdiction over national security-related research (an open-ended designation), while the shift toward market-driven research created tension between science as an engine of national prosperity and science as an international public good. American policies hardened during World War II: The United States denied French patent claims on plutonium and reactors and upset the British by supporting private claims to penicillin. Research with national security or commercial applications was kept secret and the United States ensured private ownership of research received international protection. Concern over competition complicated scientific relations: American policy was to cooperate in basic research, but fields like biotechnology muddled such distinctions. At the same time, protecting commercial research and industry shaped the American approach to global governance: International authorities served a purpose in securing property rights, less so when requiring domestic regulation.

In hindsight, American diplomacy shaped the global landscape of science, contributing to two seemingly contradictory trends after World War II: the simultaneous diffusion and concentration of science. On one hand, US assistance and funding aided the globalization of academic research: Scientists frequently publish with foreign colleagues and international networks in fields such as seismology and medicine facilitate global coordination; multiple nations, for example, participated in identifying the SARS virus genome within weeks in 2003.[2] On the other hand,

---

[2] Regarding publication, see Ludo Waltman, Robert J. W. Tijssen, and Nees Jan van Eck, "Globalisation of Science in Kilometers," *Journal of Infometrics* 5 (2011): 574–582. Regarding seismology, see "Tectonic Shifts: The Rise of Global Networks," in Caroline

US advocacy for intellectual protections allowed corporations to control access to commercial research; today, an unknown percentage of scientific knowledge is removed from public view through classification on national security or proprietary grounds.[3]

Over time, alternative "empires" of knowledge arose. One "empire" resulted from collaboration among experts, creating a science-based supranational authority, with the Intergovernmental Panel on Climate Change the most prominent example. Scholar Caroline Wagner suggested research networks function as a "New Invisible College" in the twenty-first century, but the college remains informal and largely dependent on national or corporate funding.[4] Nor has scientific agreement compelled national action – witness US policy on global warming. A second "empire" belonged to corporations able to own and limit access to scientific knowledge; its supporters highlighted the conveniences of modern life and argued intellectual protections were necessary to incentivize research. Of course, national "empires" persisted: In 2014, for example, fifteen countries performed 90 percent of global R&D and patents were equally concentrated.[5] The United States and PRC accounted for 27 percent and 20 percent of global R&D respectively, confirming American leadership and China's rapid ascension.[6]

American scientific preeminence and influence lessened after the Cold War. In the 1950s, for example, American policies for atomic energy set global standards, American funding underwrote global scientific initiatives and the country determined access to advanced science and technology through classification, export controls, and passport/visa policies. But such dominance could not last. American programs aided the development of European and Asian rivals, who recovered from World War II and competed in the global knowledge economy. Research specialized and diffused, while COCOM and export controls collapsed alongside the Soviet Union. After the Cold War, the American government continued to

S. Wagner, *The New Invisible College: Science for Development* (Washington, DC: Brookings Institution Press, 2008), 51–68. Finally, regarding SARS, see E.T. Liu, "Global Health Research Diplomacy," in Davis and Patman, eds., *Science Diplomacy: New Day or False Dawn?* (Hackensack, NJ: World Scientific Publishing Co., 2015), 219–229.

[3] Peter Galison, "Removing Knowledge: The Logic of Modern Censorship," in Proctor and Schiebinger, *Agnotology*, 37–54.

[4] Wagner, *The New Invisible College.*

[5] Regarding nations, see National Science Foundation, *Science & Engineering Indicators 2016* (Washington, DC: National Academies Press, 2016), 0–17. Regarding patents, see also Richard Florida, "The World Is Spiky," *Atlantic Monthly* (October 2005): 48–51.

[6] Ibid.

support domestic research, maintained authority over national security-related research (an open-ended category) and worked to secure intellectual property rights for commercial research. Thus, while American influence lessened, the country continued to benefit from its status as the leading scientific nation.

The election of Donald Trump revealed the significance of science to American international relations, standing and self-image. Dozens of scientific groups urged the president-elect to appoint respected scientific leaders; others worried he would roll back President Obama's science diplomacy.[7] After inauguration, the *Bulletin of Atomic Scientists* moved its Doomsday clock closer to Armageddon, arguing "Trump's statements and actions have been unsettling," especially his "growing disregard for scientific expertise."[8] When Trump ordered immigration restrictions, more than 160 scientific and academic organizations objected, while the president of the Mexican Academy of Sciences suggested it was "an opportune moment" for Mexican students to consider programs in the EU and China.[9] Europeans considered "US-proofing" the Paris Climate Accord before Trump's election; fear the United States would withdraw led to global marches only months into his term.[10]

American scientific organizations protested the administration's rejection of climate change and people around the world demonstrated in support.[11] Global protestors also highlighted issues specific to their countries: South African marchers decried AIDS misinformation, Mexican researchers demanded more federal funding, and Londoners worried about the impact of Brexit on British science and scientists.[12] But the international response revealed widespread concern the United States could renounce its scientific leadership and discontinue its postwar

---

[7] Alexandra Witze, "Trump Agenda Threatens US Legacy of Science Diplomacy," *Nature* (January 27, 2017).

[8] Rachael Lallensack, "Doomsday Clock Ticks 30 Seconds Closer to Midnight, Thanks to Trump," *Science* (Jan 26, 2017). See also David Kramer, "With Trump in Charge, Uncharted Waters Lie Ahead for Science," *Physics Today* 70 (2017).

[9] Steven T. Corneliussen, "Stakeholders Fear for Trump-Era Science," *Physics Today* (February 3, 2017). For the MAS President, see Lizzie Wade, "Mexican Scientists feel the Trump effect," *Science* 355 (February 3, 2017): 440–441, quote on 441.

[10] Luke Kemp, "US-proofing the Paris Climate Agreement," *Climate Policy* 17 (2016): 86–101. See also Lindzi Wessel, "On Eve of Science March, Planners Look Ahead," *Science* 356 (April 14, 2017): 118.

[11] Tim Appenzeller, "An Unprecedented March for Science," *Science* 356 (April 28, 2017): 356–357.

[12] Ibid.

advocacy for science (French President Macron invited American climate researchers to work in his country). Although it is too soon to tell how politicization of science will impact America's global scientific standing, the emigration of American scientists abroad for research, whether in high-energy physics or climatology, marks a departure from the post-World War II era.

Disagreement endures over how scientific cooperation fits American national interests. NGOs suggest the country advance a broader spectrum of national interests than security and commerce: Scientific and technical outreach in the developing world, for example, could be considered in the national interest (though assistance is not a panacea, previous initiatives are not a fair measurement of potential). Environmentalists contend American diplomacy relies on a limited, commercial view of nature and believe mitigating climate change is in the national interest (the DOD has agreed since the 1990s). Academics recommend the country support scientific internationalism and institutions such as UNESCO, the ICGEB, and TWAS. Finally, though cooperation in commercial and national security research remains uniquely complicated, it is possible, even in areas of commercial value and between geopolitical competitors.

Contemporary scientific relations occur on many levels and through many channels. Although China opposed many American positions at the Copenhagen climate summit (2009), the two countries agreed to work together on global warming and the transition to a low-carbon economy.[13] The United States and the PRC established a Clean Energy Research Center (CERC) to coordinate research in three areas – clean coal, clean vehicles, and more efficient construction.[14] CERC expanded even as international climate negotiations stagnated; by 2014, more than eighty Chinese and forty American organizations participated, employing approximately 1,100 researchers.[15] Representatives from both nationalities received more than a dozen patents within the first few years (without

---

[13] Björn Conrad, "China in Copenhagen: Reconciling the 'Beijing Climate Revolution' and the "Copenhagen Climate Obstinacy," *The China Quarterly* 210 (June 2012): 435–455.

[14] State Department, "Scientific Cooperation, Clean Energy Research Center: Protocol Between the United States of America and China," (November 17, 2009), available at: https://permanent.access.gpo.gov/gpo23156/190877.pdf.

[15] Joanna I. Lewis, "Managing Intellectual Property Rights in Cross-Border Clean Energy Collaboration: The Case of the U.S.-China Clean Energy Research Center," *Energy Policy* 69 (2014): 546–554, esp. 549.

conflict) and the energy sectors of each country signed a tentative agreement.[16] It is too soon to know if CERC will provide a model moving forward, but it is a reminder scientific cooperation may provide a path even as diplomacy falters.

The centrality of scientific knowledge to international affairs will only increase. Cooperation is easily politicized in such a highly charged atmosphere, whether with geopolitical rivals (Iran), economic competitors (Japan), or both (China). Yet cooperation still offers unique opportunities for international relations, especially since science is too specialized for any one country to lead all fields. Advantages are temporary, and the United States must keep abreast of foreign research, even if only for self-preservation. Scientific preeminence and cooperation serve the national interest, whether through accessing foreign research or from the positive relations engendered by sharing. More than two centuries ago, President Jefferson observed lighting another's candle did not diminish his own flame; whenever possible, America should follow this lead.

---

[16] Lewis, "Managing Intellectual Property Rights in Cross-Border Clean Energy Corroboration." See also See also Zheng Wan and Brian Craig, "Reflections on China-US Energy Cooperation: Overcoming Differences to Advance Collaboration," *Utilities Policy* 27 (2013): 93–97.

Appendix

Selected Chronology

1743   American Philosophical Society est.
1789   US Constitution commits government to "progress of science and technical arts"
1803   President Jefferson's "Secret" Message to Congress
1804   US Army Corps of Discovery surveys Louisiana ("Lewis and Clark Expedition")
1838   US Exploring Expedition of the Pacific ("Wilkes Expedition")
1863   National Academy of Sciences est.
1872   Great Britain launches the HMS *Challenger*
1882   First International Polar Year
1892   US agricultural mission to Mexico ("Nelson Mission")
1901   National Bureau of Standards est.
1915   National Advisory Committee for Aeronautics est.
1916   National Research Council est.
1938   Industrial Research Institute est.
         Committee on Scientific and Cultural Cooperation with Latin America est.
1939   Hahn and Strassman paper on nuclear fission
1940   British Tizard Mission to the United States
         Radiation Laboratory at MIT est.
1941   US/British/Canadian scientific collaboration expands
1942   Office of Scientific Research and Development est.
         Manhattan Engineering District est.
         Executive Order 9095 and the Office of the Alien Property Custodian est.

1943   ALSOS mission est.
       Food and Agricultural Organization est.
       US–Mexican Agricultural Program est.
1945   Operation Paperclip
       UN and United Nations Educational Scientific and Cultural
          Organization est.
1946   "Goushenko Affair" and revelations of Soviet Espionage
       "Baruch Plan" and the United Nations Atomic Energy Committee
       US opposition to French international laboratory plan
       UNESCO science coordination offices and UN Population
          Division est.
       World Health Organization and question of US ratification
       "KR Affair" causes tension between the United States and
          Soviet Union
1947   Atomic Bomb Casualty Commission est.
       Executive Order 9835 introduces loyalty oaths for federal
          researchers
1948   United States ratifies and joins WHO
       Soviet "Peace Offensive" regarding science and technology
       Trial of Mendelian genetics in the Soviet Academy of Agricultural
          Sciences
       State Department and MIT initiate Project Troy
       Economic Cooperation Act provides scientific and technical aid
          for the Middle East
1949   North Atlantic Treaty Organization est.
       Coordinating Committee on Multilateral Exports est.
1949   President Truman's "Point Four" speech
       UN technical aid programs est.
1950   US scientific and technical assistance to Iran begins
1953   United States sponsors Hamburg Conference on "Science and
          Freedom"
       US Atoms for Peace program
1954   *Conseil European pour la Recherche Nuclaire* est.
       Soviet Union rejoins UNESCO
       United States passes the Agricultural Trade Development and
          Assistance Act (PL-480)
1955   Malaria Eradication Program est. (US/UN partnership)
1957   International Atomic Energy Agency est.
       International Geophysical Year begins
       Soviets launch *Sputnik*

NATO science committee est.

French proposal for a Western Foundation for Scientific
   Cooperation

Soviet Union rejoins the World Health Organization

Department of Defense est. JASON

1958 Defense Research Projects Agency est.

National Aeronautics and Space Administration est.

UN Committee on the Peaceful Uses of Outer Space est.

1961 President Kennedy announces the "Alliance for Progress" for
   Latin America

The International Rice Research Institute est.

1962 Publication of Rachel Carson's *Silent Spring*

Operation Ranch Hand in Vietnam

Communication Satellite Act and Communications Satellite
   Corporation est.

1963 UN Conference for the Application of S&T for the Benefit of Less-
   Developed Areas

Beginning of World Weather Watch

International Telecommunications Satellite Consortium est.

1964 The People's Republic of China tests an atomic bomb

International Biological Program est.

UN Conference on Trade and Development and creation of G77

1965 US "Early Bird" communications satellite orbits

1966 The International Maize and Wheat Improvement Center est.

1967 Soviets establish *Interkosmos* to launch communist bloc satellites

UN Outer Space Treaty

Global Atmospheric Research Program est.

1968 US Foreign Assistance Act ties PL-480 funds to population
   control

1970 The People's Republic of China launches its first satellite

US Environmental Protection Agency est.

1972 UN Conference on the Human Environment (Stockholm
   Conference)

UN Environmental Program est.

Office of Technology Assessment est.

SALT treaty and scientific and technical détente agreements with
   Soviets

First "Landsat" satellite launched

Shanghai Communique on US/Chinese relations

United States and Israel est. Binational Science Foundation

1974   UN New International Economic Order adopted
State Department Bureau of Oceans and International
    Environmental Scientific Affairs
Federal Laboratory Consortium Technology Transfer Act
OECD est. International Energy Agency to respond to oil
    embargo
1975   US/Soviet Apollo–Soyuz Test Project (the "Handshake in Space")
NASA Satellite Instructional Television Experiment in India
United States and Saudi Arabia est. the Joint Economic
    Cooperation Office in Riyadh
UNIDO and UNESCO est. biotechnology initiatives
1976   Eight equatorial states claim part of geostationary orbital space
1977   US introduces "Budapest Treaty" to secure global genetic patent
    rights
Genex and Biogen founded
1978   US scientific and technical delegation visits China
United States and Israel est. Binational Industrial Research and
    Development Foundation
1979   United States signs the Convention on Long-Range
    Transboundary Air Pollution
UN Conference on Science and Technology for Development
Islamic Revolution in Iran ends ongoing American programs in Iran
Soviet invasion of Afghanistan ends many American détente
    exchanges
Normalization of US/Chinese relations
United States and Israel est. Binational Agricultural Research and
    Development Foundation
US Climate Program est.
International conference on genetics and computer databases
1980   *Chakrabarty* v *Diamond* ruling on patenting living organisms
Cohen–Boyer patent granted for recombinant DNA process
UN World Climate Research Program est.
1982   French propose G7 Science and Technology Initiative at Versailles
    Summit
Japan begins "Fifth Generation" computing initiative
1983   G7 scientific and technical agreement at Williamsburg Summit
President Reagan announces the Strategic Defense Initiative
United States introduces intellectual protections in the Caribbean
    Trade Act
GenBank established with federal support

1984   First European Framework Program for Research and
          Development
        US National Cooperative Research Act
        United States withdraws from UNESCO
        President Reagan invites participation in *Space Station Freedom* at
          London Summit
        NSDD 145 initiates government monitoring of
          telecommunications and data
1985   Vienna Convention acknowledges an ozone problem
1986   Chernobyl nuclear accident and explosion (USSR)
        G7 scientific and technical initiative ends
        US space shuttle *Challenger* explodes
1987   United States introduces intellectual protections in Uruguay round
          of GATT talks
        United States signs the Montreal Protocol on Substances that
          Deplete the Ozone Layer
1988   International Climate Change Panel est.
1990   International Human Genome Project est.
1992   UNEP "Earth Summit" at Rio
        United States signs the UN Framework Convention on Climate
          Change
        United States signs the UN Convention on Biological
          Diversity
        NIH and MRC (Great Britain) file first gene patents
1993   Superconducting Supercollider construction ends
1995   TRIPS agreement (Trade Related aspects of Intellectual Property
          rightS)
        Space shuttle *Atlantis* and *Mir* dock in second "Handshake
          in Space"
        WTO Agreement on Application of Sanitary and Phytosanitary
          Measures
1996   "Valentine's Day Massacre" when Intelsat 7A explodes
        "Bermuda Principles" for sharing HGP DNA sequences
          announced
1997   Byrd-Hagel Resolution in the Senate pre-emptively rejects Kyoto
          protocol
1998   Commercial Space Act
        International Space Station launched
        European Directive on the Legal Protection of Biotechnological
          Inventions

1999   UNESCO Science Action Framework adopted for developing
            countries
          United States indicts Wen Ho Lee for nuclear espionage
2000   US-China Economic Security Review Commission est.
          Cartagena Protocol to the CBD
2001   United States withdraws from the Kyoto Protocol
2003   United States rejoins UNESCO
          Iraqi International Center for Science and Industry est.
2005   United States est. Virtual Science Libraries for Iraq, Pakistan,
            Morocco & other countries
2009   US/Israeli Stuxnet computer virus infects Iranian computers
          United States and PRC est. Clean Energy Research Center
2011   US Congressional hearings on science diplomacy with China
          United States leaves UNESCO over recognition of Palestine
          United States est. Global Innovation through Science and
            Technology
2015   United States signs the Paris climate agreement
          United States signs the Iranian nuclear agreement

Bibliography

Note on Sources: A wide variety of sources contributed to this history. The journal *Science*, published by the American Association for the Advancement of Science, was an invaluable and constant resource. State Department publications, especially the *Foreign Relations of the United States (FRUS)* and *Science, Technology and American Diplomacy* (1980–1996), detailed government policies. Online databases, including the CIA's FOIA request library, the Digital National Security Archive (DNSA) and the Declassified Document Retrieval System (DDRS), proved essential as well. The National Science Foundation's *Science & Engineering Indicators* provided the statistical background. Finally, hundreds of articles from newspapers, industry publications, and scientific and academic journals added context and nuance.

CONGRESSIONAL PUBLICATIONS

Note: Titles listed by publication date and published by the US Government Printing Office.

US House of Representatives. *Science, Technology, and American Diplomacy in the Age of Interdependence* (1976).
    Science, Technology, and American Diplomacy: An Extended Study of the Interactions of Science and Technology with United States Foreign Policy (1977).
    Soviet Scientific and Technical Cooperation with Countries Other Than the United States (1979).
    United States–China Science Cooperation (1979).
    United States Scientific and Technical Exchanges with the Soviet Union (1980).
    Survey of the Science and Technology Issues Present and Future (1981).
    Status of U.S.–Saudi Arabian Joint Commission on Economic Cooperation (1983)
    Overview of International Science and Technology Policy (1986).
    National Security Export Controls Report by the National Academy of Science (1987).

The Effect of Changing Export Controls on Cooperation in Science and Technology (1991).

Science, Technology and Global Competitiveness (2005).

International Science and Technology Cooperation (2008).

The Role of Non-Governmental Organizations and Universities in International Science and Technology Cooperation (2008).

International Science and Technology Cooperation (2008).

Communist Chinese Cyber-Attacks, Cyber-Espionage and Theft of American Technology (2011).

Efforts to Transfer America's Leading-Edge Science to China (2011).

Iranian Nuclear Talks: Negotiating a Bad Deal? (2014).

US Senate. *The Convention on Biological Diversity* (1994).

Review of Alleged Chinese Espionage at Department of Energy Laboratories (2000).

The Wen Ho Lee Matter (2001).

The Peter Lee Case (2001).

Report on the Government's Handling of the Investigation and Prosecution of Dr. Wen Ho Lee (2001).

Economic Development Opportunities in Nano Commercialization (2006).

The International Space Station: A Platform for Research, Collaboration and Discovery (2012).

Joint Hearing, Committee on Science (House) and Committee on Commerce, Science and Transportation (Senate). *Science and Technology to Combat Terrorism* (2002).

DEPARTMENT OF STATE PUBLICATIONS

US Department of State. *Foreign Relations of the United States* (FRUS), multiple volumes.

The Point Four Program (1949).

Point Four Projects: July 1, 1950 through December 31, 1951 (1952)

Science and Technology for Development (1978).

Some United States Activities Using Science and Technology for Development (1979).

Science, Technology and American Diplomacy (Washington, DC: Government Printing Office, 1980–1996). Abbreviated as *STAD*.

Diane B. Bendahmane and David William McClintock, eds. *Science, Technology, and Foreign Affairs: Volume I, Global Environment, Communications, and Agriculture* (1985). Work is a product of the Department of State.

Science, Technology, and Foreign Affairs: Volume II, Climate, Scientific Dialogue and Health (1985). Work is a product of the Department of State.

NATIONAL INSTITUTIONS

Office of Public Health, International Cooperation Administration, *Technical Cooperation in Health* (Washington, DC: International Cooperation Administration, 1961).

Agency for International Development, Department of State, *The Alliance for Progress ... an American Partnership* (Washington, DC: US Government Printing Office, 1965).

National Academy of Sciences. *National Academy of Sciences: International Development Programs of the Office of the Foreign Secretary, Summary and Analysis of Activities, 1961–1971* (Washington, DC: National Academy of Sciences, 1971).

Department of Commerce, Bureau of International Commerce. *Iran: A Survey of U.S. Business Opportunities* (Washington, DC: GPO, 1977).

National Academy of Sciences. *Safety of Genetically Engineered Foods: Approaches to Assessing Unintended Health Effects* (WashingtonDC: National Academies Press, 2004).

National Research Council. *U.S. and International Perspectives on Global Science Policy and Science Diplomacy: Report of a Workshop* (Washington, DC: National Academies Press, 2011).

National Academy of Sciences. *Climate Change: Evidence and Causes* (Washington, DC: National Academies Press, 2014).

National Academies of Sciences, Engineering and Medicine. *U.S.–Iran Engagement in Science, Engineering, and Health (2010–2016): A Resilient Program but an Uncertain Future* (Washington, DC: National Academies Press, 2017).

SELECTED BOOKS

Adas, Michael. *Machines as the Measure of Men: Science, Technology and the Ideologies of Western Dominance* (Ithaca, NY: Cornell University Press, 1989).

Al-Salloom, Hamad I., ed. *Science and Technology in Saudi Arabia* (Beltsville, MD: Amana Publications, 1995).

Albers, Henry H. *Saudi Arabia: Technocrats in a Traditional Society* (New York: Peter Lang, 1989).

Andrée, Peter. *Genetically Modified Diplomacy: The Global Politics of Agricultural Biotechnology and the Environment* (Vancouver: University of British Columbia Press, 2007).

Andrew, Christopher and Vasili Mitrokhin. *The Sword and the Shield: The Mitrokhin Archive and the Secret History of the KGB* (New York: Basic Books, 1999).

Avery, Donald H. *The Science of War: Canadian Scientists and Allied Military Technology during the Second World War* (Toronto: University of Toronto Press, 1998).

Baber, Zaheer. *The Science of Empire: Scientific Knowledge, Civilization and Colonial Rule in India* (New York: State University of New York Press, 1996).

Badash, Lawrence. *Scientists and the Development of Nuclear Weapons*. Atlantic Highlands, NJ: Humanities Press, 1995.

Badash, Lawrence, J. O. Hirschfelder, and H. P. Broida. *Reminiscences of Los Alamos, 1943–45* (Boston: D. Reidel Publishing Company, 1980).

Bass, Paul William. *Point Four, Touching the Dream: A Bold, New U.S. Foreign Policy* (Stillwater, OK: New Forums Press).

Baxter 3rd., James Phinney. *Scientists against Time* (Cambridge, MA: MIT Press, 1946).

Bedini, Silvio A. *Thomas Jefferson: Statesman of Science* (New York: Macmillan, 1990).

Beling, Willard A. *King Faisal and the Modernisation of Saudi Arabia* (Boulder, CO: Westview Press, 1980).

Benedick, Richard Elliot. *Ozone Diplomacy: New Directions in Safeguarding the Planet*, enlarged edition (Cambridge, MA: Harvard University Press, 1998).

Berman, Elizabeth Popp. *Creating the Market University: How Academic Science Became an Economic Engine* (Princeton, NJ: Princeton University Press, 2012).

Bergesen, Helge Ole. and Georg Parmann, eds. *Green Globe Yearbook of International Co-Operation on Environment and Development 1995* (Oxford: Oxford University Press).

Bernstein, Jeremy. *Hitler's Uranium Club: The Secret Recordings at Farm Hall* (Woodbury, NY: AIP, 1995).

Bertsch, Gary K. and Steven Elliott-Gower, eds. *Export Controls in Transition* (Durham, NC: Duke University Press, 1992).

Beyerchen, Alan D. *Scientists under Hitler: Politics and the Physics Community in the Third Reich* (New Haven, CT: Yale University Press, 1977).

Bill, James A. *The Eagle and the Lion: The Tragedy of American–Iranian Relations* (New Haven: Yale University Press, 1989).

Bingham, Jonathan B. *Shirt-Sleeve Diplomacy: Point 4 in Action* (New York: John Day Co., 1954).

Blackwell, Robert D. and Jennifer M. Harris, *War by Other Means: Geoeconomics and Statecraft* (Cambridge, MA: Harvard University Press, 2016).

Blair, Peter D. *Congress's Own Think Tank: Learning from the Legacy of the Office of Technology Assessment (1972–1995)* (New York: Palgrave Mac-Millan, 2013).

Bond, Peter. *The Continuing Story of the International Space Station* (New York: Springer Publishing, 2002).

Boyer, Paul. *By the Bomb's Early Light: American Thought and Culture at the Dawn of the Atomic Age* (Chapel Hill: University of North Carolina Press, 1994).

Borstelmann, Thomas. *The Cold War and the Color Line: American Race Relations in the Global Arena* (Cambridge, MA: Harvard University Press, 2003).

Breuer, Georg. *Weather Modification: Prospects and Problems* (New York: Cambridge University Press, 1979).

Briggs, Laura. *Reproducing Empire: Race, Sex, Science and U.S. Imperialism in Puerto Rico* (Berkeley: University of California Press, 2002).

British Royal Society. *New Frontiers in Science Diplomacy: Navigating the Changing Balance of Power* (London: Science Policy Centre, 2010)

Brockway, Lucille. *Science and Colonial Expansion: The Role of the British Royal Botanic Garden* (New Haven: Yale University Press, 2002).

Bud, Robert. *The Uses of Life: A History of Biotechnology* (New York: Cambridge University Press, 1993).

Bulkeley, Rip. *The Sputniks Crisis and the Early United States Space Policy: A Critique of the Historiography of Space* (Bloomington: Indiana University Press, 1991).

Bush, Vannevar. *Pieces of the Action* (New York: Morrow, 1970).

Butterfield, Samuel Hale. *U.S. Development Aid: An Historic First: Achievements and Failures in the Twentieth Century* (Westport, CT: Praeger, 2004).

Cantelon, Philip L., et al., eds. *The American Atom: A Documentary history of Nuclear Policies from the Discovery of Fission to the Present*, 2nd edn. (Philadelphia: University of Pennsylvania Press, 1991).

Cantell, Kari. *The Story of Interferon: The Ups and Downs in the Life of a Scientist* (New York: World Scientific Publishing Company, 1998).

Carter, James M. *Inventing Vietnam: The United States and State Building, 1954–1968* (New York: Cambridge University Press, 2008).

Chang, Iris. *Thread of the Silkworm* (New York: 1995).

Chapman, Bert. *Export Controls: A Contemporary History* (Lanham, MD: University Press of America, 2013).

Chorev, Nitsan. *The World Health Organization between North and South* (Ithaca, NY: Cornell University Press, 2012).

Clark, Jon, Celia Modgil, and Sohan Modgil, eds. *Robert K. Merton: Consensus and Controversy* (New York: Falmer Press, 1990).

Close, Frank. *Half-Life: The Divided Life of Bruno Pontecorve, Physicist/or Spy* (New York: Basic Books, 2015).

Cohen, Avner. *Israel and the Bomb* (New York: Columbia University Press, 1998).

Coleman, Peter. *The Liberal Conspiracy: The Congress for Cultural Freedom and the Struggle for the Mind of Postwar Europe* (New York: The Free Press, 1989).

Connelly, Matthew. *Fatal Misconception: The Struggle to Control World Population* (Cambridge, MA: Harvard University Press, 2008).

Creager, Angela N.H. *Life Atomic: A History of Radioisotopes in Science and Medicine* (Chicago: University of Chicago Press, 2013),

Cueto, Marcos, ed. *Missionaries of Science: The Rockefeller Foundation & Latin America* (Bloomington: Indiana University Press, 1994).

Cullather, Nick. *The Hungry World: America's Cold War Battle against Poverty in Asia* (Cambridge, MA: Harvard University Press, 2011).

Curti, Merl. *Prelude to Point Four: American Technical Missions Overseas, 1838–1938* (Madison, WI: University of Wisconsin Press, 1954).

Dannehl, Charles R. *Politics, Trade and Development: Soviet Economic Aid to the Non-Communist Third World, 1955–89* (Brookfield, VT: Dartmouth Publishing Company, 1995).

Davis, Lloyd and Robert G. Patman. *Science Diplomacy: New Day or False Dawn?* (Hackensack, NJ: World Scientific Publishing, 2015).

Dayhoff, Margaret O. *Atlas of Protein Sequence and Structure* (Silver Springs, MD: National Biomedical Research Foundation, 1965).

de Beer, Gavin. *The Sciences were Never at War* (London: Thomas Nelson & Sons, 1960).

de Tinguy, Anne. *U.S.–Soviet Relations During the Détente* (New York: Columbia University Press, 1999).

Dejong-Lambert, William. *The Cold War Politics of Genetic Research: An Introduction to the Lysenko Affair* (New York: Springer, 2012).

Dickson, David. *New Politics of Science* (Chicago: University of Chicago Press, 1988).

Divine, Robert A. *The Sputnik Challenge: Eisenhower's Response to the Soviet Satellite* (New York: Oxford University Press, 1993).

Douglas, Heather E. *Science, Policy and the Value-Free Ideal* (Pittsburgh, PA: University of Pittsburgh Press, 2009).

Drahos, Peter and John Braithwaite. *Information Feudalism: Who Owns the Knowledge Economy?* (New York: New Press, 2007).

Dupree, Hunter. *Science in the Federal Government: A History of Policies and Activities* (Baltimore: Johns Hopkins University Press, 1986).

Ekbladh, David. *The Great American Mission: Modernization & the Construction of an American World Order* (Princeton: Princeton University Press, 2010).

Engerman, David C., Nils Gilman, Mark H. Haefele, and Michael E. Latham, eds. *Staging Growth: Modernization, Development and the Global Cold War* (Boston: University of Massachusetts Press, 2003).

Etzkowitz, Henry, and Andrew Webster, eds. *Capitalizing Knowledge: New Intersections of Industry and Academia* (Albany, NY: State University of New York, 1998).

Evans, Richard. *Deng Xiaoping and the Making of Modern China* (New York: Penguin Books, 1997).

Farley, John. *Brock Chisholm, the World Health Organization & the Cold War* (Vancouver: University of British Columbia Press, 2008).

Fermi, Laura. *Illustrious Immigrants: The Intellectual Migration from Europe, 1930–1941* (Chicago: University of Chicago Press, 1968).

Fifield, Richard. *International Research in the Antarctic* (Oxford: Oxford University Press, 1987).

Finkbeiner, Ann. *The Jasons: The Secret History of Science's Postwar Elite* (Viking, 2006).

Fleming, James Rodger. *Fixing the Sky: The Checkered History of Weather and Climate Control* (New York: Columbia University Press, 2010).

Fuhrman, Matthew. *Atomic Assistance: How "Atoms for Peace" Programs Cause Nuclear Insecurity* (Ithaca, NY: Cornell University Press, 2012).

Galison, Peter and Bruce Hevly, eds. *Big Science: The Growth of Large-Scale Research* (Stanford: Stanford University Press, 1992).

Gimbel, John. *Science, Technology and Reparations: Exploitation and Plunder in Postwar Germany* (Stanford: Stanford University Press, 1990).

Gilman, Nils. *Mandarins of the Future: Modernization Theory in Cold War America* (Baltimore: Johns Hopkins University Press, 2003).

Goliszek, Andrew. *In the Name of Science: A History of Secret Programs, Medical Research and Human Experimentation* (New York: St. Martin's Press, 2003).

Goetzmann, William H. *Exploration and Empire: The Explorer and the Scientist in the Winning of the American West* (New York: Alfred A. Knopf, 1966).

Goldman, Merle. *China's Intellectuals: Advise and Dissent* (Cambridge, MA: Harvard University Press, 1981).

Goode, James F. *The United States and Iran, 1946–51: The Diplomacy of Neglect* (New York: St. Martin's Press, 1989).

Gorbachev, Mikhail. *Perestroika: New Thinking for Our Country and the World* (New York: Harper Collins, 1987).

Goss, Andrew. *The Floracrats: State-Sponsored Science and the Failure of the Enlightenment in Indonesia* (Madison: University of Wisconsin Press, 2011).

Goudsmit, Samuel A. *ALSOS* (Woodbury, New York: American Institute of Physics Press, 1996).

Graham, Loren R. *Science in Russia and the Soviet Union: A Short History* (Cambridge: Cambridge University Press, 1993).

  Lysenko's Ghost: Epigenetics and Russia (Cambridge, MA: Harvard University Press, 2016).

Graham, Loren R. and Irina Dezhina. *Science in the New Russia: Crisis, Aid, Reform* (Bloomington: Indiana University Press, 2008).

Greenaway, Frank. *Science International: A History of the International Council of Scientific Unions* (New York: Cambridge University Press, 1996).

Greenberg, Daniel S. *The Politics of Pure Science* (Chicago: University of Chicago Press, 1999).

  Science, Money and Politics: Political Triumph and Ethical Erosion (Chicago: University of Chicago Press, 2001).

Groueff, Stephane. *Manhattan Project: The Untold Story of the Making of the Atomic Bomb* (Boston: Little, Brown & Co., 1967).

Guillemin, Jeanne. *Biological Weapons: From the Invention of State-Sponsored Programs to Contemporary Bioterrorism* (New York: Columbia University Press, 2005).

Gupta, Joyeeta. *The History of Global Climate Governance* (Cambridge: Cambridge University Press, 2014).

Haber, L. F. *The Poisonous Cloud: Chemical Warfare in the First World War* (Oxford: Clarendon Press, 1986).

Hagen, Joel B. *An Entangled Bank: The Origins of Ecosystem Ecology* (New Brunswick, NJ: Rutgers University Press, 1992).

Harris, Paul G. ed. *The Environment, International Relations and U.S. Foreign Policy* (Washington, DC: Georgetown University Press, 2001).

Harrison, Neil E. and Gary C. Bryner, eds. *Science and Politics in the International Environment* (Lanham, MD: Rowman & Littlefield, 2004).

Hartmann, Betsy. *Reproductive Rights and Wrongs: The Global Politics of Population Control* (Boston: South End Press, 1995).

Harvey, Brian. *China's Space Program: From Conception to Manned Spaceflight* (New York: Springer Praxis, 2004).

  China in Space: The Great Leap Forward (Chichester, UK: Springer-Praxis Publishing, 2013).

Harvey, David. *A Brief History of Neoliberalism* (New York: Oxford University Press, 2005).

Havens, Barrington S. *History of Project Cirrus* (Schenectady, NY: Research Publication Service, 1952).

Hayes, Samuel P. *Beauty, Health and Permanence: Environmental Politics in the United States, 1955–1985* (Cambridge: Cambridge University Press, 1989).

Heilbron, John L. *The Dilemmas of an Upright Man: Max Planck as Spokesman for German Science* (Berkeley; University of California Press, 1986).

Henry, John. *The Scientific Revolution and the Origins of Modern Science* (New York: Palgrave Macmillan, 2008).

Herken, Gregg. *Cardinal Choices: Presidential Science Advising from the Atomic Bomb to SDI* (Stanford: Stanford University Press, 2000).

Hess, David J. *Alternative Pathways in Science and Industry: Activism, Innovation and the Environment in an Era of Globalization* (Cambridge, MA: MIT Press, 2007).

Hewlett, Richard G. and Jack M. Holl, *Atoms for Peace and War, 1953–1961: Eisenhower and the Atomic Energy Commission* (Berkeley: University of California Press, 1989).

Hollinger, David A. *Science, Jews and Secular Culture: Studies in Mid-Twentieth-Century American Intellectual History* (Princeton, NJ: Princeton University Press, 1996).

Holloway, David. *Stalin and the Bomb: The Soviet Union and Atomic Energy, 1939–46* (New Haven, CT: Yale University Press, 1994).

Hoole, Francis. *Politics and Budgeting in the World Health Organization* (Bloomington: Indiana University Press, 1976).

Hopgood, Stephen. *American Foreign Environmental Policy and the Power of the State* (Oxford: Oxford University Press, 1998).

Howard-Jones, Norman. *International Public Health between the Two World Wars - the Organizational Problems* (Geneva: World Health Organization, 1976).

Hulme, Mike. *Why We Disagree about Climate Change* (New York: Cambridge University Press, 2009).

Hunt, Linda. *Secret Agenda: The United States Government, Nazi Scientists, and Project Paperclip, 1945 to 1990* (New York: St. Martin's Press, 1991).

Jacobsen, Annie. *Operation Paperclip: The Secret Intelligence Program That Brought Nazi Scientists to America* (New York: Back Bay Books, 2014).

Jasanoff, Sheila. *Science at the Bar: Law, Science and Technology in America* (Cambridge, MA: Harvard University Press, 1995).

  Designs on Nature: Science and Democracy in Europe and the United States (Princeton, NJ: Princeton University Press, 2005).

Johns, Adrian. *Piracy: The Intellectual Property Wars from Guttenberg to Gates* (Chicago: University of Chicago Press, 2009).

Johnson, Jeffrey Allan. *The Kaiser's Chemists: Science and Modernization in Imperial Germany* (Chapel Hill: University of North Carolina Press, 1990).

Joravsky, David. *The Lysenko Affair* (Chicago: University of Chicago Press, 1970).

Kennedy, John F. *The Strategy of Peace* (New York: Harper & Brothers, 1960).

Kevles, Daniel. *The Physicists: The History of a Scientific Community in Modern America* (Cambridge, MA: Harvard University Press, 1996).

Kinkela, David. *DDT & the American Century: Global Health, Environmental Politics and the Pesticide That Changed the World* (Charlotte: University of North Carolina Press, 2011).

Kippis, A. *A Narrative of the Voyages Round the World Performed By Captain James Cook, with an Account of His Life during the Previous and Intervening Periods* (Philadelphia: Porter & Coates, [N.D.]).

Kirby, William C. Robert S. Ross, and Gong Li, *Normalization of U.S.–China Relations: An International History* (Cambridge, MA: Harvard University Press, 2005).

Kistiakowsky, George B. *A Scientist at the White House: The Private Diary of President Eisenhower's Special Assistant for Science and Technology* (Cambridge, MA: Harvard University Press, 1976).

Kline, Benjamin. *First Along the River: A Brief History of the U.S. Environmental Movement*, 4th edn. (New York: Rowman & Littlefield, 2011).

Kloppenburg, Jr., Jack Ralph. *First the Seed: The Political Economy of Plant Biotechnology*, 2nd edn. (Madison: University of Wisconsin Press, 2004).

Knight, Amy. *How the Cold War Began: The Goushenko Affair and the Hunt for Soviet Spies* (Toronto: McClelland & Stewart, 2005).

Kraemer, Sylvia. *Science & Technology Policy in the United States: Open Systems in Action* (New Brunswick, NJ: Rutgers University Press, 2006).

Kramish, Arnold. *The Griffin: The Greatest Untold Espionage Story of World War II* (Boston: Houghton Mifflin, 1986).

Krattinger, A., et al., eds. *Intellectual Property Management in Health and Agricultural Innovation: A Handbook of Best Practices* (Oxford, UK: MIHR, 2007).

Krementsov, Nikolai. *The Cure: A Story of Cancer and Politics from the Annals of the Cold War* (Chicago: University of Chicago Press, 2002).

Krige, John. *American Hegemony and the Postwar Reconstruction of Science in Europe* (Cambridge, MA: MIT Press, 2008).

Krige, John and Ashok Maharaj. *NASA in the World: Fifty Years of International Collaboration in Space* (New York: Palgrave MacMillan, 2013).

Lafeber, Walter. *Inevitable Revolutions: The United States and Central America*, 2nd edn. (New York: W. W. Norton & Company,1993).

Lambright, W. Henry. *Presidential Management of Science and Technology: The Johnson Presidency* (Austin: University of Texas Press, 2012).

Lancaster, Carol. *Foreign Aid: Diplomacy, Development, Domestic Politics* (Chicago: University of Chicago Press, 2007).

Laney, Monique. *German Rocketeers in the Heart of Dixie: Making Sense of the Nazi Past during the Civil Rights Era* (New Haven, CT: Yale University Press, 2015).

Lasby, Clarence G. *Project Paperclip: German Scientists and the Cold War* (New York: Atheneum, 1971).

Latham, Michael E. *Modernization as Ideology: American Social Science and "Nation Building" in the Kennedy Era* (Chapel Hill: University of North Carolina Press, 2000).

Launius, Roger D. *NASA: A History of the U.S. Civil Space Program* (Malabar, FL: Krieger Publishing Company, 1994).

Launius, R. and J. Fleming, eds. *Globalizing Polar Science: Reconsidering the International Polar and Geophysical Years* (New York: Palgrave-MacMillan, 2011).

Laves, Walter H. C. and Charles A. Thomson, *UNESCO: Purpose, Progress, Prospects* (Bloomington: Indiana University Press, 1957).

Lee, Kelley. *The World Health Organization (WHO)*, (New York: Routledge, 2009).

Lee, Wen Ho. *My Country versus Me: The Firsthand Account by the Los Alamos Scientist Who Was Falsely Accused of Being a Spy* (New York: Hachette, 2003).

Loeppky, Rodney. *Encoding Capital: The Political Economy of the Human Genome Project* (New York: Routledge, 2005).

Leslie, Stuart R. *The Cold War and American Science: The Military-Industrial-Academic Complex at MIT and Stanford* (New York: Columbia University Press, 1993).

Lewallen, John. *Ecology of Devastation: Indochina* (Baltimore: Penguin Books, 1971).

Lewis, John Wilson and Xue Litai. *China Builds the Bomb* (Stanford: Stanford University Press, 1988).

Lindee, M. Susan. *Suffering Made Real: American Science and the Survivors at Hiroshima* (Chicago: University of Chicago Press, 1994).

Litfin, Karen T. *Ozone Discourses: Science and Politics in Global Environmental Cooperation* (New York: Columbia University Press, 1994).

Logevall, Fredrik and Andrew Preston, eds. *Nixon in the World: American Foreign Relations, 1969–1977* (New York: Oxford University Press, 2008).

Logsden, John M. *Together in Orbit: The Origins of International Participation in the Space Station* (Washington, DC: NASA History Division, 1998).

Machado, Barry. *In Search of a Usable Past: The Marshall Plan and Postwar Reconstruction Today* (Lexington, VA: George C. Marshall Foundation, 2007).

Mann, Alfred K. *For Better or for Worse: The Marriage of Science & Government in the United States* (New York: Columbia University Press, 2000).

Mann, James. *About Face: A History of America's Curious Relationship with China, from Nixon to Clinton* (New York: Vintage Books, 2000).

Masters, Dexter and Katharine Way, eds. *One World or None: A Report to the Public on the Full Meaning of the Atomic Bomb* (New York: The New Press, 2007).

Matson, Pamela A., ed. *Seeds of Sustainability: Lessons from the Birthplace of the Green Revolution in Agriculture* (Washington: Island Press, 2012).

McDougall, Walter A. *The Heavens and the Earth: A Political History of the Space Age* (Baltimore: Johns Hopkins University Press, 1985).

McNeil, J. R., ed. *Environmental Histories of the Cold War* (Cambridge: Cambridge University Press, 2010).

Mearsheimer, John J. and Stephen M. Walt. *The Israel Lobby and U.S. Foreign Policy* (New York: Farrar, Straus and Giroux, 2007).

Medvedev, Zhores A. *The Rise and Fall of T. D. Lysenko* (New York: Anchor Books, 1971).

Merton, Robert K. *The Sociology of Science: Theoretical and Empirical Investigations* (Chicago: University of Chicago Press, 1973).

Midwest Research Institute. *Solar Controlled Environment Agriculture Project, Final Report: Vol. 1, Project Summary* (Kansas City, MO: Midwest Research Institute, 1986).

Mirowski, Philip. *Science Mart: Privatizing American Science* (Cambridge, MA: Harvard University Press, 2011).

Montpetit, Eric, Christine Rothmeyer, and Frédéric Varone, eds. *The Politics of Biotechnology in North America and Europe: Policy Networks, Institutions and Internationalization* (New York: Lexington Books, 2007).

Moore, Kelly. *Disrupting Science: Social Movements, American Scientists, and the Politics of the Military, 1945–1975* (Princeton: Princeton University Press, 2008).

Neal, Homer Alfred and Tobin Smith., eds. *Beyond Sputnik: U.S. Science Policy in the 21st Century* (Ann Arbor: University of Michigan Press, 2008).

Needell, Allan A. *Science, Cold War and the American State: Lloyd V. Berkner and the Balance of Professional Ideals* (New York: Routledge, 2000).

Nelkin, Dorothy. *Selling Science: How the Press Covers Science and Technology* (New York: W. H. Freeman and Company, 1995).

Nye, Mary Jo. *Michael Polanyi and His Generation: Origins of the Social Construction of Science* (Chicago: University of Chicago Press, 2011).

Oreskes, Naomi and Erik M. Conway. *Merchants of Doubt: How a Handful of Scientists Obscured the Truth from Tobacco Smoke to Global Warming* (New York: Bloomsbury, 2011).

Orleans, Leo A. ed. *Science in Contemporary China* (Stanford: Stanford University Press, 1980).

Orleans, Leo A. *Chinese Students in America: Policies, Issues and Numbers* (Washington, DC: National Academy Press, 1988).

Osgood, Kenneth. *Total Cold War: Eisenhower's Secret Propaganda Battle at Home and Abroad* (Lawrence: University of Kansas Press, 2006).

Paár-Jákli, Gabriella. *Networked Governance and Transatlantic Relations: Building Bridges through Science Diplomacy* (New York: Routledge, 2014).

Paarlberg, Robert. *Starved for Science: How Biotechnology Is Being Kept out of Africa* (Cambridge, MA: Harvard University Press, 2008).

Pach, Chester J. *Arming the Free World: The Origins of the United States Military Assistance Program, 1945–50* (Charlotte: University of North Carolina Press, 1991).

Parsi, Trita. *A Single Roll of the Dice: Obama's Diplomacy with Iran* (New Haven, CT: Yale University Press, 2012).

Pascuizzi, Giovanni, Umberto Izzo, and Matteo Macilotti, eds. *Comparative Issues in the Governance of Research Biobanks: Property, Privacy, Intellectual Property and the Role of Technology* (New York: Springer, 2013).

Paul, Harry W. *The Sorcerer's Apprentice: The French Scientist's Image of German Science, 1814–1919* (Gainesville: University of Florida Press, 1972).

Perkins, John H. *Geopolitics and the Green Revolution: Wheat, Genes and the Cold War* (Oxford: Oxford University Press, 1997).

Peters, Toine. *Interferon: The Science and Selling of a Miracle Drug* (New York: Routledge, 2005).

Phelps, James. *The Tizard Mission: The Top-Secret Operation That Changed the Course of World War II* (London: Westholme Publishing, 2010).

Picard, Louis A. and Robert Groelsema., eds. *Foreign Aid and Foreign Policy: Lessons for the Next Half-Century* (Armonk, NY: M. E. Sharpe, 2008).

Pollock, Ethan. *Stalin and the Soviet Science Wars* (Princeton, NJ: Princeton University Press, 2006).

Prados, John. *Safe for Democracy: The Secret Wars of the CIA* (Chicago: Ivan R. Dee, 2006).

Pringle, Peter. *The Murder of Nikolai Vavilov: The Story of Stalin's Persecution of One of the Great Scientists of the Twentieth Century* (New York: Simon & Schuster, 2011).

Proctor, Robert N. and Londa Schiebinger, eds. *Agnotology: The Making and Unmaking of Ignorance* (Stanford: Stanford University Press, 2008).

Rabe, Stephen G. *The Killing Zone: The United States Wages Cold War in Latin America* (New York: Oxford University Press, 2012).

Raviv, Dan and Yossi Melman. *Spies against Armageddon: Inside Israel's Secret Wars* (Sea Cliff, NY: Levant Books, 2014).

Reed, Thomas C. and Danny B. Stillman. *The Nuclear Express: A Political History of the Bomb and its Proliferation* (Minneapolis, MN: Zenith Press, 2009).

Reingold, Nathan, ed. *The Sciences in the American Context: New Perspectives* (Washington, DC: Smithsonian Institution, 1979).

Rhodes, Catherine. *International Governance of Biotechnology: Nature, Problems and Potential* (New York: Bloomsbury Academic, 2010).

Rhodes, Richard. *The Making of the Atomic Bomb* (New York: Simon & Schuster, 2012).

Rigney, Daniel. *The Matthew Effect: How Advantage Begets Further Advantage* (New York: Columbia University Press, 2010).

Roll-Hansen, Nils. *The Lysenko Effect: The Politics of Science* (Amherst, NY: Humanity Books, 2005).

Ross, Robert S. and Jiang Changbin. *Re-Examining the Cold War: U.S.–China Diplomacy, 1954–1973* (Cambridge, MA: Harvard University Press, 2001).

Rostow, W. W. *The Stages of Economic Growth: A Non-Communist Manifesto* (London: Cambridge University Press, 1961).

Russell, Edmund. *War and Nature: Fighting Humans and Insects with Chemicals from World War I to Silent Spring* (New York: Cambridge University Press, 2001).

Saich, Tony. *China's Science Policy in the 80s* (Atlantic Highlands, NJ: Humanities Press International, 1989).

Sakharov, Andrei. *Memoirs* (New York: Knopf, 1990).

Sale, Kirkpatrick. *The Green Revolution: The American Environmental Movement 1962–1992* (New York: Hill & Wang, 1993).

Sapolsky, Harvey M. *Science and the Navy: The History of the Office of Naval Research* (Princeton, NJ: Princeton University Press, 2014).

Sarewitz, Daniel. *Frontiers of Illusion: Science, Technology and the Politics of Progress* (Philadelphia, PA: Temple University Press, 1996).

Saunders, Francis Stonor. *The Cultural Cold War: The CIA and the World of Arts and Letters* (New York: The New Press, 1999).

Schmalzer, Sigrid. *Red Revolution, Green Revolution* (Chicago: University of Chicago Press, 2016).

Schweitzer, Glenn E. *Experiments in Cooperation: Assessing U.S.–Russian Programs in Science and Technology* (New York: The Twentieth Century Fund Press, 1997).

Moscow DMZ: The Story of the International Effort to Convert Russian Weapons Science to Peaceful Purposes (Armonk, NY: M. E. Sharpe, 1996).

Scientists, Engineers, and Two-Track Diplomacy: A Half-Century of U.S. – Russian Interacademy Cooperation (Washington, DC: National Academies Press, 2004)

Selin, Henrik. *Global Governance of Hazardous Chemicals: Challenges of Multilevel Management* (Cambridge, MA: MIT Press, 2010).

Shambaugh, David, ed., *Tangled Titans: The United States and China* (New York: Rowman & Littlefield, 2013).

Sheehan, Michael. *The International Politics of Space* (New York: Routledge, 2007).

Shenin, Sergei Y. *The United States and the Third World: The Origins of Postwar Relations and the Point Four Program* (Huntington, NY: Nova Science Publishers, 2000).

Shiva, Vandana. *The Violence of the Green Revolution: Third World Agriculture, Ecology and Politics* (London: Zed Books, 1991).

Siddiqi, Javed. *World Health and World Politics: The World Health Organization and the UN System* (Columbus: University of South Carolina Press, 1995).

Simon, Denis Fred and Merle Goldman, eds. *Science and Technology in Post-Mao China* (Cambridge, MA: Harvard University Press, 1989).

Simonson, Peter *Refiguring Mass Communication: A History* (Urbana: University of Illinois Press, 2010).

Slavin, Barbara. *Bitter Friends, Bosom Enemies: Iran, the U.S. and the Twisted Path to Confrontation* (New York: St. Martin's Press, 2007).

Skolnikoff, Eugene B. *Science, Technology and American Foreign Policy* (Cambridge, MA: MIT Press, 1969).

Smith, Alice Kimball. *A Peril and a Hope: The Scientists' Movement in America, 1945–47* (Chicago: University of Chicago Press, 1965).

Spence, Jonathan D. *The Search for Modern China*, 2nd edn. (New York: W. W. Norton & Company, 1999).

Speth, James Gustave and Peter M. Haas. *Global Environmental Governance* (Washington, DC: Island Press, 2006).

Spiro, David E. *The Hidden Hand of American Hegemony: Petrodollar Recycling and International Markets* (Ithaca, NY: Cornell University Press, 1999).

Staples, Amy L. S. *The Birth of Development: How the World Bank, Food and Agriculture Organization, and World Health Organization Changed the World, 1945–1965* (Kent, Ohio: Kent State University Press, 2006).

Stern, Philip M. and Harold P. Green, *The Oppenheimer Case* (New York: Harper & Row, 1969).

Stober, Dan and Ian Hoffman. *A Convenient Spy: Wen Ho Lee and the Politics of Nuclear Espionage* (New York: Simon & Schuster, 2001).

Sutter, Robert G. *U.S. – Chinese Relations: Perilous Past, Pragmatic Present*, 2nd ed. (New York: Rowman & Littlefield, 2013).

Taffet, Jeffrey F. *Foreign Aid as Foreign Policy: The Alliance for Progress in Latin America* (New York: Routledge, 2007).

Takeshita, Chikako. *The Global Biopolitics of the IUD: How Science Constructs Contraceptive Users and Women's Bodies* (Cambridge, MA: MIT Press, 2012).

Trulock, Notra. *Code Name Kindred Spirit: Inside the Chinese Nuclear Espionage Scandal* (San Francisco: Encounter Books, 2003).

van Beukering, Pieter J. H. and Elissaios Papyrakis., eds. *Nature's Wealth: The Economics of Ecosystem Services and Poverty* (Cambridge: Cambridge University Press, 2013).

Wagner, Caroline S. *The New Invisible College: Science for Development* (Washington, DC: Brookings Institution Press, 2008).

Waldby, Catherine and Robert Mitchell, *Tissue Economies: Blood, Organs, and Cell Lines in Late Capitalism* (Durham, NC: Duke University Press, 2006).

Walker, Mark. *German National Socialism and the Quest for Nuclear Power, 1939–49.* (Cambridge: Cambridge University Press, 1989).
   Nazi Science: Myth, Truth, and the German Atomic Bomb Project (New York: Plenum, 1995).

Walker, Mark, ed. *Science and Ideology: A Comparative History* (New York: Routledge, 2003).

Wallerstein, Mitchell B., ed. *Scientific and Technological Cooperation among Industrialized Countries* (Washington: National Academy Press, 1984).

Wallerstein, Mitchel B., Robert Schoen, and Mary E. Mogee, eds. *Global Dimensions of Intellectual Property Rights in Science and Technology* (Washington, DC: National Academy Press, 1993).

Wang, Jessica. *American Science in an Age of Anxiety: Scientists, Anticommunism & the Cold War* (Chapel Hill: University of North Carolina Press, 1999).

Wang, Zuoyue. *In Sputnik's Shadow: The President's Science Advisory Committee and Cold War America* (New Brunswick, NJ: Rutgers University Press, 2008).

Warne, William E. *Mission for Peace: Point 4 in Iran* (Bethesda, MD: IBEX Publishers, 1999).

Weart, Spencer R. *Scientists in Power* (Cambridge, MA: Harvard University Press, 1979).
   The Discovery of Global Warming (Cambridge, MA: Harvard University Press, 2008).

Wei, Chunjuan Nancy and Darry E. Brock. *Mr. Science and Chairman Mao's Cultural Revolution: Science and Technology in Modern China* (New York: Lexington Books, 2014).

Weinberger, Sharon. *The Imagineers of War: The Untold Story of DARPA, the Pentagon Agency That Changed the World* (New York: Alfred A. Knopf, 2017).

Weisberg, Barry. *Ecocide in Indochina: The Ecology of War* (San Francisco, CA: Canfield Press, 1970).

West, Nigel. *Mortal Crimes: The Greatest Theft in History: The Soviet Penetration of the Manhattan Project* (New York: Enigma Books, 2004).

Westwick, Peter J. *The National Labs: Science in an American System, 1947–1974* (Cambridge, MA: Harvard University Press, 2003).

Whitaker, Reg and Steve Hewitt. *Canada and the Cold War* (Toronto: James Lorimer & Company Ltd., 2003).

Whitt, Laurelyn. *Science, Colonialism and Indigenous Peoples: The Cultural Politics of Law and Knowledge* (New York: Cambridge University Press, 2009).

Wise, David. *Tiger Trap: America's Secret Spy War with China* (Boston: Houghton Mifflin, 2011).

Wright, Susan. *Molecular Politics: Developing American and British Regulatory Policy for Genetic Engineering, 1972–1982* (Chicago: University of Chicago Press, 1994).

Wolfe, Audra J. *Competing with the Soviets: Science, Technology and the State in Cold War America* (Baltimore, MD: Johns Hopkins University Press, 2012).

Wu, Felicia and William P. Butz. *The Future of Genetically Modified Crops: Lessons from the Green Revolution* (Santa Monica, CA: Rand Corporation, 2004).

Yergin, Daniel. *The Prize: The Epic Quest for Oil, Money and Power* (New York: Simon & Schuster, 2009).

Yoshikawa, Hideo and Joanne Kauffman. *Science Has No National Borders: Harry C. Kelly and the Reconstruction of Science and Technology in Postwar Japan* (Cambridge, MA: MIT Press, 1994).

Zachary, Pascal. *Endless Frontier: Vannevar Bush, Engineer of the American Century* (Cambridge, MA: MIT Press, 1999).

Zetter, Kim. *Countdown to Zero Day: Stuxnet and the Launch of the World's First Digital Weapon* (New York: Broadway Books, 2014).

Ziman, John. *Real Science: What It Is, and What It Means* (Cambridge: Cambridge University Press, 2002).

Zimmerman, David. *Top-Secret Mission: The Tizard Mission and the Scientific War* (Montreal: McGill-Queen's University Press, 1996).

Zubok, Vladislav M. *A Failed Empire: The Soviet Union in the Cold War from Stalin to Gorbachev* (Chapel Hill: University of North Carolina Press, 2009).

Index

www.ingramcontent.com/pod-product-compliance
Ingram Content Group UK Ltd.
Pitfield, Milton Keynes, MK11 3LW, UK
UKHW010853090126
466816UK00011B/211